Building Services Engineering

Also available from Taylor & Francis

Building Services Engineering
Fifth edition

David V. Chadderton

Taylor & Francis
Taylor & Francis Group

LONDON AND NEW YORK

First edition published 1991 by E & FN Spon
Second edition published 1994
Third edition published 2000
Fourth edition published 2004

Fifth edition published 2007
by Taylor & Francis
2 Park Square, Milton Park, Abingdon, Oxon OX14 4RN

Simultaneously published in the USA and Canada
by Taylor & Francis
270 Madison Ave, New York, NY 10016

Reprinted 2008, 2009

*Taylor & Francis is an imprint of the Taylor & Francis Group,
an informa business*

© 1991, 1994, 2000, 2004, 2007 David V. Chadderton

Typeset in Frutiger Light by
Newgen Imaging Systems (P) Ltd, Chennai, India
Printed and bound in Great Britain by
The Cromwell Press Group, Trowbridge, Wiltshire

British Library Cataloguing in Publication Data
A catalogue record for this book is available from the British Library

Library of Congress Cataloging in Publication Data
Chadderton, David V. (David Vincent), 1994–
     Building services engineering / David V. Chadderton. – 5th ed.
        p. cm.
     Includes bibliographical references and index.
        1. Buildings – Mechanical equipment. 2. Buildings – Environmental
engineering. I. Title.

TH6010.C4867 2007
696–dc22   2006030849

ISBN10: 0–415–41354–0 (hbk)
ISBN10: 0–415–41355–9 (pbk)
ISBN10: 0–203–96299–0 (ebk)

ISBN13: 978–0–415–41354–1 (hbk)
ISBN13: 978–0–415–41355–8 (pbk)
ISBN13: 978–0–203–96299–2 (ebk)

Contents

Preface to fifth edition

Building Services Engineering fifth edition is an update and expansion to include web site learning resources. Mechanical transportation has been included again. Two new chapters of multiple choice questions, 'Question Bank' and 'Understanding Units', are added to provide a wide range of challenging learning resources covering the whole book. Each chapter has additional multiple choice questions for self-assessment. Readers have access to self-test questions on the publishers' website. Questions may require the reader to look up answers in additional resources or use the internet with a search engine. There is only one correct answer to each multiple choice question unless specified as having more. Incorrect answers may be partially true but not considered by the author to be the entirely correct response for the purpose of this book; these may stimulate additional study, discussion, questioning with peers or the instructor.

Instructors can download the Building Services Engineering Instructors Manual of over 1500 multiple choice questions covering every chapter. All the multiple choice questions in this book are also in the Manual so that instructors can cut and paste test material rapidly. Questions are in chapters corresponding to the book chapter subjects to facilitate easy selection of questions for class quizzes, online tests, assignments and examinations. The author is well aware of the constant requirement for instructors to generate teaching resources, assignments and tests, having done so for many years.

The spreadsheet software file for Chapter 14, 'Room Acoustics', can be downloaded from the website http://www.tandfbuiltenvironment.com/0419257403.asp. Users usually have to adjust the screen display to optimize the viewed pages. It is expected that the reader can use the spreadsheet software that is available on their own computer, or is provided on a network system for their use. If this is not the case, introductory training in spreadsheet software use is needed. The reader can make use of Chapter 1, 'Computer and Spreadsheet Use', in *Building Services Engineering Spreadsheets* (Chadderton, 1997b) where sufficient introductory training in computer and spreadsheet use is provided.

Building Services Engineering fifth edition is intended to be a broad introduction to the range of subjects involved. The engineering content and calculation methods are sufficiently rigorous to match most of what is done within the industry during the design of many building services applications. The subjects covered and the depth to which they are analysed and calculated are more than sufficient to meet the syllabus requirements of higher technician, undergraduate and some postgraduate courses in building services engineering, heating, ventilating and air conditioning, energy management, architecture, building and quantity surveying, housing management, estate management and property facility management. Those preparing for clerk of works examinations will also find the book useful. The advanced user will need to progress to specialized text books and the standard references.

The reader is challenged to become actively engaged in the design calculations carried out by design engineers, through step-by-step introduction of each stage. A standard of numerical competence is expected that some lecturers may consider being higher than is necessary for some courses. This was deemed appropriate in order to broaden the potential readership and provide an adequate basis for a deeper design study.

Readers are encouraged to make use of the internet as a learning resource. Graphics included in a printed book are there to explain a basic principle. Use a search engine to view real plant items such as steel panel heating radiators, pumps, air-handling units, fire extinguishers or lifts, as needed. For example, a search for Wartsila takes the enquirer straight to the manufacturer of the geothermal heating CHP equipment. Happy surfing!

Acknowledgements

I am particularly grateful to the publishers for their investment in much of my life's work. Such a production only becomes possible through the efforts of a team of highly professional people. An enthusiastic, harmonious and efficient working relationship has always existed, in my experience, with Taylor & Francis. All those involved are sincerely thanked for their efforts and the result. My wife Maureen is thanked for her encouragement and understanding while I have been engrossed in keyboard work, on the drawing board and shuffling through piles of proofs. I would specifically like to thank those who have refereed this work. Their efforts to ensure that the book has comprehensive coverage, introductory work, adequate depth of study, valid examples of design, good-structured worked examples and exercises are all appreciated. Users and recommenders of the book are all thanked for their support; without them, it would not exist.

The psychrometric chart, Fig. 10.1, has been reproduced by permission of the Chartered Institution of Building Services Engineers. Pads of charts, for calculation purposes, may be obtained from CIBSE, Delta House, 222 Balham High Road, London SW12 9BS, UK.

Units and constants

Système International units are used and Table 1 gives the basic and derived units employed, their symbols and some common equalities.

Table 1 Units.

Quantity	Unit	Symbol	Equality
Mass	kilogram	kg	
	tonne	tonne	1 tonne $= 10^3$ kg
Length	metre	m	
Time	second	s	
	hour	h	1 h $=$ 3600 s
Energy, work, heat	joule	J	1 J $=$ 1 Nm
Force	newton	N	1 N $=$ 1 kgm/s^2
Power, heat flow	watt	W	1 W $=$ 1 J/s
			1 W $=$ 1 Nm/s
			1 W $=$ 1 VA
Pressure	pascal	Pa	1 Pa $=$ 1 N/m^2
	newton/m^2	N/m^2	1 b $= 10^5$ N/m^2
	bar	b	l b $= 10^3$ mb
Frequency	hertz	Hz	1 Hz $=$ 1 cycle/s
Electrical resistance	ohm	R, Ω	
Electrical potential	volt	V	
Electrical current	ampere	I, A	I $=$ V/R
Absolute temperature	kelvin	K	K $= (°C + 273)$
Temperature	degree Celsius	°C	
Luminous flux	lumen	lm	
Illuminance	lux	lx	1 lx $=$ 1 lm/m^2
Area	square metre	m^2	
Volume	litre	l	
	cubic metre	m^3	1 m$^3 = 10^3$ l

Table 2 Multiples and submultiples.

Quantity	Name	Symbol
10^{12}	tera	T
10^{9}	giga	G
10^{6}	mega	M
10^{3}	kilo	k
10^{-3}	milli	m
10^{-6}	micro	μ

Table 3 Physical constants.

Gravitational acceleration	g	9.807 m/s^2
Specific heat capacity of air	SHC	1.012 kJ/kgK
Specific heat capacity of water	SHC	4.186 kJ/kgK
Stefan–Boltzmann constant	σ	$5.67 \times 10^{-8} \text{ W/m}^2\text{K}^4$
Density of air at 20°C, 1013.25 rnb	ρ	1.205 kg/m^3
Density of water at 4°C	ρ	10^3kg/m^3
Exponential	e	2.718

Symbols

Symbol	Description	Units
A	area	m^2
	electrical current	A
A_f	floor area	m^2
	physical constant	dB
A_g	cross-sectional area of gutter	mm^2
A_o	area of water flow at gutter outlet	mm^2
A_r	roof area	m^2
A_w	walling area	m^2
α (alpha)	electrical temperature coefficient of resistance	$\Omega/\Omega°C$
	percentage depreciation and interest charge	%
	absorption coefficient	dimensionless
$\bar{\alpha}$	mean absorption coefficient	dimensionless
AET	allowed exposure time	min
B	building envelope number	
	sound reduction index	dB
B_f	physical constant	dB
b	barometric pressure	bar, b
β (beta)	angle	degree
C	fuel cost per appropriate unit	
	carbon	
C_1, C_2	constant	
C_i	interior air pollution	decipol
C_o	outdoor air pollution	decipol
C_r	room concentration	%
C_s	supply air concentration	%
C_T	concentration after time T	%
C_3	electrical load	W/m^2
clo	clothing thermal insulation	
C_v	ventilation coefficient	
CO_2	carbon dioxide	%, ppm
D	gutter depth	mm
DI	directivity index	dB
DU	demand or discharge unit	
d	pipe diameter	m or mm
	distance	m

Symbol	Description	Units
Δt (delta)	difference of temperature	°C
Δp	difference of pressure	N/m^2
d.b.	dry-bulb air temperature	°C d.b.
decipol	air pollution from one standard person	
E	emissivity	
E_{max}	maximum available evaporative cooling	W
E_{req}	required evaporative cooling	W
EWCT	equivalent wind chill temperature	°C
EL	equivalent length	m
EUPF	energy use performance factor	
e	exponential	
η (eta)	efficiency	%
F	radiation configuration factor	
F_G	fractional area	
F_s	factor of safety	
f	frequency	Hz
G	gradient	
	moisture mass flow rate	kg/m^2
	pollution load	olf/m^2
GCV	gross calorific value	MJ/kg
GJ	energy	gigajoule
g	gravitational acceleration	m/s^2
	air moisture content	kg water/kg dry air
H	height	m
	body internal heat generation	W/m^2
HSI	heat stress index	
Hz	frequency	cycle/s
h	time	hour
I	cost of installed thermal insulation	$£/m^3$
	electrical current	ampere
J	energy	joule
K	absolute temperature	kelvin
K_1, K_2	constant	
kg	mass	kilogram
kJ	energy	kilojoule
kW	power	kilowatt
kWh	energy	kilowatt-hour
L	load factor	
LDL	lighting design lumens	lumen
LH	latent heat of evaporation	kW
l	length	m
λ (lambda)	thermal conductivity	W/mK
LPG	liquefied petroleum gas	
M	metabolic rate	W/m^2
MF	maintenance factor	
MJ	energy	megajoule
MW	power	megawatt
m	length	metre
mm	length	millimeter
μ (mu)	diffusion resistance factor	
N	air change rate	h^{-1}
	force	Newton
	number of occupants	

(continued)

Continued

Symbol	Description	Units
NR	noise rating	dimensionless
n_f	number of storeys	
olf	concentration of odorous pollutants	
Ω (omega)	electrical resistance	ohm
P	pressure	pascal
	permeance	kg/Ns
P	carbon dioxide production	
Pa	pressure	pascal
P_1, P_2	area fraction	
PD	percentage of occupants dissatisfied	%
p_s	vapour pressure	pascal
ϕ (phi)	angle	degree
Q	fluid flow rate	m^3/s or l/s
	power	kW
	geometric directivity factor	dimensionless
Q_c	convection heat transfer	W
Q_e	extract air flow rate	m^3/s
Q_{ex}	exhaust air flow rate	m^3/s
Q_f	fresh air flow rate	m^3/s
Q_f	fabric heat loss	W
Q_{HWS}	hot water service power	kW
Q_L	leakage air flow rate	m^3/s
Q_p	total heat requirement	W
Q_r	radiation heat transfer	W
	recirculation air flow rate	m^3/s
Q_u	heat flow through fabric	W
Q_v	ventilation heat loss	W
q	water flow rate	kg/s
R	resistance, electrical	Ω
	thermal	$m^2 K/W$
	room sound absorption constant	m^2
R_A	combined resistance of pitched roof	$m^2 K/W$
R_a	air space thermal resistance	$m^2 K/W$
R_B	ceiling thermal resistance	$m^2 K/W$
R_n	new thermal resistance	$m^2 K/W$
R_{si}	internal surface thermal resistance	$m^2 K/W$
R_{so}	outside surface thermal resistance	$m^2 K/W$
R_R	thermal resistance of roof void	$m^2 K/W$
R_v	vapour resistance	Ns/kg
r	distance	m
r_v	vapour resistivity	GN s/kgm
		MN s/gm
ρ (rho)	density	kg/m^3
	specific electrical resistance	Ωm
	soil electrical resistivity	Ωm
S	spacing	m
	length of heating season	days
	surface area	m^2
s	time	second
SC	quarterly standing charge	
SE	specific enthalpy	kJ/kg
SG	specific gravity	
SH	sensible heat transfer	kW

Symbol	Description	Units
SPL	sound pressure level	dB
SWL	sound power level	dB
SRI	sound reduction index	dB
SHC	specific heat capacity	kJ/kgK
$\sum$	summation	
T	absolute temperature	kelvin
	total demand target	
	reverberation time	s
τ (tau)	time interval	
T_E	electrical demand target	
T_T	thermal demand target	
t_a	air temperature	°C
t_{ai}	inside air temperature	°C
t_{ao}	outside air temperature	°C
t_b	base temperature	°C
t_c	operative temperature	°C
t_{dp}	dew-point temperature	°C
t_e	environmental temperature	°C
t_{ei}	internal environmental temperature	°C
t_{eo}	outside environmental temperature	°C
t_f	water flow temperature	°C
t_g	globe temperature	°C
t_{HWS}	hot-water storage temperature	°C
t_m	mean water temperature	°C
	area-weighted average room surface temperature	°C
t_{max}	maximum air temperature	°C
t_{min}	minimum air temperature	°C
t_{15}	air temperature at time $\theta = 15$ h	°C
t_r	mean radiant temperature	°C
	return water temperature	°C
t_{res}	resultant temperature	°C
t_s	supply air temperature	°C
	surface temperature	°C
θ (theta)	angle	degree
	time	h
U	thermal transmittance	W/m²K
U_e	economic thermal transmittance	W/m²K
U_n	new thermal transmittance	W/m²K
U_w	wall thermal transmittance	W/m²K
UC	useful cost of a fuel	£/GJ, p/kWh
UF	utilization factor	
V	volume	m³
V	electrical potential	volt
v	velocity	m/s
v_s	specific volume	m³/kg
W	width	m or mm
W	power	watt
w.b.	wet-bulb air temperature	°C w.b.
WCI	wind chill index	
Y	admittance factor	
	annual degree days	

1 Built environment

Learning objectives

Study of this chapter will enable the reader to:

1. relate human physiological needs to the internal and external environment;
2. understand the ways in which heat exchange between the body and its surroundings takes place;
3. calculate indoor and outdoor thermal comfort equations;
4. identify essential instruments for measuring the environment;
5. understand the thermal environment terminology used by design engineers;
6. recognize the problems of experimental work;
7. make reliable technical reports based on his or her own work and not to copy the work of others;
8. understand and use the factors that influence indoor air quality;
9. calculate fresh air ventilation rate;
10. know the instrumentation used for indoor environmental monitoring.

Key terms and concepts

air velocity 9; allowed exposure time 14; atmospheric pollutants 4; *clo* value 8; comfort equation 8; conduction, convection, radiation and evaporation 3; data logger 11; decipol 4; dry-bulb air temperature 15; dry resultant temperature 20; environmental temperature 19; equivalent wind chill temperature 13; globe temperature 16; heat stress index 14; humidity 3; infrared scanner 19; Kata thermometer 17; mean radiant temperature 9; metabolic rate 8; odorants 4; olf 4; olfactory 4; operative temperature 20; percentage of people dissatisfied 11; pitot-static tube 18; predicted mean vote 11; sling psychrometer 15; thermistor anemometer 17; thermocouple 18; thermohygrograph 17; trend 11; vane anemometer 17; vapour pressure 8; wet-bulb globe temperature 15; wet-bulb temperature 15; wind chill index 13.

Introduction

The building is an enclosure for the benefit of human habitation, work or recreation; in some cases, the ruling criteria are those demanded by an industrial need, such as machinery, products stored or computer equipment. Much construction work is undertaken outdoors, where the climate influences human working effectiveness and can lead to health risk.

The way in which the design engineer calculates measurement scales and interaction with human requirements is investigated.

World energy supply and demand

Many countries have become rich in material possessions by rapidly exploiting technological advances in energy supply and use. The human ability to extract enormous amounts of natural energy from the earth has enabled rapid travel to all parts of the globe, and space, and has kept habitable structures warm or cool as desired. Buildings are maintained with a standard of comfort, in all aspects of the word, at high energy cost. Large parts of the globe are much poorer in terms of their standard of living, and it is incumbent upon those nations possessing knowledge to act responsibly with regard to the resources they use now, how they plan for future development, and what technology is sold to developing nations. The building industry as a whole can learn from the history of energy usage to promote only those systems making the best use of available power. Economical use of energy with services installations is crucially important in this respect, as is designing new buildings that are energy-efficient. The designer's responsibility for energy use ceases after construction is completed and the defects liability period ends. The building owner, or user, pays for energy use in continuum. The energy cost, whether this be low or high, of using the facility, has been welded into place. Many years pass before anyone becomes brave enough to consider retrofit energy-saving measures. Demand-side minimization of energy use becomes the responsibility of all of us in the construction industry as well as us being personal consumers in our homes and transportation. The alternative is the ever-increasing requirement for supply-side public power supply installations. The word sustainability is recently popular; evaluating what it means to the building services industry becomes a real challenge to us all.

The building as an environmental filter

One of people's basic needs is to maintain a constant body temperature while the metabolism regulates heat flows from the body to compensate for changes in the environment. We have become expert in fine-tuning the environmental conditions produced by the climate in relation to the properties of the building envelope to avoid discomfort. A simple tent or cave may be sufficient to filter out the worst of adverse weather conditions, but the ability of this type of shelter to respond to favourable heat gains or cooling breezes may be too fast or too slow to maintain comfort.

Outside of the tropics, latitudes beyond 23.5° from the equator, houses may be advantageously oriented towards the sun to take advantage of solar heat gains, which will be stored in the dense parts of the structure and later released into the rooms to help offset heat losses to the cool external air during winter. Buildings within the tropical zone require large overhanging roofs and shutters over the windows to exclude as much solar radiation as possible and to shade the rooms. Thus the building envelope acts to moderate extremes of climate, and by suitable design of illumination and ventilation openings, together with heating, cooling and humidity controls, a stable internal environment can be matched to the use of the building.

Basic needs for human comfort

The building services engineer is involved with every part of the interface between the building and its occupant. Visually, colours rendered by natural and artificial illumination are produced by combinations of decor and windows. The acoustic environment is largely attributed to the success achieved in producing the required temperatures with quiet services equipment, all of which is part of the thermal control and transportation arrangements. Energy consumption for thermally based systems is the main concern, and close coordination between client, architect and engineer is vitally important.

Heat transfer between the human body and its surroundings can be summarized as follows.

Conduction

Points of contact with the structure are made with furniture and the floor. Clothing normally has a substantial thermal insulation value and discomfort should be avoided.

Convection

Heat removed from the body by natural convection currents in the room air, or fast-moving airstreams produced by ventilation fans or external wind pressure, is a major source of cooling. The body's response to a cool air environment is to restrict blood circulation to the skin to conserve deep tissue temperature, involuntary reflex action, shivering, if necessary, and in extreme cases inevitable lowering of body temperature. This last state of hypothermia can lead to loss of life and is a particular concern in relation to elderly people.

Radiation

Radiation heat transfer takes place between the body and its surroundings. The direction of heat transfer may be either way, but normally a minor part of the total body heat loss takes place by this method. Radiation between skin and clothing surfaces and the room depends on the fourth power of the absolute surface temperature, the emissivity, the surface area and the geometric configuration of the emitting and receiving areas. Thus a moving person will experience changes in comfort level depending on the location of the hot and cold surfaces in the room, even though air temperature and speed may be constant.

Some source of radiant heat is essential for comfort, particularly for sedentary occupations, and hot-water central heating radiators, direct fuel-fired appliances and most electrical heaters provide this. The elderly find particular difficulty in keeping warm when they are relatively immobile, and convective heating alone is unlikely to be satisfactory. A source of radiant heat provides rapid heat transfer and a focal point, easy manual control and quick heat-up periods. Severe cases of underheating can be counteracted by placing aluminium foil screens in positions where they can reflect radiation onto the rear of the chair.

Overheating from sunshine can also cause discomfort and glare, and tolerance levels for radiant heating systems have been established.

Evaporation

Humid air is exhaled, and further transfer of moisture from the body takes place by evaporation from the skin and through clothing. Maintenance of a steady rate of moisture removal from the body is essential, and this is a mass transfer process depending on air humidity, temperature and speed as well as variables such as clothing and activity.

Ventilation

The quality of the air in a building depends upon the quantity, type and dispersal of atmospheric pollutants (Awbi, 1991, p. 27). Some of these, odorants, can be detected by the olfactory receptors in the nose. These are the odours, vapours and gases that ingress from the outdoor environment and are released from humans, animals, flora, furnishings and the structural components of the building. Solid particles of dust, pollen and other contaminants often have little or no smell. These might be seen in occasional shafts of sunlight, and become visible when they have settled. Cleaning fluids such as ammonia, cigarette smoke, hair spray, deodorants and perfumes can be most noticeable. The inflow of diesel exhaust fumes, road tar, paint vapours and creosote creates unpleasantly noticeable pollution, even when of short duration. The presence of harmful pollutants such as carbon monoxide and radon gases is not detectable by the occupant. Indoor air quality may be said to be acceptable when not more than 50% of the occupants can detect any odour. Pollutants may still be present even if not noticeable by most occupants. The nasal cavity also reacts to pollutants with the general chemical sense of irritation. Olfactory response adapts to pollutants over time making people less sensitive to them while irritation increases with time (Chartered Institution of Building Services Engineers, CIBSE Guide A).

Professor Ole Fanger has introduced units of subjective assessment for odorants only. The olf quantifies the concentration of odorous pollutants. The decipol is the evaluation of the pollutant as determined by the recipient through the olfactory sensations from the nose. One olf is the emission rate of biological effluents from one standard person, or the equivalent from other sources. One decipol is the pollution caused by one standard person when ventilated with 10 l/s of unpolluted air. The number of olfs corresponding to different levels of human activity is shown in Table 1.1 (Fanger, 1988).

Office accommodation normally has one person for each 10 m² of floor area, so the biological effluent pollution load produced by normal occupancy is 0.1 olf/m². Smokers, building and furnishing materials and ventilation systems add to the pollution load. The average pollution in an existing building that has 40% of the occupants as smokers produces a load G of 0.7 olf/m². A low-pollution building with an absence of smoking has a load G of 0.2 olf/m². When there is complete mixing of the ventilation air with the air in the room, the rate of supply of outdoor air that is necessary to maintain the required standard of air quality is found from

$$Q = \frac{10 \times G}{C_i - C_o} \text{ l/s}$$

where C_i is the perceived air pollution within the enclosure (decipol), C_o is the perceived air pollution of outdoor air, usually 0.05 decipol but which may rise to 0.3 in a city with moderate pollution, and G is the concentration of pollution in the enclosure and the ventilation system (olf).

Table 1.1 Olf values for human activities.

Human activity	Number of olfs
Sedentary	1
Active	5
Highly active	11
Average for a smoker	6
During smoking	25

The perceived air pollution C_i within the enclosure is found from the percentage of the occupants who are dissatisfied with the conditions, PD, from

$$C_i = \frac{112}{(5.98 - \ln PD)^4} \text{ decipol}$$

where PD is the percentage of the occupants who are dissatisfied (ASHRAE (1985) recommend 20%), and $\ln = $ logarithm to base e ($\log_e$).

EXAMPLE 1.1

Calculate the outdoor air ventilation rate, from the Fanger method, that is required to satisfy 75% of the occupants of a commercial building where none of them are smokers. The building is located on the edge of a large town in the UK.

To satisfy 75% of the occupants, 25% are dissatisfied, so PD is 25%.

$$C_i = \frac{112}{(5.98 - \ln 25)^4} \text{ decipol}$$

$$= \frac{112}{(5.98 - 3.219)^4} \text{ decipol}$$

$$= \frac{112}{(2.761)^4} \text{ decipol}$$

$$= 1.93 \text{ decipol}$$

For an existing building with no smokers, $G = 0.2$ olf/m^2. $C_o = 0.05$ decipol for clean suburban air. For 1 m^2 of office floor area

$$Q = \frac{10 \times G}{C_i - C_o} \text{ l/s m}^2$$

$$= \frac{10 \times 0.2}{1.93 - 0.05} \text{ l/s m}^2$$

$$= 1.07 \text{ l/s m}^2$$

EXAMPLE 1.2

A new aerobics gymnasium is being designed for the basement of a commercial building in the City of Manchester. The room is to be 15 m long, 12 m wide and 3 m high. It has no exterior windows. There will be between 10 and 40 simultaneous users of the facility. There will not be any smoking and all the furnishings and building materials will have the low-pollution emission of 0.1 olf/m^2. At peak usage, all the occupants will be highly active. Calculate the outdoor air ventilation rate, from the Fanger method, that will be required so that 75% of the occupants will be satisfied. Recommend how the outdoor air ventilation system could be controlled for energy economy and comfort. Recommend an acceptable solution for the client.

(a) At full occupancy (40 people), to satisfy 75% of the occupants, 25% (10 people) are dissatisfied, so $PD = 25\%$.

$$C_i = \frac{112}{(5.98 - \ln 25)^4} \text{ decipol}$$

$$= 1.93 \text{ decipol}$$

The low-pollution building produces 0.1 olf/m^2.
At peak usage, there will be 40 people each producing 11 olf, from Table 1.1.

$$\text{gymnasium floor area} = 15 \text{ m} \times 12 \text{ m} = 180 \text{ m}^2$$

$$\text{pollution load } G = 0.1 \text{ olf/m}^2 + \frac{40 \text{ people} \times 11 \text{ olf/person}}{180 \text{ m}^2}$$

$$= 0.1 + 2.44 \text{ olf/m}^2$$

$$= 2.54 \text{ olf/m}^2$$

$C_o = 0.3$ decipol for vitiated city air that enters the basement from street level. For 1 m^2 of gymnasium floor area,

$$Q = \frac{10 \times G}{C_i - C_o} \text{ l/s m}^2$$

$$= \frac{10 \times 2.54}{1.93 - 0.3} \text{ l/s m}^2$$

$$= 15.6 \text{ l/s m}^2$$

For the whole gymnasium,

$$Q = 180 \text{ m}^2 \times 15.61 \text{ l/s m}^2$$

$$= 2808 \text{ l/s}$$

the room volume

$$V = 15 \text{ m} \times 12 \text{ m} \times 3 \text{ m}$$

$$= 540 \text{ m}^3$$

air change rate

$$N = 2808 \frac{l}{s} \times \frac{1 \text{ m}^3}{10^3 \text{ l}} \times \frac{1 \text{ air change}}{540 \text{ m}^3} \times \frac{3600 \text{ s}}{1 \text{ h}}$$

$$= \frac{2808 \times 3600}{10^3 \times 540} \text{ air changes/h}$$

$$= 18.7 \text{ air changes/h}$$

This is a high air change rate and will result in large heating and cooling costs during winter and summer if the thermal conditions are to remain within acceptable ranges for the activity. The number of room occupants who can be satisfied with the air quality will need to be reduced. Air quality, here, refers to maintaining a low enough level of odours so that the occupants are satisfied. Alternative solutions to supplying 100% outdoor air into the basement are available.

(b) When the gymnasium has 10 occupants with 75% satisfaction, 3 people are dissatisfied, so PD is 25%, C_i is 1.93 decipol, G is 0.71 olf/m^2 when C_o is 0.3 decipol, Q becomes 4.4 l/s m^2, 792 l/s for the whole room and N is 5.3 air change/h.

(c) The supply of outside air into the gymnasium is required to comply with the minimum of 10 l/s per person (CIBSE Guide A).

$$\text{Minimum outside air supply } Q_o = 40 \text{ people} \times \frac{10}{\text{person}} \frac{l}{s}$$

$$= 400 \text{ l/s}$$

$$\text{Air change rate } N = 400 \frac{l}{s} \times \frac{1 \text{ m}^3}{10^3 \text{ l}} \times \frac{1 \text{ air change}}{540 \text{ m}^3} \times \frac{3600 \text{ s}}{1 \text{ h}}$$

$$= \frac{400 \times 3600}{10^3 \times 540} \text{ air changes/h}$$

$$= 2.7 \text{ air changes/h}$$

(d) The air change rates of outdoor air that are, apparently, needed to satisfy the Fanger odour air quality criteria vary from zero, when the gymnasium is unoccupied, to 5.3 per hour when 10 people use the gymnasium, to 18.7 per hour when 40 people use the room. The minimum required outdoor air supply is 400 l/s; this corresponds to a room air change rate of 2.7 per hour when 40 people are using the gymnasium. Mechanical ventilation will be needed owing to the location of the room. The design engineer will recommend a minimum energy use by recirculating the maximum amount of room air and minimizing the use of outdoor air to match the room occupancy. All the incoming outdoor air can be passed through a flat-plate heat exchanger to recover heat that is being discharged to the atmosphere through the exhaust air duct, if this is possible for the duct installation. When the room is unoccupied, the outside air supply can be closed with a motorized damper. The number of occupants in the gymnasium can be detected from a carbon dioxide sensor in the extract air duct, which can be used to control the opening of the outside air motorized damper as well as the supply air and return air fan speeds. The varying air flow rates can be provided with a system of supply-and-extract air fans that have variable performance. Each fan can have a variable-speed electric motor which has its supply frequency controlled between 0 and 50 Hz from an inverter drive. The outdoor air that is supplied into the gymnasium can be heated to around 16°C in winter and cooled to 20°C in summer, for thermal comfort. When lower than the Fanger-recommended outside air intake quantities are used, that will normally be always, the odour air quality can be improved by passing the return air through an activated carbon filter or by injecting a deodorant spray into the supply air duct. Whichever alternative is selected, the energy-conscious design engineer will minimize the use of outside air to ventilate the room. Those using the gymnasium are fairly unlikely to find the presence of some body odour totally unacceptable. Significant energy savings will be achieved by the use of carbon dioxide sensing of the occupants in a gymnasium that is used intermittently each day. The outside air damper will remain closed whenever the room is unoccupied and it will only open in response to the actual occupancy level.

Table 1.2 Comparison of ventilation rates for offices.

Data source	Ventilation rate	
	$(l/s\ m^2)$	(l/s per person)
Fanger equation	5	50
ASHRAE	0.7	7
BS5925	1.3	13
DIN 1946	1.9	19
CIBSE, office	1.3	10

Awbi (1991, p. 31) shows a comparison between the current standards for ventilation with outdoor air. They are based upon one person for each 10 m² of office floor area, the standard person and activity plus some smoking. To these, recommendations (CIBSE Guide A) have been added, as shown in Table 1.2. The general recommendation is 10 l/s per person. Where less is attempted, measurement of return air CO_2 content may be applied to ensure adequate air quality.

Other applications are taken on their merits: for example, toilets 5 air changes/h, bathrooms 15 l/s and kitchen 60 l/s in dwellings (CIBSE Guide A). The designer needs to evaluate all the aspects of the need for outdoor air ventilation. These include the heating and cooling plant loads that will be generated, the potential for energy recovery between the incoming fresh air and outgoing air at room temperature, avoidance of draughts in the occupied rooms, the variation of load with the occupancy level and the ability to utilize outdoor air to provide free cooling to the building when the outdoor air is between 10°C and 20°C (Chadderton, 1997a, p. 5).

Comfort equation

The fundamental purpose of heating or cooling an occupied space is to maintain constant body temperature. Regulation of personal comfort is achieved by clothing choice, and a successful building can be said to be one where 5% or less of its occupants complain. Definitive work was done by Fanger (1972) to produce a complete mathematical statement of the heat transfer between the body and its surroundings, and then to conduct subjective tests and produce guidelines for building and engineering designers. Fanger's comfort equation is of the form

$$f(H, clo, v, t_r, t_a, p_s) = 0$$

that is, a balance, where f represents a mathematical function connecting all the variables contained in the brackets, H is the internal heat production in the human body, clo is the thermal insulation value of clothing, v is the air velocity, t_r is the mean radiant temperature, t_a is the air temperature and p_s the atmospheric vapour pressure.

The energy released per unit time in the human body by oxidation processes is known as the metabolic rate M. Some of this can be converted into useful mechanical work W, so that

$$M = H + W$$

The mechanical efficiency of the body is given by

$$\frac{W}{M} \times 100\%$$

and this varies from zero while reading this book to a maximum of 20% during heavy physical work.

Internal heat production H varies from 35 W/m² (watts per square metre of body surface) while sleeping to 440 W/m² during maximum exertion. The thermal insulation value of clothing is expressed in *clo* units. One *clo* unit is equal to the total thermal resistance from the skin to the outer surface divided by 0.18. It is worth noting that 1 *tog* = 0.645 *clo*. Values of *clo* vary from zero when nude through 1 when wearing a normal business suit to 4 for a polar suit.

Moving-air velocity in normally occupied rooms will be between 0 and 2 m/s (metres per second), where the upper figure relates to a significantly uncomfortable hot or cold draught. Still air conditions are most unlikely to occur, as convection currents from people and warmed surfaces will promote some circulation. Room air movement patterns should be variable rather than monotonous, and ventilation of every part of the space is most important.

The mean radiant temperature is a measure of radiation heat transfers taking place between various surfaces and has an important bearing on thermal comfort. Heat transfer Q_r by radiation from a warm emitting surface to a receiving plane is given by

$$Q_r = \frac{5.67}{10^8} A_1 F_1 E_1 E_2 (T_1^4 - T_2^4) \text{ W}$$

where E_1 and E_2 are surface emissivities, which range from 0.04 for polished aluminium to 0.96 for matt black paint, and T_1 and T_2 are the absolute temperatures of the emitting and receiving surfaces in Kelvin (°C + 273). F_1 is the configuration factor denoting the orientation of and distance between the two surfaces. Flat parallel surfaces are evaluated using

$$A_1 F_1 = \frac{A_1 A_2}{\pi l^2}$$

where A_1 and A_2 are the surface areas (m²) and l is the distance between the surfaces (m). This only fits the simplest cases, such as radiation across the cavity in a wall. Other real problems become geometrically rigorous and complex.

EXAMPLE 1.3

A hot-water central heating system has a ceiling-mounted radiant panel of surface area 8.0 m² and a surface temperature of 70°C. It faces a person 1.1 m away whose upward projected body surface area and temperature are 0.3 m² and 21°C respectively. The surface emissivities of the radiator and the person are 0.85 and 0.90 respectively. Calculate the radiant heat transfer.

$$A_1 F_1 = \frac{8.0 \text{ m}^2 \times 0.3 \text{ m}^2}{\pi (1.1^2 \text{ m}^2)} = 0.63 \text{ m}^2$$

$$T_1 = (273 + 70) \text{K} = 343 \text{ K}$$

$$T_2 = (273 + 21) \text{K} = 294 \text{ K}$$

$$Q_r = \frac{5.67}{10^8} \times 0.63 \times 0.85 \times 0.90 \times (343^4 - 294^4) \text{W}$$

$$= 174 \text{ W}$$

Because of the complexity of such calculations for a heated room, an approximation to the mean radiant temperature t_r is often made that estimates its value for the geometric centre of the room volume:

$$t_r \cong t_m = \frac{A_1 t_{s1} + A_2 t_{s2} + \cdots + A_n t_{sn}}{A_1 + A_2 + \cdots + A_n}$$

Thus t_m is seen to be the average room surface temperature weighted in proportion to the surface area, room geometry and point of measurement; n is the number of surfaces.

EXAMPLE 1.4

A room is heated with under floor low-temperature hot-water pipework and the winter data shown in Table 1.3 were found during site measurements.

Table 1.3 Data for Example 1.4.

Surface	Area (m²)	Average surface temperature (°C)
Window	12	2
Outside wall	24	12
Inner walls	30	16
Ceiling	22	24
Floor	22	30

Estimate the mean radiant temperature of the room at the centre of its volume.

$\sum A$ and $\sum At_s$ are calculated as shown in Table 1.4. Then

$$t_m \cong \frac{\sum (At_s)}{\sum A}$$
$$= \frac{1980}{110}$$
$$= 18°C$$

Air temperature t_a is that recorded by a mercury-in-glass dry-bulb thermometer freely exposed to the air stream; thermometer is usually in a sling psychrometer.

Table 1.4 Calculations for Example 1.4.

Surface	Area A (m²)	t_s (°C)	At_s
Window	12	2	24
Outside wall	24	12	288
Inner walls	30	16	480
Ceiling	22	24	528
Floor	22	30	660
	$\sum A = 110$		$\sum (At_s) = 1980$

Atmospheric vapour pressure p is that part of the barometric pressure produced by the water vapour in humid air. Standard atmospheric pressure at sea level is 1013.25 mb (millibars) and this comprises about 993.0 mb from the weight of dry gases and 20.25 mb from the water vapour, depending on the values of barometric pressure, air temperature and humidity.

The comfort equation balances to zero when heat transfers between the body and surroundings are stable. This is the thermally neutral condition when there are no feelings of discomfort. It is unlikely to be satisfied for a group of people, and a comfort zone is used to specify the range of acceptable levels for the majority.

Comfort measurement

Figure 1.1 shows the percentage of people dissatisfied (*PPD*) and predicted mean vote (*PMV*) that are used for the assessment of indoor thermal environments, from Fanger. The chart demonstrates that dissatisfaction among less than one-third of the occupants is achieved in indoor environments within the slightly cool to slightly warm categories. Personal variation in clothing, sedentary position and control over the microclimate being the final control mechanism.

A thermal comfort analogue computer may be used to make this assessment with an ellipsoidal-shaped sensor that is around the size of a straight banana. This shape has the same mathematical relationship between its heat loss by radiation and convection as the human body.

To obtain values of temperature, velocity, carbon dioxide and humidity in the atmosphere, each is measured separately as 0–10 V or 0–20 mA analogue signals, converted into digital bytes with an analogue-to-digital converter and then stored in a data logger in random access memory. Numerous readings are easily retained in memory for later access through a desktop or portable computer. Lists of data numbers and their time of occurrence can be stored on hard and floppy disks for processing with dedicated software or any spreadsheet program. Graphical output can show the history of the recorded temperature, or other variable, during a period of time. Such graphs are termed trends. The software can be programmed to take readings at, say, intervals of 2 s or longer, or only whenever a significant change takes place. Figure 1.2 shows the schematic arrangement of a data-logging system with a trend display of room air temperature. Figure 1.3 is a photograph of a portable data-logging cart created to assess existing building conditions.

1.1 *PPD* versus *PMV*.

Visual display
screens showing
same data trend

240 V
power

RS232
ports

Portable computer

Digital
communication
cable

Data logger, A/D, RAM and battery

Desktop computer

240 V
power
supply

0–10 V, 2-wire cable

t_a°C d.b.

Lux

t_a°C w.b.

CO_2

v m/s

t_{ao} °C

t_g°C globe

Range of detectors

Occupancy sensor

1.2 Multi-channel data logger linked to desktop and portable computers.

The sensors, from left to right, are globe temperature, relative humidity, air velocity and air dry-bulb temperature.

Research results indicate that nationality, geographic origin, location, age, sex and body build have no significant effect on basic comfort requirements. The achievement of acceptable comfort levels for the variety of activities that are conducted in offices, factories, schools, hospitals and homes can be very expensive in terms of building and services construction and energy costs. In practice, compromise solutions are made that satisfy most requirements for the majority of the occupants.

1.3 Thermal comfort data logger (reproduced by courtesy of Mobile Architecture and Built Environment Laboratory, Deakin University, Geelong).

External environments

Extremes of external climate are mainly of concern to construction workers in severe environments. There are two main indices: wind chill index (*WCI*) and heat stress index (*HSI*). Wind chill index measures the cooling effect on the body of a moving air stream, given by:

$$WCI = (10.45 + 10\sqrt{v} - v)(33 - t_a)$$

Wind speed *v* is in metres per second; the equation can be used for values of up to 22.0 m/s (79 km/h (kilometres per hour)). Frostbite should be avoided if the wind chill index is less than 1400 at an air temperature of $-10.0°C$ d.b. during a maximum exposure of 30 min. A convenient use of the wind chill index is to calculate the equivalent wind chill temperature (*EWCT*), which is the air temperature, calculated from measured data, that would provide the same chilling effect but at an air velocity of 1.78 m/s (6.4 km/h):

$$EWCT = (-0.045WCI + 33)°C$$

EXAMPLE 1.5

A winter day in northern England has an outdoor air velocity measured as 25.2 km/h (7 m/s) and the sling psychrometer air temperature showed $-2.5°C$ d.b. Find the wind chill index and the equivalent wind chill temperature.

$$v = 7 \text{ m/s} \quad t_a = -2.5°C$$

$$WCI = (10.45 + 10\sqrt{7} - 7)[33 - (-2.5)]$$

$$WCI = 1061.7$$

$$EWCT = (-0.045 \times 1061.7 + 33)°C$$

$$EWCT = -14.8°C$$

Construction workers in hot climates are exposed to heat hazards, and severe cases will lead to an inevitable increase in body temperature. This is the zone of unacceptable body heating. In its attempt to dissipate the metabolic heat produced, which cannot be transferred to the high-temperature environment, the body raises its temperature to compensate. The symptoms are fatigue, headache, dizziness, vomiting, irritability, fainting and failure of normal blood circulation to cope with the problem. Heat exhaustion of this sort can normally be counteracted by removal to a cool place. Cramp may occur as a result of loss of some body salts. Salt tablets can be taken to redress the balance. Heatstroke occurs if the body temperature rises to 40.6°C (normal body temperature is 36.9°C), and in this condition sweating ceases, the body enters a comatose state, brain damage can occur and death is imminent.

The heat stress index is a measure of the maximum elevation of sweat rate, body temperature and heart rate that can be tolerated (Belding and Hatch, 1955):

$$HSI = \frac{E_{req}}{E_{max}} \times 100 \quad \text{(this is a ratio and not a percentage)}$$

where $E_{req} = M - Q_r - Q_c$ is the required evaporative cooling, $Q_r = 4.4(35 - t_r)$ is the heat lost from the clothed body by radiation, $Q_c = 4.6 \, v^{0.60}(35 - t_a)$ is the heat lost from the clothed body by convection and $E_{max} = 7.6 \, v^{0.60}(56 - p_a)$ is the maximum available evaporative cooling.

The numerical value of the heat stress index can be related to practical circumstances as shown in Table 1.5. The allowed exposure time (AET) is given by

$$AET = \frac{2440}{E_{req} - E_{max}} \text{ min}$$

When the heat stress index is less than 100.0, the exposure time is unrestricted. The heat stress index reduces and the allowed exposure time increases when a prevailing wind or fans produce extra cooling.

Table 1.5 Evaluation of heat stress index.

Heat stress index	Response to 8-h exposure
−20	Mild cold strain
0	Neutral
10–30	Mild strain
40–60	Unsuitable for mental effort
70–90	Selected personnel
100	Maximum possible for a fit acclimatized young man

EXAMPLE 1.6

A construction worker is exposed to conditions of 38°C d.b. and 24% saturation in the shade in city centre Melbourne. Metabolic rate is 320 W/m^2 while lifting and placing structural formwork. Local air movement is 0.25 m/s. Find the heat stress index and the allowed exposure time. Assume that $t_r = t_a$. The atmospheric vapour pressure p_a is 16.74 mb.

$$E_{req} = 320 - 4.4(35 - 38) - 4.6 \times 0.25^{0.60}(35 - 38)$$

$$= 339.2 \text{ W/m}^2$$

$$E_{max} = 7.6 \times 0.25^{0.60}(56 - 16.74)$$

$$= 129.9 \text{ W/m}^2$$

$$HSI = \frac{339.2}{129.9} \times 100$$

$$= 261.1$$

$$AET = \frac{2440}{339.2 - 129.9} \text{ min}$$

$$= 11.7 \text{ min}$$

By providing increased air movement around the construction worker with a portable fan, an elevation of v to 1.5 m/s would reduce the heat stress index to below 100 and make exposure time unrestricted.

Environmental measurements

Measurement of air temperature and humidity is accurately made by using both dry- and wet-bulb mercury-in-glass thermometers in a sling psychrometer, otherwise called whirling hygrometer, Figure 1.4. The dry-bulb thermometer defines the air temperature t_a°C d.b. and evaporation of water from the cotton wick cools the wet-bulb thermometer; its reading is known as the wet-bulb temperature t_a°C w.b. The difference between dry- and wet-bulb temperatures is the wet-bulb depression. Temperatures are sometimes symbolized by θ (theta) (CIBSE Guide A).

In order to find the mean radiant temperature t_r, the dry-bulb temperature t_a and the air velocity v in m/s are measured, and the empirical relationship

$$t_r = t_g(1 + 2.35\sqrt{v}) - 2.35t_a\sqrt{v}$$

where t_g is the globe temperature, is used. The globe temperature is measured at the centre of a blackened globe of diameter 150 mm suspended at the measurement location. Figure 1.5 shows the large and small globe thermometers used. The resultant temperature, found using a globe of diameter 100 mm, can be used, and t_r can be calculated by replacing 2.35 with 3.17 in the previous formula.

Wet-bulb globe temperature, WBGT, is used for assessment of warm humid environments for health and safety at work conditions. It is found from air wet-bulb t_a°C w.b., globe t_g°C and air dry-bulb t_a°C d.b. in weighted amounts to give greater emphasis to humidity. It is humidity that causes significant heat stress, while hot dry air flow is efficient at removing bodily heat

1.4 Sling psychrometer or whirling hygrometer (reproduced by courtesy of Casella CEL Ltd).

38 mm diameter miniature globe

150 mm diameter globe

1.5 Large and small globe thermometers.

production. When local air temperature approaches core body temperature, the sole remaining mechanism for bodily heat removal is evaporation from breath and skin; that is, when there is no handy swimming pool.

$$WBGT = 0.7t_a°C \text{ w.b.} + 0.2t_g°C + 0.1t_a°C \text{ d.b.}$$

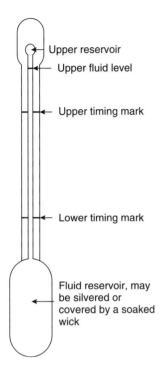

Upper reservoir
Upper fluid level

Upper timing mark

Lower timing mark

Fluid reservoir, may
be silvered or
covered by a soaked
wick

1.6 Kata thermometer.

The Kata thermometer shown in Fig. 1.6 is used to measure the cooling power of room air movements and air velocity. The thermometer is heated by immersing the large bulb in hot water until the alcohol enters the upper reservoir. The bulb is dried, and the Kata thermometer is suspended at the location being investigated. The time taken for the meniscus to fall between the two marks on the stem is noted, the operation is repeated twice and the average time of cooling is found. A Kata factor, which is the amount of heat lost from the bulb surface during cooling (typically 475 mcal/cm^2 of bulb surface as the thermometer cools through 2.8°C), is inscribed on the stem. The cooling power is defined as the Kata factor divided by the average cooling time in seconds. A cooling power of around 10 is suitable for sedentary occupation. Wet-bulb Kata readings can also be taken, and these include humidity effects. A nomogram or formula can be used to find the air velocity at that location.

Direct-reading air velocity measuring instruments are shown in Figs. 1.7 and 1.8. Rotating vane anemometers are used to measure airstreams through ventilation grilles, where the rotational speed of the blades is magnetically counted. Thermistor and hot-wire anemometers utilize the air stream cooling effect on the probe; the latter type is mainly used for research work. Duct air flow rates are found by inserting a pitot-static tube into the airway, taking up to 48 velocity readings at locations specified in BS 848: Part 1: 1980 and evaluating the air volume flow rate from the average air velocity found from all the readings.

The term for air humidity is percentage saturation, and the most reliable method of measurement is to take dry- and wet-bulb air temperature readings using a sling psychrometer and referring to a psychrometric chart (Chapter 5). Hair hygrometers can be used to display percentage saturation. A combined clockwork-driven thermohygrograph can be used to monitor air dry-bulb and humidity room conditions over a period of days, but electronic data loggers are currently preferred. Permanent monitoring and control requires an electronic sensor utilizing a

1.7 Thermistor anemometer.

1.8 Pitot-static tube and U-tube manometer.

hygroscopic salt covering a coil of wire. The output signal from this unit depends on the amount of moisture absorbed by the sensor.

Thermocouple temperature-sensing enables supervision of large numbers of locations by manual or automatic scanning. Inaccessible areas can be reached by installing thermocouple wires during construction or by drilling. Figure 1.9 shows the basic construction of a copper–constantan thermocouple circuit. The small electrical voltages caused between the junctions of dissimilar metals at different temperatures produce an electron flow, and thus the potential difference between the wires can be measured and calibrated into temperature. Surface temperature measurements can be made using portable instruments to check the integrity of thermal insulation (Chapter 3) or in experimental investigations into the thermal transmittance of building components using multi-channel recording systems.

1.9 Basic thermocouple circuit and calibration graph.

Non-touch temperature sensors receive infrared heat radiation emitted from all surfaces that are above absolute zero, and they are calibrated to give the surface temperature. Small areas can be surveyed with portable gun instruments and large scans can be performed by mobile television camera equipment, which produces temperature contour maps. Ground-level or aerial surveys are used to detect energy waste from non-existent, inadequate or damaged thermal insulation in homes, in factories or where there are buried pipes in district heating systems.

Environmental temperature

Heat losses from buildings and the conditions required for thermal comfort both depend on the mean radiant temperature and the air temperature. The environmental temperature t_{ei} combines these two measures:

$$t_{ei} = 0.667t_r + 0.333t_{ai}$$

EXAMPLE 1.7

An open-plan office is designed for sedentary occupation and is to have general air movement not exceeding 0.25 m/s and an air temperature of 22°C d.b., in winter. It is expected that a globe temperature of 20°C would be found at the centre of the room volume. What would be the mean radiant, resultant and environmental temperatures?

$$t_r = t_g(1 + 2.35\sqrt{v}) - 2.35t_a\sqrt{v}$$

$$= 20(1 + 2.35 \times \sqrt{0.25}) - 2.35 \times 22 \times \sqrt{0.25}$$

$$= 17.6°C$$

$$t_{res} = t_g(1 + 3.17\sqrt{v}) - 3.17t_a\sqrt{v}$$

$$= 20(1 + 3.17 \times \sqrt{0.25}) - 3.17 \times 22 \times \sqrt{0.25}$$

$$= 16.8°C$$

$$t_{ei} = 0.667t_r + 0.333t_{ai}$$

$$= 0.667 \times 17.6 + 0.333 \times 22$$

$$= 19.1°C$$

Operative temperature

Operative temperature, t_c, is that recorded in the centre of a 40 mm diameter blackened globe freely exposed to room air. Sufficient time is needed for stabilization of the radiant and convective heat transfers to take place with the thermometer, thermocouple or thermistor temperature sensor in the globe. When taking readings, realize that room conditions may be varying due to movement of people, solar heat gains, air movement or the room temperature control system. During normal room conditions of low air velocity and where mean radiant and air temperatures are very close, operative, air dry-bulb and dry resultant temperatures, t_{res}, are substantially equal. Well-insulated modern buildings having small areas of glazing have the effect of raising room mean radiant temperature close to the air dry bulb. Dry resultant temperature is the temperature recorded by a thermometer at the centre of a 100 mm diameter blackened globe. Operative temperature is now recommended for use as that specified for comfort conditions in the internal environment. Comfortable values range from 21°C for a residential living room through 18°C for a bedroom or lecture room to 16°C in passageways. It is related to the mean radiant, air and resultant temperatures and the air velocity v m/s by:

$$t_c = t_{res} = \frac{t_r + t_{ai}\sqrt{10v}}{1 + \sqrt{10v}}$$

In normally occupied rooms the air temperature and the mean radiant temperature should be within a few degrees of each other. The air velocity within a habitable space should be barely discernible, in the region of 0.1 m/s. Under these conditions the dry resultant temperature at the centre of the room is

$$t_c = t_{res} = 0.5t_{ai} + 0.5t_m$$

There should be no significant effect upon comfort conditions when t_c is within 1.5°C of its design value; this allows for some flexibility in the actual value of the mean radiant and air temperatures and the air velocity due to the occupants changing position within the room, or to weather variations.

Comfort criteria

The main comfort criteria for sedentary occupants in buildings in climates similar to that of the British Isles are as follows.

1. Operative temperature should be in the range 19–23°C depending on room use.
2. A feeling of freshness is produced when the mean radiant temperature is slightly above air temperature. A significant amount of radiant heating is needed in order to achieve this.

1.10 Sedentary comfort zone.

3. Air temperature and the mean radiant temperature should be approximately the same. Large differences cause either radiant overheating or excessive heat loss from the body to the environment, as would be experienced during occupation of a glasshouse through seasonal variations.
4. Percentage saturation should be in the range 40–70%.
5. Maximum air velocity at the neck should be 0.1 m/s for a moving-air temperature of 20°C d.b. Both hot and cold draughts are to be avoided. Data are available for other temperature and velocity combinations.
6. Variable air velocity and direction are preferable to unchanging values of these variables. This is achieved by changes in natural ventilation from prevailing wind, movement of people around the building, on–off or high–low thermostatic operation of fan-assisted heaters or variable-volume air-conditioning systems.
7. The minimum quantity of fresh air for room use that will remove probable contamination is 10 l/s per person.
8. Mechanical ventilation systems should provide at least 4 air changes/h to avoid stagnant pockets and ensure good air circulation.
9. Incoming fresh air can be filtered to maintain a clean dust-free internal environment.
10. The difference between room air temperatures at head and foot levels should be no more than 1°C.
11. Ventilation air quantity can be determined by some other controlling parameter: for example, removal of smoke, fumes or dust, solar or other heat gains and dilution of noxious fumes.

Figure 1.10 shows a comfort zone, and it can be seen that the data for the previous example produce an acceptable office environment.

Experimental work

Various measurements of the thermal environment can be made using relatively simple equipment, and some suggestions for practical projects are listed here. It is most important to take great care with observation of the conditions being measured. Record all the relevant details: time of day, date, external weather, temperatures, air speeds, solar radiation, number of people, clothing, activity level, heating and cooling room terminal units and outlets and, most important of all, where the data measurements are being taken. Take digital photographs of the measuring instruments and everything done, if the occupants allow; permission is very important. Do not take all the readings near a draughty window and then claim that the results show average room conditions.

1. Effects of room air temperature gradients on comfort: measure floor-to-ceiling air temperature gradients with a multi-channel thermocouple probe and discuss their effects on the comfort conditions encountered.
2. Room air velocity distribution: produce velocity contours for different parts of the room and compare with observed comfort conditions.
3. Measure the Kata cooling power and air velocity, the globe temperature and the dry-bulb air temperature. Compute the other temperature scales and compare with subjective assessments of the conditions.
4. Make and test a thermocouple temperature-measuring circuit with copper–constantan wire and an accurate potentiometer. Build junctions into walls and building structures, plot their temperature profiles and calculate their thermal transmittances.

Questions

Descriptive answers must be made in your own words based on a thorough understanding of the subject. Copied work displays a noticeable discontinuity of style between your own production and the reference material. Also, the work has not been fully comprehended if it can only be answered by copying. The ability to pass on reliable and concise information is a vitally important part of business and government work, and you should realize that as much practice and experience as possible is needed to become an effective communicator.

1. List and discuss the factors affecting thermal comfort.
2. List the factors involved in making buildings work in terms of sustainability.
3. Define sustainability in relation to the construction and use of habitable buildings.
4. Write notes on what you understand by the word sustainability for building services engineers and users of services installations.
5. Discuss the following statement: 'a building acts as an environmental filter'.
6. State how extremes of heat and cold affect the workers on a site, what environmental measurements can be taken, and the corrective actions possible to ensure safe and healthy working conditions.
7. Describe how an investigation into the thermal comfort provided in a building can be measured. State which instruments would be used and give reasons.
8. Describe with the aid of sketches how each of the following instruments functions: dry-bulb thermometer, wet-bulb thermometer, globe thermometer, Kata thermometer, vane anemometer, thermocouple and infrared scanner.
9. Describe how heating, cooling and humidity control systems interact with the building to provide a comfortable environment.
10. A survey of a heated room revealed the information given in Table 1.6. The dry-bulb air temperature is 19°C. Estimate the mean radiant and environmental temperatures and assess the thermal comfort conditions for sedentary occupation. State any remedial measures that may need to be taken.

Table 1.6 Data for Question 10.

Surface	Area (m²)	Average surface temperature (°C)
Window	7	0
Outside	20	10
Inner wall	28	17
Ceiling	30	23
Floor	30	12

11. A hot-water central heating system has a radiator of surface area 4.5 m² and a surface temperature of 76°C. It faces a person 2.25 m away whose body surface area and temperature are 1 m² and 21°C respectively. The surface emissivities of the radiator and the person are 0.85 and 0.90 respectively. Calculate the radiant heat transfer.

12. A radiant panel at high level in an industrial facility has a surface area of 20.0 m² and an average surface temperature of 165°C. It faces a steel work bench 3.2 m directly below whose surface area and temperature are 8 m² and 26°C respectively. The surface emissivities of the radiator and the bench are 0.85 and 0.90 respectively. Calculate the radiant heat transfer. What danger may be produced?

13. A gas-fired radiant tube overhead heating system has a surface area 20.0 m² and a surface temperature of 225°C. Workers are 3.5 m below. Each person has an upward projected surface area and temperature of 0.25 m² and 25°C respectively. The surface emissivities of the radiant tube and the person are 0.90. Calculate the radiant heat transfer.

14. A room is heated with a hot-water panel radiator and the data shown in Table 1.7 were found during site measurements. Estimate the mean radiant temperature of the room at the centre of its volume.

Table 1.7 Data for Question 14.

Surface	Average surface temperature (°C)	Area (m²)
Window	8	8
Radiator	70	4
Outside wall	12	20
Inner walls	17	24
Ceiling	24	20
Floor	11	20

15. A refrigerated food cold room is maintained at −4°C with ammonia refrigerant fan coil units. Site measurements found the data given in Table 1.8. Estimate the mean radiant temperature of the room at the centre of its volume.

Table 1.8 Data for Question 15.

Surface	Average surface temperature (°C)	Area (m²)
Doors	5	10
Walls	−1	310
Ceiling	3	450
Floor	2	400

16. An hotel lounge remains at 24°C with air conditioning. Measurements found the data given in Table 1.9. Estimate the mean radiant temperature of the room at the centre of its volume.

Table 1.9 Data for Question 16.

Surface	Average surface temperature (°C)	Area (m²)
Window	12	20
Outside wall	14	90
Inner walls	18	110
Ceiling	26	150
Floor	12	150

17. A building site in Manchester was exposed to a wind velocity measured at 5.2 m/s and the sling psychrometer air temperature showed –2°C d.b. Find the wind chill index and the equivalent wind chill temperature.

18. An exposed hill top has the air velocity measured at 15 km/h, 4.2 m/s and the sling psychrometer air temperature showed -1.5°C d.b. Find the wind chill index and the equivalent wind chill temperature.

19. An underground tunnel has an air velocity of 45 km/h (12.5 m/s) during full traffic flow in a European city in winter. The sling psychrometer air temperature showed -1.2°C d.b. Find the wind chill index and the equivalent wind chill temperature.

20. Compare the wind chill index and the equivalent wind chill temperature on two sites operated by workers wearing similar clothes.
Site A: air velocity, 5 m/s; air temperature, 2.0°C d.b.
Site B: air velocity, 0.6 m/s; air temperature, -10°C d.b.

21. A scaffold erector works in air at -5°C d.b. and 20% saturation with a metabolic rate of 300 W/m². The wind velocity gusts up to 2.5 m/s. Given that the atmospheric vapour pressure is 0.8 mb, calculate the heat stress index produced. Assume that $t_r = t_a$.

22. Site work is being conducted in bright sunshine when the air peaks at 35°C d.b. and 20% saturation at 3 p.m. A worker's metabolic rate is 280 W/m². Atmospheric vapour pressure is 12 mb and the air speed is 0.8 m/s. Calculate the heat stress index produced and comment on the conditions. Assume that $t_r = t_a$.

23. A construction surveyor is exposed to conditions of 35°C d.b., 22°C w.b. and 22% saturation in the sunshine in the Middle East. His metabolic rate is 300 W/m² while moving and using instruments. Local air movement is 0.2 m/s. Find the heat stress index and the allowed exposure time. Assume that $t_r = t_a$. Atmospheric vapour pressure p_a is 17.56 mb.

24. An air-conditioning unit maintenance technician works on roof-mounted fan coil units in a hot climate. Outside air conditions are 40°C d.b., 25.7°C w.b. and 30% saturation in the sunshine. His metabolic rate is 340 W/m² while working and climbing external ladders. Local air movement is 15 km/h (4.2 m/s). Find the heat stress index and the allowed exposure time. Assume that $t_r = t_a$. Atmospheric vapour pressure p_a is 23.3 mb.

25. An air-conditioning ductwork technician works indoors on a building site in a hot humid climate. There is no local cooling or mechanical ventilation available during construction work. Outside air conditions are 36°C d.b., 31.2°C w.b. and 70% saturation. Indoor conditions are about the same as outdoor air as walls are incomplete. His metabolic rate is 360 W/m² while working out of the sunshine. Local air movement is 0.3 m/s. Find the

heat stress index and the allowed exposure time. Assume that $t_r = t_a$. Atmospheric vapour pressure p_a is 42.34 mb.

26. Measurements in an office showed that the general air movement amounted to 0.25 m/s at a temperature of 23°C d.b. The globe temperature was 19°C. Calculate the mean radiant and environmental temperatures, and discover whether room conditions are within the comfort zone.

27. An open-plan office is designed for sedentary occupation and is to have general air movement not exceeding 0.2 m/s and an air temperature of 22°C d.b., in winter. It is expected that a globe temperature of 20°C would be found at the centre of the room volume. What would be the mean radiant, resultant, environmental and operative temperatures?

28. A lecture theatre is designed for sedentary occupation and is to have general air movement not exceeding 0.5 m/s and an air temperature of 21°C d.b., in winter. It is expected that a globe temperature of 18°C would be found at the centre of the room volume. What would be the mean radiant, resultant, environmental and operative temperatures?

29. A conference room is designed for sedentary occupation and is to have general air movement not exceeding 0.35 m/s and an air temperature of 24°C d.b., in summer. It is expected that a globe temperature of 21°C would be found at the centre of the room volume. What would be the mean radiant, resultant, environmental and operative temperatures?

30. List the temperatures used to describe comfort conditions and explain how they are relevant for room users.

31. Survey the factors affecting thermal comfort and explain what they mean.

32. Explain how a building filters the external climate.

33. Explain the meaning of the terms olf and decipol and state how they are used in the design of ventilation systems.

34. Which of these is where indoor odours, vapours and gases come from? More than one correct answer.

    1. The air-conditioning systems.
    2. Ingress from outdoor environment.
    3. The basement car park of the building.
    4. Humans, animals, plants and furnishings within the building.
    5. Dust, pollen and materials in waste paper bins.

35. Which of these is where indoor odours, vapours and gases come from? More than one correct answer.

    1. Cleaning fluids used overnight.
    2. New furniture, carpets, floor coverings, sealants and adhesives.
    3. Old furniture, carpets and floor coverings.
    4. Personal hygiene products.
    5. Cigarette smoke, diesel engine exhaust, road tar, painting work being done and creosote used on roofing.

36. Which of these is where indoor odours, vapours and gases come from? More than one correct answer.

    1. Radon gas emanating from the ground beneath the building.
    2. Carbon monoxide from traffic.
    3. People, our clothes and what we put on our skin.
    4. Passively acquired cigarette smoke prior to entry into the office building.
    5. Last night's spicy meal.

37. Are any of these correct for biological effluent? More than one correct answer.

    1. Is too complicated to be measured.
    2. Comes from many sources within the working environment.
    3. From one office worker in a 10.0 m²working space is standardized at 1.0 olf.
    4. Is counteracted by plants within the occupied building, particularly with open atria.
    5. We walk into the building with odours on our clothes.

38. Are any of these incorrect for biological loading? More than one correct answer.

    1. From one office worker is around 0.10 olf/m².
    2. Of a smoker while smoking is around 25.0 olf/m².
    3. Of a smoker when not smoking is around 6.0 olf/m².
    4. Within a gymnasium in use is around 11.0 olf/m².
    5. Within a low-pollution office building with an absence of smoking is around 0.20 olf/m².

39. Satisfactory air quality may be deemed when:

    1. 100% of the full-time occupants are satisfied.
    2. 85% of the full-time occupants are satisfied.
    3. 50% of the full-time occupants are satisfied.
    4. Complaints cease.
    5. Odours have been eliminated.

40. Which of these are correct for excellent air quality in a building? More than one correct answer.

    1. May need very high room air change rates.
    2. May need outside air to be collected from the roof of a tall city centre building.
    3. May be unachievable where the building is located in a polluted outdoor industrial environment.
    4. Can be improved with air filtering equipment.
    5. Is mainly impractical due to its high cost.

41. State two uses for the Kata thermometer.
42. Where are thermocouples used? What are their advantages?
43. How can heat leakages due to inadequate thermal insulation and damaged pipes or cables be detected?
44. What is the function of environmental temperature?
45. A high-temperature gas-fired radiant heater is used to provide warmth for site workers. The heater has a red-hot area of 300 mm × 500 mm at a temperature of 700°C. A worker of surface area 1 m² and temperature 12.0°C is 2.5 m from the heater. Surface emissivities are 1.0 for the heater and 0.9 for the worker. Calculate the radiant heat transfer to the worker.
46. State the factors that are taken into account when designing for the provision of ventilation with outdoor air.
47. Write a technical report on the aspects of the provision of outdoor air for the ventilation of the following:

    1. commercial building that has offices, retail shops and a pedestrian atrium;
    2. an underground high-security manufacturing and storage facility for nuclear materials;
    3. a large open-plan metal-fabrication factory;
    4. a college or university;

5. an hotel;
6. a sports centre that has swimming, weight training, aerobics, racquet courts and restaurant facilities.

48. Explain the meaning of the terms olf and decipol and state how they are used in the design of ventilation systems.

49. List the atmospheric pollutants that are likely to be present within normally occupied buildings. Identify those pollutants that are used for the design of the ventilation system, the filtration equipment, acoustic insulation and general maintenance during occupation.

50. A 10-storey office building is located in Birmingham city centre alongside a highway that has continual, heavy traffic density. The building has 30 m of road frontage, is 20 m deep and each floor is 3 m high. The occupancy averages one person for each 10 m^2 of floor area. Calculate the outdoor air ventilation rate, from the Fanger method, that is required to satisfy 85% of the occupants of an office where it is expected that none of them are smokers. State the recommended location for the fresh air intake and discharge louvres. The office occupants can be taken as mainly sedentary. The interior has a low-pollution load of around 0.20 olf/m^2.

51. Calculate the outdoor air ventilation rate, from the Fanger method, that is required to satisfy 80% of the occupants of an office where it is expected that none of them are smokers. It is located in a county area close to a rural town in the UK.

52. Calculate the outdoor air ventilation rate, from the Fanger method, that is required to satisfy 95% of the occupants of an office where it is expected that none of them are smokers. It is located in a city in the UK.

53. Calculate the outdoor air ventilation rate, from the Fanger method, that is required to satisfy 50% of the occupants of an office where it is expected that none of them are smokers. It is located in a county area close to a rural town in the UK.

54. A new conference room is being designed for an inner city in the UK. The room is to be 20 m long, 12 m wide and 4 m high. Windows are not openable. The conference room is to seat 30 adults. There will not be any smoking and all the furnishings and building materials will have the low-pollution emission of 0.1 olf/m^2. Calculate the outdoor air ventilation rate, from the Fanger method, that will be required so that 90% of the occupants will be satisfied.

55. A new office area for 8 workstations in a commercial building in the City of Southampton is to be 10 m long, 8 m wide and 3 m high. There will not be any smoking and all the furnishings and building materials will have the low-pollution emission of 0.1 olf/m^2. Calculate the outdoor air ventilation rate, from the Fanger method, that will be required so that 80% of the occupants will be satisfied.

56. A lecture theatre is being designed for a rural campus. The theatre is to be 22 m long, 16 m wide and 4.5 m high. It has no exterior windows. Potential occupancy is 100. There will not be any smoking and all the furnishings and building materials will have the low-pollution emission of 0.2 olf/m^2. Calculate the outdoor air ventilation rate, from the Fanger method, that will be required so that 50% of the occupants will not complain. Is this a suitable design?

57. An open-plan office is being designed for a city centre development. Each floor is 50 m long, 25 m wide and 3.5 m high and is to have 100 workstations for the occupants. Use the data from Table 1.2 to evaluate the outdoor ventilation air requirement from the different authorities for this application for one floor. What is the maximum room air change rate that could be produced?

58. A new lecture theatre is being designed for the Bournemouth City University. The theatre is to be 25 m long, 12 m wide and 5 m high. It has no exterior windows. The peak occupancy can be 125 and can be considered to be sedentary. There will not be any smoking, the

building has a low-pollution load of 0.1 olf/m^2 and each occupant creates a load of 1 olf. The external atmosphere is considered to have a low-pollution count. Calculate the outdoor air ventilation rate, from the Fanger method, that will be required so that 90% of the occupants will be satisfied. Recommend how the outdoor air ventilation system could be controlled for energy economy and comfort.

59. Discuss the difference between the Fanger method of assessing the provision of outdoor air with that based on the supply of a fixed quantity, such as 10 l/s per person. State the implications for the energy used in heating and cooling the supply of outdoor air for both systems.

60. Explain, with the aid of sketches, how the ventilation design can be arranged to minimize the use of outdoor air in the provision of acceptable indoor air quality and temperature control for human thermal comfort. The reader can make use of Chapter 5 in answering this question.

61. New instruments for measuring and recording environmental variables such as air temperature, velocity, relative humidity, air quality and thermal comfort become increasingly available. Conduct a search for such devices and write a brief description for the benefit of assessing their relative usefulness to your own work. You may wish to limit the range of equipment to just one type, such as air temperature measurement, in compiling the information for reporting. Make a recommendation for purchase, with reasons, as if it were for your own organization to use.

62. Explain with practical examples why continuous logging of environmental variables is useful to the engineer. You may decide to concentrate on one variable and provide data from a logging instrument to demonstrate its usefulness.

63. State why continuous logging is of value to the energy audit engineer, environmental system design engineer, building designer and building occupants, giving reasons for your statements.

64. Which of these correctly describes how heat transfers within buildings? More than one answer is correct.

    1. Radiation through the concrete floor.
    2. Convection currents within room air.
    3. Conduction between the occupants and the surfaces of the building.
    4. Conduction through solid building materials.
    5. Radiation across a wall cavity when there is aluminium foil-faced sisalization.

65. What is included in *WBGT*?

    1. It does not exist.
    2. Water basic (demand) and gross (heat) transfers for a building.
    3. Wet-bulb temperature, globe temperature and dry-bulb air temperature.
    4. Wet-bulb gradient temperature.
    5. Wet-bulb ground temperature

66. Where does poor indoor air quality come from? More than one correct answer.

    1. Toxic substances that occupants bring into the building by hand or on their clothing.
    2. Inward leakage of outdoor air.
    3. Insufficient fresh air ventilation quantity.
    4. Volatile organic compounds released into the building air from furnishings, cleaning fluids and electro-mechanical equipment.
    5. Lack of adequate air filtration systems.

67. Thermal comfort zone is:

    1. Where everyone is satisfied.
    2. All data falls within one standard deviation of the ideal conditions.
    3. Where the predicted mean vote of all occupants fall within the slightly cool to slightly warm band.
    4. When the environmental temperature is correct for the application.
    5. When nobody complains.

68. Which are correct about cold outdoor climates? More than one correct answer.

    1. Wind chill index found from local air velocity and dry-bulb air temperature.
    2. Frost bite occurs at a wind chill index of 900.
    3. Frost bite can happen at any negative Celsius air temperature.
    4. Frost bite should be avoided if wind chill index, WCI, is less than 1400 at an air temperature of –10°C d.b. during a 30 min exposure.
    5. Adequate clothing always avoids frost bite.

69. Which does not apply to heat stroke in a hot climate?

    1. To avoid it, get into a swimming pool.
    2. Occurs at a body temperature of 40.6°C.
    3. Sweating ceases.
    4. Body becomes involuntarily hyperactive.
    5. Body becomes comatose, brain damage from reduced blood supply and death is imminent.

70. Which is a correct meaning for comfort criteria?

    1. 95% of occupants are satisfied with the dry-bulb air temperature.
    2. Mean radiant temperature just below the air dry-bulb temperature in winter.
    3. Dry resultant temperature in the range 19–23°C.
    4. Dry resultant temperature in the range 24–26°C in summer.
    5. Dry-bulb air temperature in the range 15–26°C depending upon room application and season.

71. Which is not an appropriate statement for moving air comfort criteria?

    1. Variable air velocity and temperature is preferred.
    2. Varying the air velocity during the occupied day is impossible.
    3. Varying air velocity during the working day may require the supply air fan to have a variable-speed control.
    4. Low-energy buildings with natural ventilation systems have air velocity variations due to changes in prevailing winds.
    5. Low-energy buildings may have active systems to vary air movement around occupants.

72. Which is the function of the human body thermoregulatory system?

    1. Maintains comfort when heat gains to the body exceed its ability to lose heat.
    2. Provides alarm signals to prompt appropriate response.
    3. Stops blood temperature rising too high.
    4. Averages heat gains and losses between extremities to maintain comfort.
    5. Attempts to maintain energy balance and maintain 37°C core temperature.

73. What was the earlier name for operative temperature regarding comfort assessment?

    1. Wet resultant temperature.
    2. Wet-bulb globe temperature.
    3. Dry resultant temperature.
    4. Environmental temperature.
    5. Dry-bulb air temperature.

74. In well-insulated buildings having modest glazing areas and little air movement, which will operative temperature be closest to?

    1. Globe temperature.
    2. Mean radiant temperature.
    3. Wet-bulb temperature.
    4. Dry-bulb air temperature.
    5. Environmental temperature.

75. Which of these working environments may lead to construction worker heat stress?

    1. High radiant heat load, outdoor work, outdoor air 25°C d.b. and low humidity.
    2. Indoor work in confined space, outdoor air 24°C d.b., high humidity and adequate ventilation.
    3. Clear blue sky, outdoor air 35°C d.b., 20°C w.b., little wind, all work conducted in a deep underground mining tunnel.
    4. Clear blue sky, outdoor air 29°C d.b., high humidity, little wind, all work conducted beneath concrete floor slabs with no perimeter walling.
    5. Completely cloudy sky, high air dry-bulb temperature, strong wind from inland and high humidity.

2 Energy economics

Learning objectives

Study of this chapter will enable the reader to:

1. understand the basis of energy auditing and design an energy audit;
2. be able to calculate energy costs for all applications and different types of measuring unit within SI (Système International);
3. use multiples and submultiples of SI units;
4. identify usefully employed energy;
5. evaluate energy costs in pounds per useful gigajoule;
6. understand degree days and their use;
7. compare the energy efficiency of different buildings;
8. calculate the economic thickness of thermal insulation for both hot surfaces and building structures;
9. analyse financial return on investment in energy-saving measures;
10. calculate energy use targets for buildings;
11. calculate greenhouse gas carbon emission.

Key terms and concepts

annual energy 39; capital repayment period 49; carbon emission 42; cash flow 52; cost per useful therm and gigajoule 37; degree day 44; economic thickness 49; electrical target 53; energy audit 32; energy target 53; energy use performance factor 34; greenhouse gas 42; gross calorific value 36; liquefied petroleum gas 40; load factor 46; multiples and submultiples of units 35; overall efficiency 38; percentage return on investment 52; therm 35; thermal storage of a building 38; thermal target 53; total energy demand target 53; useful energy 37; use of gigajoules (GJ), kilowatts (kW), kilojoules (kJ), kilowatt-hours (kWh), megajoules (MJ) 35.

Introduction

Buildings are such major consumers of primary energy, that is, coal, oil, gas, nuclear energy and renewable sources such as hydroelectric, wind, solar and wave energy, that accounting for its use and calculating the consequent financial implications are of paramount importance to all who are involved in building design and operation.

Logical methods of dealing with the calculations are introduced to enable the new user to cope with the complex conversion equations and calculation of energy costs per standard unit, the annual energy cost and the economic thickness of thermal insulation.

Degree days are explained and their use as an accounting tool is explored. The financial implications of purchasing or leasing energy-saving hardware equipment are investigated. Energy demand targeting is outlined. Greenhouse gas emissions to the atmosphere are explained and calculated.

Energy audit

Management of the energy that is used for buildings has three major components:

1. initial design;
2. retrofitting energy-saving measures;
3. maintenance practices.

Design engineers should provide heating, ventilating and air-conditioning systems that consume the minimum amount of fossil fuel energy in satisfying the needs of the site. The initial installation is designed in conjunction with the architecture and the client's requirements, in accordance with statutory legislation and in compliance with the standards of good engineering practice. The design includes the means of controlling the use of energy. Control can mean anything from switching building services plant on and off manually, up to a fully automated computer-based system that gives audible and visual alarms when something goes wrong. The best efforts of the design engineer are limited by the initial construction cost of the new installation, which is usually minimized. The installation of energy-saving systems often increases this cost. Some buildings are designed to be low-energy users. Most building services engineering systems are designed to provide thermal comfort for the conditions that are found in the building. For example, perimeter heating to overcome the heat loss through large areas of glazing and thermostatically controlled ventilation louvres in a naturally ventilated building in a cool climate.

Energy-saving measures that are installed after the first few years of use of a new building, or during a major upgrade, can be justified for two primary reasons. Either the owner of the building has decided to refurbish it, and has found the capital funding that is needed for all the work, or the operating cost of the site is significantly greater than comparable facilities, and the owner or tenant is prepared to invest in measures that will reduce annual outgoings. The owner may be forced to provide lettable office space that has competitive energy and maintenance costs. A building that has a labour-intensive maintenance and supervision workload, from steam boilers, manual switching of mechanical plant and lighting systems, unreliable water chillers, poorly maintained closed-circuit water conditions and highly stressed belt drives on fans, is not attractive to a new user.

Many sites have maintenance practices that encourage the provision of breakdown repairs and replacements, rather than by preventing breakdowns through good-quality methods. The financial controller of the business may view the annual maintenance budget as expendable, through a lack of understanding about engineering equipment. This is understandable. The maintenance engineer has only to ask the finance director whether it is preferable for a company car to be taken for regular servicing or to wait for the car to break down on a motorway during inclement weather, because the engine has run out of oil, the engine cooling system has boiled dry and the brake pads have worn down to the metal! This is how the maintenance budget of some sites is managed, that is, breakdown maintenance only. Many building services are critical to the life safety of the users. These life-safety systems are not just the emergency exit lighting, smoke spill ventilation fans, stairway air pressurization fans and electrical earth leakage circuit-breakers, but also include the air conditioning to hospital operating theatres, lifts, outside air ventilation dampers, domestic hot-water and cooling-tower bacteria controls. Proficient maintenance practice helps to prevent breakdowns by:

1. monitoring the condition of plant;
2. optimizing the maintenance activity to replace items only when they are needed;
3. keeping the maintenance team well motivated;
4. planning expenditure;
5. comprehensive maintenance record keeping;
6. enabling a quick response to problems, such as the failure of a fan motor, before the tenants complain of experiencing poor quality air conditions. The building maintenance manager usually has about half an hour from when an air-conditioning fan ceases to function to when the tenants complain on the telephone. If the plant failure has been monitored through the building management system computer with audible and visual alarms, and an automatically sent message to the engineer's pager or mobile phone, the corrective response can be made within 5 min and the tenants provided with a briefing.

The energy audit engineer assesses the practical and financial viability of energy-saving measures for each site, as is appropriate. The purpose of the energy-saving analysis is to identify suitable investments in capital equipment that will reduce the use of energy and labour, so that the savings will provide a payback on the investment in a reasonable period. This period will vary from 1 year, for those only interested in this year's profits, to 3 years for those who rely upon their bank for capital funding, to 5 years for those who can source capital funds from an equity performance contracting partner, to the longer terms of 10–25 years when the user is a government department and is to retain ownership of the public buildings indefinitely. The retrofit energy-saving measures that are usually considered include the following:

1. thermal insulation of the building;
2. solar shading;
3. changing the fuel source for heating and cooling;
4. heat pumps;
5. heat reclaim;
6. cogeneration of electricity with heating or cooling;
7. computer-based building management system;
8. digital control refrigerant circuit of the water chiller;
9. hot-water, chilled-water or ice thermal storage;
10. load shedding large electrical loads at critical times for short periods;
11. energy tariff change;
12. reducing the lighting system power usage;
13. variable-speed drives of fan and pump motors;
14. reducing the usage of water by taps and in toilets;
15. economy air recycling ductwork and motorized damper controls;

16. air-to-air heat exchange between exhaust and incoming outside air ducts;
17. occupancy-sensing with infrared, acoustic or carbon dioxide sensing to control lighting and the supply of outside air;
18. air curtains at doorways;
19. oxygen sensing in the boiler flue gas to modulate the combustion air supplied to the burner;
20. replacement of old inefficient boilers and heating systems;
21. distribution of domestic hot water at 45°C with a mixing valve and temperature control;
22. replacement of steam-to-water heat exchangers and calorifiers with local gas-fired heating and domestic hot-water systems;
23. thermal insulation of heating, cooling and steam pipework and heat exchangers;
24. recovery of the maximum quantity of condensate in a steam distribution system;
25. replacement and overhaul of steam traps and condensate pumping.

An energy audit of an existing building or a new development is carried out in a similar manner to a financial audit but it is not only money that is accounted. All energy use is monitored and regular statements are prepared showing final uses, costs and energy quantities consumed per unit of production or per square metre of floor area as appropriate. Weather data are used to assess the performance of heating systems. Monthly intervals between audits are most practical for building use, and in addition an annual statement can be incorporated into a company's accounts.

Payne (1978) is useful for further reading. An initial energy audit has certain basic aims. To:

1. establish total costs of energy purchased;
2. locate the principal energy-consuming areas;
3. notice any obvious losses or inefficient uses of heat, fuel and electricity;
4. take overall data to gain initial results quickly, which can be refined later and broken down into greater detail;
5. find where additional metering is needed;
6. take priority action to correct wastage;
7. survey buildings and plant use at night and weekends as well as during normal working hours;
8. initiate formal records monitored by the energy manager;
9. compare all energy used on a common basis (kilowatt-hours, therms or megajoules);
10. list energy inputs and outputs to particular buildings or departments.

A vital part of auditing is enlisting the cooperation of all employee groups, and explaining the problem not just in financial terms but also in quantities of energy. A joint effort by all staff is needed. Posters, stickers and prizes for ideas can be used to stimulate interest.

An overall energy audit will list each fuel, the annual quantity used and the cost for the year, including standing charges and maintenance; then a comparison is made with other fuels by converting to a common unit of measurement.

Energy use performance factors (*EUPF*) enable comparisons to be made between similar buildings or items of equipment. These can be litres of heating oil per degree day, kilowatt-hours of electricity consumption per square metre of floor area, megajoules of energy per person per hour of building use, or other accounting ratios as appropriate. For example, car manufacturers may analyse energy used per car. As experience is gained in auditing a particular building, data can be refined to monthly energy use in conjunction with degree day figures for this period.

This detailed analysis can be made for each building or department of a large site, each large room or factory area, each type of heating, air-conditioning or power-using system, each industrial process and each item of plant. The most serious deficiency in the acquisition of data is likely to be the lack of sufficient metering stations. Electricity, gas and other fuels are metered by the supply authority at the point of entry to the building or site; further metering is the responsibility of the site user. Frequently, no further meters are installed and capital expenditure is needed to obtain data. A careful cost benefit approach is required to assess the viability of this equipment (Moss, 1997).

Unity brackets

In order to deal with the numbers involved, a degree of familiarity with the units and conversion between the various types is needed. A handling technique known as the unity bracket helps to avoid errors being made when dealing with unfamiliar combinations of units. Suppose that we wish to convert 1260 mm into metres.

$$\text{length} = 1260 \text{ mm}$$

Now, 1000 mm $\equiv$ 1 m. Divide each side by 1 m: thus

$$\frac{1000 \text{ mm}}{1 \text{ m}} \equiv 1$$

The left-hand side is now a unity bracket exactly equal to 1 (or unity). Similar unity brackets can be formed for any suitable conversion problem. Now, multiply length by the unity bracket:

$$\text{length} = 1260 \text{ mm} \times \frac{1 \text{ m}}{1000 \text{ mm}}$$

$$= 1.26 \text{ m}$$

Notice that the unity bracket is arranged so that its denominator units cancel the original units. A long chain of conversions can easily be handled and the method avoids errors of logic that can occur if an attempt is made to cope with the problem using mental arithmetic.

EXAMPLE 2.1

British Gas sold gas by the therm up to 1992. Harmonization of the units of measurement in Europe caused the change to kWh units. If 1 therm is equal to 105.5 MJ, how many kilowatt-hours are there in 1 therm?

$$1 \text{ therm} = 105.5 \text{ MJ}$$

$$1 \text{ MJ} = 10^6 \text{ J}$$

$$1 \text{ h} = 3600 \text{ s}$$

$$1 \text{ Watt} = 1 \text{ J/s}$$

$$1 \text{ kWh} = 1 \text{ kWh} \times \frac{3600 \text{ s}}{1 \text{ h}} \times \frac{10^3 \text{ W}}{1 \text{ kW}} \times \frac{1 \text{ J}}{1 \text{ Ws}} \times \frac{1 \text{ MJ}}{10^6 \text{ J}} \times \frac{1 \text{ therm}}{105.5 \text{ MJ}}$$

$$= \frac{3600 \text{ s} \times 10^3}{10 \times 105.5}$$

$$= 0.0341 \text{ therms}$$

Therefore,

$$1 \text{ therm} = \frac{1}{0.0341} \text{ kWh}$$

$$= 29.3056 \text{ kWh}$$

Gross calorific value of a fuel

The total heat energy content of a fuel is known as the gross calorific value (GCV) and is usually expressed in megajoules per kilogram (MJ/kg).

EXAMPLE 2.2

Heating oil has a specific gravity of 0.84 and a GCV of 44.8 MJ/kg. Find its heat content in kWh per litre.

$$GCV = 44.8 \frac{\text{MJ}}{\text{kg}} \times \frac{0.84 \text{ kg}}{1 \text{ l}} \times \frac{10^3 \text{ kJ}}{1 \text{ MJ}} \times \frac{1 \text{ kWs}}{1 \text{ kJ}} \times \frac{1 \text{ h}}{3600 \text{ s}}$$

$$= \frac{44.8 \times 0.84 \times 10^3}{3600} \text{ kWh/l}$$

$$= 10.453 \text{ kWh/l}$$

EXAMPLE 2.3

A site uses 48 000 l of oil of GCV 44.0 MJ/kg and specific gravity 0.84 costing £22 500 and 2×10^6 kWh of electricity costing 7.2 p per kWh. The fixed charge for the electrical installation is £3500 and the servicing cost for the oil-fired heating system is £8500. The period of use being considered is 1 year. Draw up an overall energy audit for the year.

$$\text{oil } GCV = 44.0 \frac{\text{MJ}}{\text{kg}} \times \frac{0.84 \text{ kg}}{1 \text{ l}} \times \frac{10^3 \text{ kJ}}{1 \text{ MJ}} \times \frac{1 \text{ kWs}}{1 \text{ kJ}} \times \frac{1 \text{ h}}{3600 \text{ s}}$$

$$= 10.27 \text{ kWh/l}$$

$$\text{total cost of oil} = £22 500 + £8500$$

$$= £31 000$$

Table 2.1 Fuel cost data for Example 2.3.

Fuel	Quantity	Total cost (£)	kWh	Cost per kWh
Oil	48 000 l	31 000	492 960	6.29
Electricity	2×10^6 kWh	147 500	2 000 000	7.38
Total		178 500	2 492 960	7.16 (average)

$$\text{kWh of oil} = 48\,000 \text{ l} \times 10.27 \, \frac{\text{kWh}}{\text{l}}$$

$$= 492\,960 \text{ kWh}$$

$$\text{cost per kWh for oil} = \frac{£31\,000}{492\,960} \times \frac{100 \text{ p}}{£1}$$

$$= 6.29 \text{ p/kWh}$$

$$1 \text{ kWh} = 1 \text{ unit of electricity}$$

$$\text{total cost of electricity} = 2 \times 10^6 \text{ kWh} \times \frac{7.2 \text{ p}}{\text{kWh}} \times \frac{£1}{100 \text{ p}} + £3500$$

$$= £147\,500$$

$$\text{cost per kWh of electricity} = \frac{£147\,500}{2\,000\,000 \text{ kWh}} \times \frac{100 \text{ p}}{£1}$$

$$= 7.38 \text{ p/kWh}$$

$$\text{Average cost of all energy used} = \frac{£178\,500}{2\,492\,960 \text{ kWh}} \times \frac{100 \text{ p}}{£1}$$

$$= 7.16 \text{ p/kWh}$$

The data are shown in Table 2.1.

Energy cost per useful gigajoule

The supplying authority or company quotes energy costs in the unitary system most convenient to their industry, and there is no obvious method of comparing the real cost of providing a specific amount of useful heat or power in the building. A decision can be made to reduce all costs to a common base unit and this may be the therm, the kilowatt-hour or the gigajoule, where,

$$1 \text{ GJ} = 10^9 \text{ J} = 10^6 \text{ kJ} = 10^3 \text{ MJ}$$

If the overall efficiency of the energy conversion process is included in the unit cost, then the incurred cost of using that system of heat or power can be realistically assessed. The cost per useful kWh is the cost of providing 1 kWh of useful heat, or energy, at the place of use. The overall efficiency of a central heating system will include the following.

1. Combustion efficiency of the fuel. Regular maintenance is necessary with all fuel-burning appliances to ensure that the correct fuel-to-air ratio is maintained.

2. Heat transfer efficiency of the appliance. Both flue gas and water-side surfaces must be kept clean.
3. Heat losses from distribution pipework. All hot-water pipes and surfaces must be adequately insulated unless they provide a useful heating surface in rooms. It is not good practice to allow heating system pipes to be bare metal as the uncontrolled heat transfer will lead to high fuel costs.
4. Ability of the final heat emitter to transfer warmth to the occupants. A hot-water central heating radiator placed under a window-sill should counteract down-draughts and provide a reasonably adequate air temperature at the window and at the inner surface of the outside wall; this has the effect of increasing heat flows through the window and wall. Placing the radiator on a warm internal wall will improve its useful heat output but at the loss of some warm usable space in the room as the window region will be colder.
5. Thermal storage capacity of the heating system and building. Large amounts of heat are stored in the water in heating systems and the dense fabric of buildings. An insensitive automatic control system, or the lack of such a system, will lead to wild swings above and below the desired resultant temperature and cause excessive fuel use.

The estimated overall system efficiencies are listed in Table 2.2 from Uglow (1981).

The cost in use of a fuel or source of energy, UC, can be calculated from the basic price C and the overall efficiency η. For gas,

$$UC = C\,\frac{p}{kWh} \times \frac{100}{\eta} + \frac{\text{annual standing charge}}{\text{annual kWh}}$$

Thus once the total kWh used during the year are assessed, the annual standing charge can be apportioned to each kWh. This is not necessary if the standing charge has already been incurred by another use, for example, cooking and water heating, and UC is being evaluated for an additional heating system.

Table 2.2 Overall efficiencies of heating systems.

Fuel	Appliances	Overall efficiency η (%)
Electricity	Individual appliances	100
	Storage radiators	90
	Storage warm air	90
	Storage under floor	90
Gas	Individual appliances	55
	Boiler and hot-water radiators	65–70
	Ducted warm air	70–75
Solid fuel	Open grate fire	35
	Closed stove	60
	Boiler and hot-water radiators	60
Oil	Boiler and hot-water radiators	65–70
	Ducted warm air	70–75

Note: The higher figures relate to intermittent heating system operation.

EXAMPLE 2.4

A gas-fired central heating and hot-water system is to be installed in a residential property. The gas tariff is 1.8 p/kWh plus a standing charge of 10 p per day. The estimated annual heat energy that will be used by the occupants is 165 000 kWh. The annual maintenance works cost £125.Find the total energy bill and the average cost per kWh, for the year.

From Table 2.2 the overall efficiency of the heating system, η, is 70%.

$$\text{energy usefully consumed} = 165\,000 \text{ kWh/year}$$

$$\text{energy paid for at the gas meter} = 165\,000 \times \frac{100}{70} \text{ kWh/year}$$

$$= 235\,714 \text{ kWh/year}$$

The nearest whole number of kWh is the significant number. Do not waste time with too many decimal places. Fractions of a millimetre, Watt, Pascal or kWh are of little or no significance to the reality of the calculation.

$$\text{total cost of energy} = 1.8\,\frac{\text{p}}{\text{kWh}} \times 235\,714\,\frac{\text{kWh}}{\text{year}} \times \frac{£1}{100\,\text{p}}$$

$$= £4243$$

$$\text{annual standing charge} = 365\,\frac{\text{days}}{\text{year}} \times 10\,\frac{\text{p}}{\text{day}} \times \frac{£1}{100\,\text{p}}$$

$$= £37$$

$$\text{total annual energy bill} = £4243 + £36 + £125$$

$$= £4404$$

$$\text{overall average cost of energy} = \frac{£4404}{235\,714\text{ kWh}} \times \frac{100\,\text{p}}{£1}$$

$$= 1.87 \text{ p/kWh}$$

The cost in use, UC, for other fuel or energy sources is calculated from the basic price. These are pence per litre for oils, £ per tonne for solid fuels and £ per kg refill for liquefied petroleum gas. The specific gravity of liquid fuel, SG, is used to convert volume to mass measurements. When the specific gravity of an oil is 0.83, 1 l of the oil weighs 0.83 kg. One litre of water at 4°C weighs 1 kg. Oil and paraffin is sold by the litre but its heat content, gross calorific value GCV, is usually listed as around 45 MJ/kg. For oil

$$UC = C\,\frac{\text{p}}{1} \times \frac{1\,\text{l}}{SG\,\text{kg}} \times \frac{\text{kg}}{GCV\,\text{MJ}} \times \frac{100}{\eta} \times \frac{1\,\text{kJ}}{1\,\text{kWs}} \times \frac{3600\,\text{s}}{1\,\text{h}}$$

$$= \frac{C \times 100 \times 3600}{SG \times GCV \times \eta \times 10^3} \text{ p/kWh}$$

For solid fuel

$$UC = C \, \frac{p}{kg} \times \frac{kg}{GCV \ MJ} \times \frac{100}{\eta} \times \frac{1 \ MJ}{10^3 \ kJ} \times \frac{1 \ kJ}{1 \ kWs} \times \frac{3600 \ s}{1 \ h}$$

$$= \frac{C \times 100 \times 3600}{GCV \times \eta \times 10^3} \ p/kWh$$

For electricity

$$UC = C \, \frac{p}{kWh} \times \frac{100}{\eta}$$

$$= \frac{C \times 100}{\eta} \ p/kWh$$

Liquefied petroleum gas (LPG) (butane or propane) can be used from refillable cylinders on sites that are remote from the mains gas distribution of methane. LPG is sold at a refill charge per number of kilograms, depending upon the size of the cylinder. Convert the refill cost into a price per kilogram of LPG by dividing by its weight. For example, a 25 kg refill might cost £18.50:

$$UC = C \, \frac{p}{kg} \times \frac{kg}{GCV \ MJ} \times \frac{100}{\eta} \times \frac{1 \ MJ}{10^3 \ kJ} \times \frac{1 \ kJ}{1 \ kWs} \times \frac{3600}{1 \ h}$$

$$= \frac{C \times 100 \times 3600}{GCV \times \eta \times 10^3} \ p/kWh$$

The calculation for paraffin is the same as that for oils:

$$UC = \frac{C \times 100 \times 3600}{SG \times GCV \times \eta \times 10^3} \ p/kWh$$

EXAMPLE 2.5

Construct a table of fuel cost per useful kWh from the data given below, assuming that standing charges have already been allocated to other services and need not be included here.

Gas: 1.8 p/kWh, $\eta = 75\%$, $GCV = 38.5$ MJ/m^3
Heating oil: $SG = 0.84$, $GCV = 45.8$ MJ/kg, $\eta = 70\%$, 42.0 p/l
Anthracite: $GCV = 26.7$ MJ/kg, $\eta = 60\%$, 9.0 p/kg
Electricity: daytime 10.0 p/kWh, $\eta = 100\%$; night-time 4.5 p/kWh, $\eta = 90\%$
LPG (propane): 32.0 p/kg, $\eta = 70\%$, $GCV = 50$ MJ/kg
Paraffin: 29.0 p/l, $\eta = 80\%$, $GCV = 46.4$ MJ/kg, $SG = 0.79$

$$\text{gas cost} = 1.80 \, \frac{p}{kWh} \times \frac{100}{75}$$

$$= 2.4 \ p/kWh$$

$$\text{oil cost} = \frac{42.0 \times 100 \times 3600}{70 \times 0.84 \times 45.8 \times 1000}$$

$$= 5.6 \ p/kWh$$

$$\text{anthracite cost} = \frac{9.0 \times 100 \times 3600}{60 \times 26.7 \times 1000} \text{ p/kWh}$$

$$= 2.02 \text{ p/kWh}$$

$$\text{electricity (day)} = 10.0 \times \frac{100}{100} \text{ p/kWh}$$

$$= 10.0 \text{ p/kWh}$$

$$\text{electricity (night)} = 4.50 \times \frac{100}{90} \text{ p/kWh}$$

$$= 5.0 \text{ p/kWh}$$

$$\textit{LPG} \text{ (propane) cost} = \frac{32.0 \times 100 \times 3600}{70 \times 50.0 \times 1000} \text{ p/kWh}$$

$$= 3.3 \text{ p/kWh}$$

$$\text{paraffin cost} = \frac{29.0 \times 100 \times 3600}{80 \times 0.79 \times 46.4 \times 1000} \text{ p/kWh}$$

$$= 3.6 \text{ p/kWh}$$

Notice that the cost of energy in £ per useful GJ is likely to be within the range from £2 to £25 for the foreseeable future. £3 per GJ may apply to the lowest grades of solid and liquid fuels, which are used in large power-generating stations. £25 per GJ will be the upper limit for electricity consumed during the day by household consumers.

Deregulation of the electricity in the UK and Australia has led to falling prices for peak electricity. Natural gas, where it is available, remains the most popular means of generating electrical power, heating and cooling, owing to its cleanliness, convenience and lack of site storage requirement.

EXAMPLE 2.6

Evaluate the fuel cost in pounds per useful gigajoule and add them to Table 2.3.

Table 2.3 Summary of fuel costs.

Fuel	Cost per useful unit	
	Pence/kWh	£/GJ
Gas	2.4	7.14
Electricity (night)	5.0	13.9
Paraffin	3.6	9.9
Oil	5.6	15.6
Anthracite	2.02	5.61
LPG (propane)	3.3	9.14
Electricity (day)	10.0	27.78

$$\text{gas cost} = 1.80 \, \frac{p}{kWh} \times \frac{100}{70} \times \frac{1 \, kWs}{1 \, kJ} \times \frac{1 \, h}{3600 \, s} \times \frac{10^6 \, kJ}{1 \, kJ} \times \frac{£1}{100 \, p}$$

$$= 1.80 \times \frac{100}{70} \times \frac{10^6}{3600 \times 100} \, £/GJ$$

$$= 2.57 \times 2.778 \, £/GJ$$

$$= £7.14 \, \text{per GJ}$$

The 2.778 is the constant which, when multiplied by the fuel cost in pence per useful kWh, will give the equivalent in £/GJ.

$$\text{oil cost} = \frac{42.0 \times 100 \times 10^3}{70 \times 0.84 \times 45.8 \times 100} \, £/GJ$$

$$= £15.6 \, \text{per GJ}$$

$$\text{anthracite cost} = \frac{9.0 \times 100 \times 10^3}{60 \times 26.7 \times 100} \, £/GJ$$

$$= £5.61 \, \text{per GJ}$$

$$\text{electricity (day)} = \frac{10.0 \times 100 \times 10^6}{100 \times 3600 \times 100} \, £/GJ$$

$$= £27.78 \, \text{per GJ}$$

$$\text{electricity (night)} = \frac{4.5 \times 100 \times 10^6}{90 \times 3600 \times 100}$$

$$= £13.9 \, \text{per GJ}$$

$$\text{LPG (propane) cost} = \frac{32.0 \times 100 \times 10^3}{70 \times 50.0 \times 100}$$

$$= £9.14 \, \text{per GJ}$$

$$\text{paraffin cost} = \frac{29.0 \times 100 \times 10^3}{80 \times 0.79 \times 46.4 \times 100} \, £/GJ$$

$$= £9.9 \, \text{per GJ}$$

Greenhouse gas

The energy used by building services systems is one of the global sources of greenhouse gas production. Greenhouse gases in the atmosphere include the carbon dioxide emitted from combustion of fuels. Fuel is consumed within the building and in the provision of electricity for the building. Electricity is generated at power stations from a variety of primary energy sources: coal, oil, natural gas, nuclear fuel and renewables such as hydroelectric systems, wind turbines, wave-driven turbines and solar-powered photo-voltaic cells. There is not an electricity generation system that is free of greenhouse gas production. Where does nuclear fuel come from? Uranium 235-rich rock is mined from the ground in Canada and Australia by diesel-driven excavators, the rock is processed to separate out the silvery-white metal that is processed into fuel and transported, all of which consumes diesel fuel and electrical energy. Disposal of the spent nuclear fuel rods takes further energy for transportation, processing, mining, hundreds of years of storage with manual supervision and maintenance. Renewable energy systems require diesel energy for construction, concrete, steel, electrical energy when the wind or sunshine is not available, manual work for supervision, maintenance and replacement parts.

Table 2.4 Energy–CO_2 conversion factors.

Energy source	CO_2 conversion factor (kg C/kWh)
Natural gas	0.055
Oil	0.079
Coal	0.093
Electricity	0.142

The greenhouse gas, carbon dioxide, conversion factors for energy use are shown in Table 2.4 (Action Energy publication EEB006 Offices, appendix 1, figure A1.1). These are the quantity of carbon, supplied from the energy source, combusted and converted into carbon dioxide that is discharged into the atmosphere when the energy is used. These kilograms of carbon per kilowatt-hour of energy source consumed (kg C/kWh) factors usually are for the full cycle involved in acquiring the energy source, the energy used in its processing and the losses in distribution to the final user. For example, natural gas is heated at the land terminal from undersea gas fields, pumped to overcome friction in hundreds of kilometres of pipelines and some gas leaks occur. Electricity comes from a mixture of raw energy sources in the UK such as coal, oil, natural gas, nuclear and hydro schemes; the overall CO_2 conversion factor has to accommodate the mix of sources used. Electrical distribution between the power stations and the final user requires cables having resistance and leakage losses. Oil and coal are processed and then distributed by diesel engine-driven transport by road, rail and ship.

EXAMPLE 2.7

A small commercial building has a predicted energy consumption of 250 000 kWh per year for only the space heating system. The design engineer is to recommend the energy source and system type to be used on the basis of minimizing the greenhouse gas emissions. The usage efficiency of the alternative systems are 100% for electrical individual appliances, 70% for gas-fired radiator heating system, 60% for coal-fired radiator heating system and 75% for an oil-fired ducted warm-air heating system. Use the carbon conversion factors in Table 2.4 and make a suitable recommendation to the client.

The analysis is shown in Table 2.5.

$$\text{For natural gas, carbon emission} = \frac{250\,000 \text{ kWh}}{70\%} \times 0.055 \, \frac{\text{kg C}}{\text{kWh}}$$
$$= 19\,643 \text{ kg C p.a.}$$

$$\text{For heating oil, carbon emission} = \frac{250\,000 \text{ kWh}}{75\%} \times 0.079 \, \frac{\text{kg C}}{\text{kWh}}$$
$$= 26\,333 \text{ kg C p.a.}$$

$$\text{For coal-fired, carbon emission} = \frac{250\,000 \text{ kWh}}{60\%} \times 0.093 \, \frac{\text{kg C}}{\text{kWh}}$$
$$= 38\,750 \text{ kg C p.a.}$$

$$\text{For electrical heating, carbon emission} = \frac{250\,000 \text{ kWh}}{100\%} \times 0.142 \, \frac{\text{kg C}}{\text{kWh}}$$
$$= 35\,500 \text{ kg C p.a.}$$

Table 2.5 Carbon emissions in Example 2.7.

Energy source	Useful energy (kWh)	System efficiency (%)	CO_2 factor (kg C/kWh)	Carbon emission (kg p.a.)
Natural gas	250 000	70	0.055	19 643
Oil	250 000	75	0.079	26 333
Coal	250 000	60	0.093	38 750
Electricity	250 000	100	0.142	35 500

Table 2.6 Degree day data showing 20-year averages.

Region	Month											
	Sep.	Oct.	Nov.	Dec.	Jan.	Feb.	Mar.	Apr.	May	June	July	Aug.
Thames	56	129	252	333	349	306	281	200	113	49	25	27
NE Scotland	125	196	321	382	399	362	340	274	203	112	86	86

The client will be advised that a natural gas-fired heating system provides the lowest green-house gas emissions and that using a condensing water heater would further reduce carbon emissions by, around, 10% as the water heater efficiency would rise from a seasonal average of around 75% to at least 85%.

Annual energy costs

Annual fuel costs can be estimated in advance of their occurrence from a knowledge of the following:

1. energy cost per useful unit;
2. length of heating season;
3. operational hours of the system;
4. mean internal building temperature;
5. design external temperature;
6. degree days for the locality.

The design steady-state building heat loss is known as the design external air temperature (Chapter 3) and ranges from −1°C to −5°C. Throughout the heating season, the heat loss will fluctuate with the cyclic variations in ambient temperature. Fortuitous heat gains will reduce fuel consumption provided that the automatic controls can reduce heating system performance sufficiently and avoid overshooting the desired room temperatures. Weather variations are evaluated by using the degree day data issued monthly by the Department of Energy. Table 2.6 shows degree days for 2 of the 17 geographical regions covering the UK (Moss, 1997).

Degree days are recorded temperature data that facilitate the production of a climatic correction or load factor for calculation of heating system operational costs and efficiency over months or yearly time intervals. They are applied to normally occupied buildings where the heat loss from the warm interior is balanced by gains of heat from the sun, occupants, lighting, cooking and hot-water usage at an external air temperature of 15.5°C; this is known as the base temperature t_b. The value taken for the base temperature is an estimate of the conditions under which there will be no net heat loss from a traditionally constructed residence; thus no fuel will be consumed at this and higher outside temperatures.

Calculation of the actual base temperature for a particular building may reveal another value; consequently, care is needed in the application of degree day data, and correction factors may be included for other than traditionally constructed dwellings: for example, highly insulated or commercial structures and where internal heat gains from electrical equipment are high (CIBSE, 1986).

The standard method of use is to assess the daily difference between the base temperature and the mean value of the external air temperature during each 24-h period. A modified calculation is made when the base temperature is below the external mean temperature, as this would indicate a net heat gain to the building. Degree day data are not used for air-conditioning cooling-load calculations as they are not appropriate. Figure 2.1 shows a typical fluctuation in external air temperature relative to base temperature. As the maximum and minimum air temperatures are 10°C and 6°C respectively, the 24-h mean is 8°C. Therefore, as there is a difference of 7.5°C per day, 7.5 degree days are added to the accumulated total for that month.

The maximum possible number of degree days for a particular location and period of heating system operation can be found as shown in the following example.

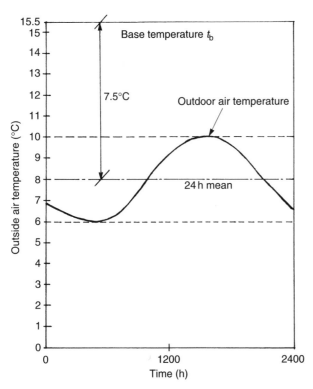

2.1 Method of calculating degree days during a 24-h period.

EXAMPLE 2.8

A house is continuously occupied during a 30-week heating season. The design external air temperature is $-1.0°C$. Find the maximum possible number of degree days.

$$\text{days} = 30 \text{ weeks} \times \frac{7 \text{ days}}{1 \text{ week}}$$

$$= 210 \text{ days}$$

$$\text{maximum temperature difference} = [15.5 - (-1.0)]°C$$

$$= 16.5°C$$

$$\text{maximum degree days} = 210 \text{ days} \times 16.5°C$$

$$= 3465 \text{ degree days}$$

The load factor L is the ratio of actual to maximum degree days and is used to find the average rate of heat loss from a building over the heating season:

$$L = \frac{\text{degree days for locality}}{\text{maximum possible degree days}}$$

EXAMPLE 2.9

Find the average rate of boiler power used during the heating season when there were 2460 degree days, and steady-state heat losses were calculated to be 24.5 kW at an outside air temperature of $-1°C$.

$$L = \frac{2460}{3465}$$

$$= 0.71$$

$$\text{seasonal average rate of heat loss} = \text{design heat loss} \times \text{load factor}$$

$$= 24.5 \text{ kW} \times 0.71$$

$$= 17.4 \text{ kW}$$

The boiler will have an average heat output of 17.4 kW over the heating season, that is, in addition to the hot-water service requirement and heat losses from pipework.

Degree days can be used to monitor fuel consumption and check that it is not being used wastefully. Incorrectly serviced fuel-burning appliances would show an increasing use of energy per degree day rather than a constant rate. Deterioration of the performance of an automatic control system or lack of proper manual regulation of ventilation openings would also result in a departure from expected ratios. A graph of energy consumption against degree days should be linear for a building, and any major divergence will show that corrective action is needed. Figure 2.2 shows an example.

The calculation of expected annual fuel costs is now a matter of finding the number of gigajoules or therms consumed in the building and then the cost of providing this useful amount of heat.

EXAMPLE 2.10

A hospital is heated for 24 h per day, 7 days per week for 30 weeks a year. Its steady-state heat loss at $-1°C$ outside is 2850 kW, and a gas-fired boiler with hot-water radiator central heating system is used. Estimate the annual fuel cost for heating the building if there are likely to be 2240 degree days in that locality. Maximum degree days are 3465. Gas costs £7.20 per useful gigajoule.

$$\text{load factor} = \frac{2240}{3465} = 0.65$$

$$\text{annual energy} = \text{heat loss} \times \text{load factor} \times \text{operating time}$$

$$= 2850 \text{ kW} \times 0.65 \times \frac{24 \text{ h}}{\text{day}} \times \frac{210 \text{ days}}{\text{year}} \times \frac{3600 \text{ s}}{1 \text{ h}}$$

$$\times \frac{1 \text{ kJ}}{1 \text{ kWs}} \times \frac{1 \text{ GJ}}{10^6 \text{ kJ}}$$

$$= 2850 \times 0.65 \times 24 \times 210 \times 3600 \times 10^{-6} \text{ GJ}$$

$$= 33\,612 \text{ GJ/year}$$

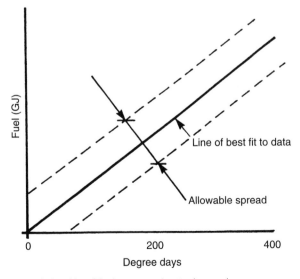

2.2 Relationship of fuel consumption to degree days.

$$\text{annual cost} = \text{useful energy required} \frac{GJ}{\text{year}} \times \frac{\text{cost £}}{\text{useful GJ}}$$

$$= 33\,612\,\frac{GJ}{\text{year}} \times \frac{£7.20}{GJ}$$

$$= £242\,006 \text{ p.a.}$$

EXAMPLE 2.11

An initial energy audit of a hospital revealed the data shown in Table 2.7. The data were for a month that had 260 degree days, and the energy manager required energy use performance factors of total cost per square metre of floor area, heating system energy used per degree day and electrical energy used per person per hour. All gas consumed was for the heating system and cost 1.85 p/kWh plus £750 per month standing charge and £550 per month for maintenance.

Electricity cost was 10 p/kWh and maintenance costs amounted to £550 per month.

Table 2.7 Energy audit data for Example 2.11.

Location	Electricity (kWh)	Gas (kWh)	Floor (m^2)	Usage (h)	Occupants
Medical	12 000 000	13 000 000	45 000	670	2300
Administration	1 500 000	1 000 000	3500	350	220
Engineering	150 000	250 000	1000	400	23
Totals	13 650 000	14 250 000	49 500	1420	2543

Energy use performance factor of total cost per square metre of floor area for the month, $EUPF_1$:

$$\text{electricity cost} = 13\,650\,000 \text{ kWh} \times 10\,\frac{p}{kWh} \times \frac{£1}{100\,p} + £550$$

$$= £1\,365\,550$$

$$\text{gas cost} = 14\,250\,000 \times 1.85\,\frac{p}{kWh} \times \frac{£1}{100\,p} + £750 + £550$$

$$= £264\,925$$

$$EUPF_1 = \frac{£1\,365\,550 + £264\,925}{49\,500 \text{ m}^2}$$

$$= £32.94 \text{ per m}^2 \text{ floor area}$$

$EUPF_2$ gas heating system energy used per degree day:

$$EUPF_2 = \frac{14\,250\,000 \text{ kWh}}{260 \text{ degree days}}$$

$$= 54\,808 \text{ kWh/degree day}$$

$EUPF_3$, electrical energy used per person per hour:

$$\text{total occupation} = \text{sum of (occupants} \times \text{usage hours)}$$

$$= (2300 \times 670) + (220 \times 350) + (23 \times 400)$$

$$= 1\,627\,200 \text{ person-hours}$$

$$EUPF_3 = \frac{14\,250\,000 \text{ kWh}}{1\,627\,200 \text{ person-hours}}$$

$$= 8.76 \text{ kWh/person/h}$$

You may wish to evaluate the energy use performance factors for various locations for comparison.

Economic thickness of thermal insulation

A balance needs to be made between the capital cost of thermal insulation of buildings or hot surfaces and the potential reduction in fuel costs in order to obtain the lowest total cost combination of these two cash flows. Capital cost is normally expected to be recovered from fuel cost savings during the first 2–3 years of use; however, longer periods than this are needed for major structural items, such as cavity fill and double glazing, and there will be additional benefits, such as improved thermal storage capacity, reduced external noise transmission, fewer draughts and added value to the property, that do not fit easily into a financial treatment of their worth.

For a flat surface, the cost of heat loss per square metre through the structure can be represented as follows:

$$\text{Fuel cost} = U\,\frac{W}{m^2 K} \times (t_{ai} - t_{ao})\,K \times L \times S\,\frac{h}{\text{year}} \times \frac{3600\,s}{1\,h} \times \frac{1\,J}{1\,Ws} \times \frac{1\,GJ}{10^9\,J} \times \frac{£C}{GJ}$$

$$= £[U(t_{ai} - t_{ao})LS \times 3.6C \times 10^{-6}] \text{ per } m^2 \text{year}$$

The cost of fuel usage for a range of thermal transmittances U can be calculated for a particular structure. This is usually a decreasing curve for increasing insulation thickness as each additional layer reduces the thermal transmittance by progressively smaller amounts.

If the cost $£/m^3$ of the thermal insulation as installed is known, then the cost for each thickness per square metre of surface area can be found from

$$\text{insulation cost} = \frac{£}{m^3} \times \frac{\text{thickness m}}{\text{repayment time years}}$$

Data from these equations can be drawn on a graph. Addition of the two curves produces a total cost curve. The lowest point on this curve gives the optimum insulation thickness; if its lower part is fairly flat, then any one of a number of commercially available thicknesses will be economic.

EXAMPLE 2.12

Expanded polystyrene board is to be added to the internal face of a wall having an initial thermal transmittance of 3.30 W/m^{2}K in thicknesses of 25, 50, 75, 100, 125 and 150 mm. Insulated wall thermal transmittances will be 0.96, 0.56, 0.40, 0.31, 0.25 and 0.21 W/m^{2}K. The insulation costs £48 per cubic metre fitted and the capital recovery period is to be 3 years. Fuel costs £8.93 per useful gigajoule. Internal and external design temperatures are 21°C and −1°C respectively, the load factor is 0.608 and the building is to be heated for 3000 h per year. Use the information provided to find the economic thickness of insulation.

$$\text{fuel cost} = U[21 - (-1)] \times 0.608 \times 3000 \times 3.6 \times 8.93 \times 10^{-6} \text{ £/m}^2 \text{ year}$$

$$= 1.29U \text{ per m}^2 \text{ year}$$

$$\text{insulation cost} = \frac{£48}{\text{m}^3} \times \frac{\text{thickness } l \text{ m}}{3 \text{ years}}$$

$$= 16 \, l \text{ per m}^2 \text{year}$$

The results are shown in Table 2.8. Figure 2.3 shows that the total cost curve can be drawn by adding the fuel and insulation cost curves for each insulation thickness. The economic thickness is 50 mm.

The economic thermal transmittance of a structure with a flat surface can be evaluated from the following equation (Diamant, 1977):

$$U_e = \left[\frac{\lambda \alpha l}{8.64 C (Y + S \Delta t)} \right]^{1/2}$$

where U_e is the economic U value (W/m^{2}K), λ is the thermal conductivity (W/mK), α is the depreciation and interest charges (%), l is the total cost of thermal insulation fitted into the building (£/m^3), C is the cost of useful heat (£/MJ), Y is the annual degree days for a base temperature of 18°C, S is the length of the heating season (days) and Δt is the difference between the average internal air temperature and 18°C. Once U_e has been found, the thermal insulation thickness can be found from

$$l = \lambda \left(\frac{1}{U_e} - \frac{1}{U} \right) \text{m}$$

where U is the uninsulated thermal transmittance (W/m^{2}K).

Table 2.8 Cost data for Example 2.12.

Thickness l (m)	0.025	0.050	0.075	0.100	0.125	0.150
U (W/m^{2}K)	0.96	0.56	0.40	0.31	0.25	0.21
Insulation cost (£)	0.40	0.80	1.20	1.60	2.00	2.40
Fuel cost (£)	1.24	0.72	0.52	0.40	0.32	0.27
Total cost (£)	1.64	1.52	1.72	2.00	2.32	2.67

2.3 Economic thickness of the wall insulation in Example 2.12.

EXAMPLE 2.13

The roof of a factory in Lancashire is insulated with expanded polystyrene (EPS) slab. The locality has 3070 degree days for a base temperature of 18°C and a heating season of 235 days. The average internal air temperature is 20°C. The EPS costs £65 per cubic metre fitted, interest is charged at 7% and the life expectancy of the insulation is 45 years. Heat costs £8.50 per gigajoule. The thermal conductivity of EPS is 0.035 W/mK. Find the economic insulation thickness.

The original thermal transmittance of the single sheet roofing is 6.0 W/m²K.

$\alpha = 7\%$ interest per annum $+ 2\%$ depreciation per annum

$\quad = 9\%$

$$C = \frac{£8.50}{GJ} \times \frac{1\ GJ}{10^3\ MJ}$$

$$= £8.50 \times 10^{-3}\ per\ MJ$$

$$U_e = \left[\frac{0.035 \times 9 \times 65}{8.64 \times 8.5 \times 10^{-3} \times (3070 + 235 \times (20 - 18))} \right]^{1/2}\ W/m^2K$$

$$= 0.28\ W/m^2K$$

$$I = 0.035 \left(\frac{1}{0.28} - \frac{1}{U} \right)$$

$U = 6.5$ W/m^{2}K for the uninsulated roof.

Therefore

$$I = 0.035 \left(\frac{1}{0.28} - \frac{1}{6.5} \right)$$

$$= 0.120 \text{ m}$$

Thus the economic thickness of insulation for this roof is 120 mm. As EPS slab is manufactured in 50 mm thickness, the insulated U value for 150 mm thickness will be 0.225 W/m^{2}K.

Accounting for energy-economizing systems

Once the capital cost and fuel cost savings have been assessed for thermal insulation, fuel-saving equipment or automatic controls, the capital repayment period or return on capital investment can be calculated in simple terms:

$$\text{capital repayment period} = \frac{\text{capital cost}}{\text{energy savings per year}}$$

$$\text{percentage return on investment} = \frac{\text{energy savings per year}}{\text{capital cost}} \times 100$$

Further refinements such as discounted cash flow, loan interest charges, tax allowances and grants can be included to improve accuracy.

Cash flow statements for limited companies are handled differently from those for home-owners, as allowances for capital expenditure and taxation on increased profitability due to energy economies can markedly improve estimates of payback times. A purchase costing £500 000, which would save £200 000 in the first year's energy bill, would appear to take 2.5 years for capital recovery, but the cash flow projection may be as shown in Table 2.9. Figures in parentheses are outward cash flows from the business. Energy costs are indicated to increase by 5% per year, so savings increase by the same amount.

This investment commences cash generation during the second year of equipment use and would start to provide funds for further investment. Certain items of equipment can be leased

Table 2.9 Cash flow forecast for the purchase of an energy-economizing system.

		Year 1	Year 2	Year 3
A	Cash balance brought forward	0	(315 000)	(53 000)
B	Capital purchase	(500 000)	0	0
C	Energy saving	200 000	210 000	220 500
D	Capital allowance, 1 year in arrears 25% × B	0	125 000	0
E	Cash balance (A − B + C + D)	(300 000)	20 000	167 500
F	Interest, say 10% × E × 0.5	(15 000)	1000	8375
G	Tax, 1 year in arrears, 40% × (C + F)	0	(74 000)	0
H	Net cash flow (C + D + F − B − C)	(315 000)	262 000	228 875
I	Cash balance (A + H)	(315 000)	(53 000)	175 875

Table 2.10 Cash flow forecast for the leasing of an energy-economizing equipment.

		Year 1	Year 2	Year 3
A	Cash balance brought forward	0	157 500	270 375
B	Leasing payment, 10% × cost	(50 000)	(50 000)	(50 000)
C	Energy saving	200 000	210 000	220 500
D	Capital allowance	0	0	0
F	Cash balance (A − B + C + D)	150 000	317 500	440 875
F	Interest, say 10% × E × 0.5	7500	15 875	22 044
G	Tax, 40% × (C + F − B), 1 year in arrears	0	(63 000)	(70 350)
H	Net cash flow (C + D + F − B − G)	157 500	112 875	122 194
I	Cash balance (A + H)	157 500	270 375	392 569

rather than purchased, and this releases cash earlier but calls for continuous payments to the leasing company. Table 2.10 shows a sample cash flow forecast. Cash flow is always positive to the company, but leasing payments are made for 10 years and then at a reduced rate after that period. Self-contained items of plant such as heat pumps or electricity generators may be leased.

Low-energy buildings

Low-energy buildings are those that utilize energy efficiently to maintain a comfortable thermal environment suitable for the purpose of the building. Design software for steady-state and dynamic use of the building are available from many sources such as the Building Regulations, Building Research Establishment Environmental Assessment Method, BREEAM, Leadership in Energy and Environmental Design, LEED, Green Star and Greenhouse Rating schemes, and are applied variously to the design stage and also during the service period of the building to meet agreed standards. Energy design targets may be proposed that include uses for heating, ventilation, hot-water services, lighting and electrical power. The total demand target T for a building is assessed by adding the thermal demand target T_T to the electrical demand target T_E in the CIBSE *Building Energy Code* (CIBSE, 1981).

In heated and naturally ventilated buildings, the rate of heat loss is related to the floor area by the dimensionless building envelope number B:

$$B = \frac{A_w}{A_f} + \frac{K_1}{n_f} + K_2 H$$

where A_w is the gross external walling surface area (m^2), A_f is the total floor area (m^2), n_f is the number of storeys and H is the floor-to-ceiling height (m). When hot-water services are included:

$$T_T = C_1 B + C_2 \text{ W/m}^2$$

where K_1, K_2, C_1 and C_2 are constants given in Table 2.11, C_3 is the mean electrical power requirement for the lighting system (W/m^2) and,

$$T_E = C_3 + 0.10 T_T \text{ W/m}^2$$

which shows that the electrical power consumption associated with the heating services is expected to be 10% of the thermal target.

Table 2.11 Values of constants for demand target.

Building type	K_1	K_2	C_1	C_2	C_3
Office, 5 days/week	0.5	0.1	13	−5	24
Shop, 6 days/week	0.5	0.1	16	−6	27
Factory, 5 days/week, single shift	1.1	0.2	6	3	8
Hotel	1.0	0.1	15	−3	15
Warehouse	1.1	0.2	6	−2	6
Hospital	1.0	0.1	17	+12	15
Institutional residence	1.0	0.1	15	±4	15
Educational	0.5	0.2	16	−4	13

EXAMPLE 2.14

Find the total demand target for a proposed 10-storey hospital medical building 50 m long, 30 m wide and 3 m floor-to-ceiling height, and compare it with an alternative design having the same floor area but of single-storey design, 172 m long and 87 m wide. Comment on the relative energy use of these alternative configurations.

For the 10-storey block

$$A_f = 50 \times 30 \times 10 \text{ m}^2$$

$$= 15\,000 \text{ m}^2$$

$$A_w = 10 \times 3 \times 2 \times (50 + 30) \text{ m}^2$$

$$= 4800 \text{ m}^2$$

$$n_f = 10$$

$$H = 3 \text{ m}$$

From Table 2.11, $K_1 = 1$ and $K_2 = 0.1$. Then:

$$B = \frac{4800}{15\,000} + \frac{1}{10} + (0.1 \times 3)$$

$$= 0.72$$

Also from Table 2.11, $C_1 = 17$, $C_2 = +12$ and $C_3 = 15$ W/m^2. Then:

$$T_T = (17 \times 0.72 + 12) \text{ W/m}^2$$

$$= 24.24 \text{ W/m}^2$$

$$T_E = (15 + 0.1 \times 24.24) \text{ W/m}^2$$

$$= 17.4 \text{ W/m}^2$$

Thus:

$$T = T_T + T_E$$

$$= (24.24 + 17.4) \text{ W/m}^2$$

$$= 41.6 \text{ W/m}^2$$

Similarly, for the single-storey building:

$B = 1.4$

$T_T = 35.8$ W/m^2

$T_E = 18.6$ W/m^2

$T = 54.4$ W/m^2

The single-storey building has a better chance of being ventilated by assisted natural systems, rather than the mechanical plant needed in multi-storey designs. The walls and perimeter glazing are more easily shaded from solar heat gains; floor usage is more efficient as vertical service shafts, lifts and stairways are unnecessary; wind exposure is reduced; and maintenance of the external surfaces of the building is less costly.

Both buildings have the same floor area; the lower fatter configuration has less external wall surface area and consequently a higher total demand target.

Compliance with the demand target is achieved when the total demand of the proposed building does not exceed its target figure; the thermal demand may be up to 10% greater than its target value, but the total demand target must not be surpassed.

The effect on gas consumption of thermal insulation in houses

When the design steady-state heat loss from a dwelling is reduced by the addition of thermal insulation and draught-proofing, increased standards of thermal comfort are provided. However, the full potential saving due to the extra insulation may not be reflected in the fuel bills as expected. Field measurements (British Gas, 1980) have shown a correlation between domestic gas consumption, design heat loss, occupancy and degree days for the locality:

$$\text{annual therms} = 61 + \frac{70YQ}{2222} + 59N$$

where Y denotes annual degree days, N is the number of persons in the household and Q is the design heat loss in kilowatts. This relationship permits an assessment of anticipated gas consumption for heating and hot-water services in housing and quantification of thermal insulation savings.

Therms are no longer used, so they are converted into kWh by multiplying by 29.3056:

$$\text{annual gas consumption} = 29.3056 \times \left(61 + \frac{70 \times 2120 \times 36}{2222} + 59N \right) \text{ kWh}$$

EXAMPLE 2.15

A house in the Thames region has a design heat loss of 28 kW and up to six occupants. Added thermal insulation reduces the design heat loss to 23 kW. Estimate the probable energy, greenhouse gas and cost savings for the gas-fired central heating and hot-water system.

Natural gas costs 2.8 p/kWh.

From Table 2.4 the total degree days for the year are 2230.

$$\text{Energy before insulation} = 29.3056 \times \left(61 + \frac{70 \times 2230 \times 28}{2222} + 59 \times 6\right) \text{ kWh}$$

$$= 69\,808 \text{ kWh}$$

$$\text{Energy after insulation} = 29.3056 \times \left(61 + \frac{70 \times 2230 \times 23}{2222} + 59 \times 6\right) \text{ kWh}$$

$$= 59\,514 \text{ kWh}$$

$$\text{Energy saving} = (69\,808 - 59\,514) \text{ kWh}$$

$$= 10\,294 \text{ kWh}$$

$$\text{Carbon emission saving} = 10\,294 \text{ kWh} \times \frac{0.055 \text{ kg C}}{\text{kWh}}$$

$$= 566 \text{ kg}$$

$$\text{gas cost saving} = 10\,294 \text{ kWh} \times \frac{2.8}{\text{kWh}} \times \frac{£1}{100 \text{ p}}$$

$$= £288.00 \text{ during the first year}$$

Questions

1. State the function of an energy audit. What data are collected? How are the data presented? What is likely to be the most serious barrier to data collection?
2. Explain the uses of energy use performance factors.
3. How can the costs of different fuels be compared with each other?
4. Ascertain current energy prices and update Table 2.3.
5. Explain the term 'degree day' and state its use.
6. How is the load factor calculated and how is it used?
7. A factory uses 20 000 l of oil for its heating and hot-water systems, 160 000 kWh of electrical power and 300 000 kWh of gas for furnaces in a year. Fixed charges are £800 for the oil, £700 for the electrical equipment and £1200 for gas equipment. Use the data provided in this chapter and current energy prices to produce an overall energy audit based on the gigajoule unit and find the average cost of all the energy used.
8. Calculate the annual cost of a gas-fired heating system in a house with a design heat loss of 30 kW at −2°C for 16 h per day, 7 days per week for 30 weeks in the year. Use the data provided in this chapter and the current fuel price.
9. Find the total annual cost of running a gas-fired heating and hot-water system in a house with four occupants if its design heat loss is 32 kW. Maintenance charges amount to £160 per year.
10. Determine the following energy use performance factors for offices A and B:

    (a) total energy cost per square metre of floor area;
    (b) heating system energy used per degree day;
    (c) total energy used per person per occupation hour.

    The data required are given in Table 2.12. They refer to a year having 2150 degree days for office A and 2310 degree days for office B. Maintenance costs are £800 for the gas system and £750 for the electrical installation in each building. All the gas is used for heating systems. Compare the performance of the two office buildings.

Table 2.12 Buildings A and B for Question 10.

Building	Electricity (kWh)	Gas (kWh)	Floor (m²)	Usage (h)	Occupants
Office A	115 000	720 000	2400	2600	160
Office B	100 000	850 000	2600	2300	210

11. Calculate the economic thickness of rock wool thermal insulation which is to be applied to a wall with an uninsulated thermal transmittance of 1.6 W/m²K. The thermal conductivity of rock wool is 0.04 W/mK. The locality has 2900 degree days for a base temperature of 18°C and a heating season of 240 days. The average internal temperature is 20°C. Rock wool costs £50 per cubic metre fitted. Interest is charged at 9% and depreciation at 10%. Fuel oil for heating costs £5.00 per useful gigajoule.

12. Find the total demand target for an eight-storey office 22 m long, 15 m wide, 3 m floor-to-ceiling height. How will compliance with this target be achieved?

13. A building has a predicted energy consumption of 1 500 000 kWh/year for only the space heating system. The design engineer is to recommend the energy source and system type to be used on the basis of minimizing the greenhouse gas emissions. The average seasonal usage efficiency of the alternative systems are 95% for electrical heating systems of various types, 75% for gas-fired radiator heating system, 65% for coal-fired radiator heating system and 75% for an oil-fired ducted warm-air heating system. Use the carbon conversion factors in Table 2.4 to calculate the carbon emission in tonnes per year and make a suitable recommendation to the client.

14. Which of these adequately describe an energy audit of a building?

    1. Points out what building operators could do to save energy.
    2. Are only for cosmetic appearance of doing something to reduce greenhouse emissions.
    3. Identify and quantify viable energy-saving investments.
    4. Concentrate on finding almost zero cost short-term payback energy-saving opportunities.
    5. Only analyses technical projects and not financial investment criteria.

15. Which of these is not a correct multiple?

    1. $kJ = 10^3$ J.
    2. MWh $= 1000$ W $\times$ 1 h.
    3. 1 GJ $= 10^6$ kJ.
    4. 1 mm $= 10^{-3}$ m.
    5. 1 GW $= 1000$ MW.

16. Which is not correct for CO_2 greenhouse gas?

    1. Produced by ruminating animals.
    2. Continuously converted back into O_2 by photosynthesis.
    3. It is burnt carbon from fuel combined with atmospheric oxygen.
    4. Easily reversed to solid carbon plus oxygen gas into the atmosphere.
    5. Combusted hydrocarbons are not the sole source of greenhouse gases.

17. Which is degree day load factor not relevant to?

    1. Calculation of heating system kW load for design.
    2. Ratio of degree days from meteorological data.

3. Minimum outside air temperature for design.
4. Seasonal weather variability.
5. Maximum possible degree days for the locality.

18. Which is correct about energy use in buildings?

1. Design energy use accurately predicts actual consumption.
2. Design energy use predictions rely on input data from the owner.
3. Usage of a new building always complies with design prediction patterns.
4. New building users never find design inadequacies.
5. Predicted energy use for new buildings is often exceeded in reality.

19. Which of these is not included in the annual financial accounts for energy-saving projects?

1. Energy cost savings due to the investment.
2. Cash balance of the proposed investment prior to the commencement of the energy-saving project.
3. Cash balance of the proposed investment at the end of each year of the project.
4. Financial capability of the contractor who is to undertake the energy-saving installation.
5. Capital allowance from taxation system against costs of the energy-saving installation.

20. Which of these describes low-energy buildings?

1. Design predictions of energy use mean nothing to the user of the building.
2. Always uncomfortable for occupants.
3. Cannot maintain low-energy use after three years of use.
4. Must comply with an energy rating mandatory standard upon construction.
5. Cannot have large areas of glazing.

21. What does sustainability mean for low-energy buildings?

1. Brick, steel and concrete walls save more energy consumption while in use than they cost to produce and construct.
2. All windows have low emissivity glass.
3. Aluminium window frames are not used in this building.
4. Double-glazed windows always used in cold climates.
5. All thermal insulation, glass, aluminium, steel, timber and concrete in this building were harvested from renewable resources.

22. What does sustainability mean for low-energy buildings?

1. The concrete for this building came from a very large quarry that can supply the national need beyond the lifetime of the present occupants.
2. Heat loss through the windows in winter are exceeded by the heat gains in summer.
3. All the timber used in concrete formwork were recycled from previous projects and are to be passed on to our next construction site.
4. All timber in this building came from managed forestry.
5. Electricity used to crush and melt primary ground resources into glass and aluminium used in this building all came from renewable energy sources.

23. Which of these has the correct units?

1. 1 Newton $= 1$ kg $\times 1$ m^2.
2. 1 Joule $= 1$ kg $\times 1$ m.
3. 1 Watt $= 1$ kg $\times$ g m/s^2.

4. 10^3 Joules $= 3600$ kN/m^2.
5. 1 Joule $= 1$ N/m^2.

24. Which of these has the correct units?

    1. 1 atmosphere $= 10^3$ b.
    2. 1 Pascal $= 1$ N/m^2.
    3. Pascal is a unit of radiation measurement.
    4. 1 kN/m$^2 = 1$ b.
    5. 1 mb $= 10^3$ N/m^2.

25. Which of these does not apply to low-energy buildings?

    1. Should be audited and re-accredited as complying with nationally mandated standard at regular intervals.
    2. Never need re-accreditation.
    3. Energy rating accreditation maintains credibility of design.
    4. Must be accredited by licensed raters complying with a professional code of conduct.
    5. Highly insulated.

26. What does BREEAM stand for?

    1. Building Rehabilitation Electrical Energy Alternative Methodology.
    2. Building Research Establishment Energy Audit Methodology.
    3. Building Recycling Energy Effectiveness Association Member.
    4. Brick Recycling Energy and Environment Assessment Method.
    5. Building Research Establishment Environmental Assessment Method.

27. What does LEED stand for?

    1. Low Energy Environmental Design.
    2. Leadership in Energy and Environmental Design.
    3. Low Electrical Energy Demand.
    4. Leading Electrical Energy Demonstration.
    5. Leader in Energy Environment and Design.

28. What does Green Star stand for?

    1. This building only uses renewable energy sources.
    2. No such thing as a green star.
    3. A low mould-growth building.
    4. Zero condensation risk.
    5. Standard for environmental performance of the building.

29. What does sustainability mean for low-energy buildings?

    1. All the glass in this building comes from self-sustaining resources.
    2. All the aluminium in this building comes from self-sustaining resources.
    3. All the primary energy used by this building comes from self-sustaining resources.
    4. All the concrete and reinforcing steel in this building comes from self-sustaining resources and recycled materials.
    5. None of these answers.

3   Heat loss calculations

Learning objectives

Study of this chapter will enable the reader to:

1. identify and use the thermal conductivity and resistivity of building materials;
2. calculate the thermal resistance of a composite structure;
3. use building exposure categories;
4. use surface and cavity thermal resistance values;
5. identify high- and low-emissivity building materials;
6. calculate, or find, the thermal transmittance of walls, flat and pitched roofs, floors and windows;
7. use the proportional area method to calculate the average U value of a thermally bridged wall or other structure;
8. calculate the fabric and ventilation building heat loss components for a building;
9. calculate air and environmental temperatures produced in a room from a specified resultant temperature;
10. identify the use of admittance Y values;
11. hot-water storage boiler power requirements;
12. calculate total boiler power.

Key terms and concepts

boiler power 71; edge insulation 66; exposure: sheltered, normal and severe 63; hot-water heat load 71; intermittent heat load 70; proportion area method 64; steady-state heat loss 67; surface emissivity 63; surface resistance 63; thermal conductivity 61; thermal resistance 61; thermal storage of structure 67; thermal transmittance 64; U value 64; Y value 70.

Introduction

The terms and techniques for handling thermal properties of building materials are introduced to enable calculation of the thermal resistance R, thermal transmittance U and use of admittance

Y of composite building elements. Chartered Institution of Building Services Engineers data are used throughout and representative values are given. Calculation of building heat loss allows load estimation for heating equipment.

Thermal resistance of materials

The thermal resistance of a slab of homogeneous material is calculated by dividing its thickness by its thermal conductivity:

$$R = \frac{l}{\lambda}$$

where R is the thermal resistance (m^{2}K/W), l is the thickness of the slab (m) and λ is the thermal conductivity (W/mK). Resistance to heat flow by a material depends on its thickness, density, water content and temperature. The latter two parameters result from the material's location within the structure. Insulating materials are usually protected from moisture and the possibility of physical damage as they are of low density and strength. The thermal conductivity of masonry can be found from the bulk dry density and the moisture content, which depends on whether it is exposed to the climate or is in a protected position. Table 3.1 shows data for building materials taken from the *CIBSE Guide*, to which further reference can be made.

EXAMPLE 3.1

Find the thermal resistance of a 110 mm thickness of brickwork inner leaf.

From Table 3.1, $\lambda = 0.62$ W/mK and $l = 0.11$ m. Therefore:

$$R = \frac{0.110}{0.62} \text{ m}^2\text{K/W}$$
$$= 0.1774 \text{ m}^2\text{K/W}$$

EXAMPLE 3.2

A designer wishes to replace 200 mm thick heavyweight concrete blocks in the design of a wall with fibreboard having the same thermal resistance. What thickness of fibreboard could be used?

The values of λ are 1.63 W/mK for the heavyweight concrete block and 0.06 W/mK for the fibreboard.

$$R \text{ (concrete)} = \frac{0.200}{1.63}$$
$$R \text{ (fibreboard)} = \frac{l}{0.06}$$

Table 3.1 Thermal conductivities of materials.

Material	Density (kg/m³)	Thermal conductivity λ (W/mK)	Specific heat capacity (J/kgK)
Walls (external and internal)			
Asbestos cement sheet	700	0.36	1050
Asbestos cement decking	1500	0.36	1050
Brickwork (outer leaf)	1700	0.84	800
Brickwork (inner leaf)	1700	0.62	800
Cast concrete (dense)	2100	1.40	840
Cast concrete (lightweight)	1200	0.38	1000
Concrete block (heavyweight)	2300	1.63	1000
Concrete block (medium weight)	1400	0.51	1000
Concrete block (lightweight)	600	0.19	1000
Fibreboard	300	0.06	1000
Plasterboard	950	0.16	840
Tile hanging	1900	0.84	800
Surface finishes			
External rendering	1300	0.50	1000
Plaster (dense)	1300	0.50	1000
Plaster (lightweight)	600	0.16	1000
Roofs			
Aerated concrete slab	500	0.16	840
Asphalt	1700	0.50	1000
Felt–bitumen layers	1700	0.50	1000
Screed	1200	0.41	840
Stone chippings	1800	0.96	1000
Tile	1900	0.84	800
Wood–wool slab	500	0.10	1000
Floors			
Cast concrete	2000	1.13	1000
Metal tray	7800	50.00	480
Screed	1200	0.41	840
Timber flooring	650	0.14	1200
Wood blocks	650	0.14	1200
Insulation			
Expanded polystyrene (EPS) slab	25	0.035	1400
Glass fibre quilt	12	0.040	840
Glass fibre slab	25	0.035	1000
Mineral fibre slab	30	0.035	1000
Phenolic foam	30	0.040	1400
Polyurethane board	30	0.025	1400
Urea formaldehyde (UF) foam	10	0.040	1400

Therefore for the same resistance,

$$\frac{0.200}{1.63} = \frac{l}{0.06}$$

Hence,

$$l = 0.06 \times \frac{0.200}{1.63} \text{ m} \times \frac{10^3 \text{ mm}}{1 \text{ m}}$$

$$= 7.4 \text{ mm}$$

Thermal transmittance (*U* value)

Thermal transmittance is found by adding the thermal resistances of adjacent material layers, boundary layers of air and air cavities, and then taking the reciprocal. Boundary layer or surface film thermal resistances result from the near-stationary air layer surrounding each part of a building, with an allowance for the radiant heat transfer at the surface. Heat transmission across cavities depends upon their width, ventilation and surface emissivities. The external surface resistance depends upon the building's exposure.

Sheltered: up to the third floor of buildings in city centres
Normal: most suburban and rural buildings; fourth to eighth floors of buildings in city centres
Severe: buildings on coastal or hill sites; floors above the fifth in suburban or rural districts; floors above the ninth in city centres.

Surface resistances are shown in Tables 3.2–3.5.

Table 3.2 Inside surface resistance R_{si}.

Building element	Heat flow	R_{si} (m^2 K/W)
Wall	Horizontal	0.12
Ceiling, floor	Upward	0.10
Ceiling, floor	Downward	0.14

Note: These values are for the high-emissivity surfaces ($E = 0.90$) common to most building components.

Table 3.3 Outside surface resistance R_{so}.

Builing element	Surface emissivity	R_{so} (m^2 K/W)		
		Sheltered	Normal	Severe
Wall	High	0.08	0.06	0.03
Wall	Low	0.11	0.07	0.03
Roof	High	0.07	0.04	0.02
Roof	Low	0.09	0.05	0.02

Table 3.4 Thermal resistance R_a of ventilated air spaces.

Air space of 25 mm or more	R_a(m^2 K/W)
Loft space between flat plaster ceiling and pitched roof with tiles on felt	0.18
Air space behind tiles on tile hung wall	0.12
Air space in cavity wall	0.18
Air space between high- and low-emissivity surfaces	0.30

Table 3.5 Thermal resistances R_a for unventilated air spaces.

Air space thickness (mm)	Surface emissivity	R_a(m^2 K/W)		
		Horizontal	Upward	Downward
5	High	0.10	0.10	0.10
5	Low	0.18	0.18	0.18
25 or more	High	0.18	0.17	0.22
25 or more	Low	0.35	0.35	1.06

Table 3.6 Data for Example 3.3.

Element	Length l (m)	λ (W/mK)	R ($m^2 K/W$)
R_{so}			0.030
Brick	0.105	0.84	0.125
R_a			0.180
Brick	0.105	0.84	0.125
Plaster	0.013	0.5	0.026
R_{si}			0.120
			$\sum R = 0.606$

EXAMPLE 3.3

An external wall consisting of 105 mm brick, 50 mm unventilated cavity, 105 mm brick and 13 mm dense plaster has a severe exposure. Find its U value.

The calculation of $\sum R$ is shown in Table 3.6. The thermal transmittance U is calculated as follows:

$$U = \frac{1}{\sum R}$$

$$= \frac{1}{0.606} \ W/m^2 K$$

$$= 1.65 \ W/m^2 K$$

EXAMPLE 3.4

Calculate the thermal transmittance of the wall in Example 3.3 if the cavity is filled with urea formaldehyde (UF).

The calculation of $\sum R$ is shown in Table 3.7. Then the new value of U is 0.6 W/m²K.

Elements of buildings that are bridged by a material of noticeably different thermal conductivity, such as a dense concrete or steel lintel in a lightweight concrete wall, can be handled by combining the U values of the two constructions using the proportional area method. If U_1 and P_1 are the thermal transmittance and the unbridged proportion respectively of the gross wall area, and U_2 and P_2 are the same parameters for the bridging material, the overall U value is given by

$$U = P_1 U_1 + P_2 U_2$$

Table 3.7 Data for Example 3.4.

Element	Length l (m)	λ (W/mK)	R (m²K/W)
Previous $\sum R$			0.606
Less R_a			−0.180
Net			0.426
UF foam	0.050	0.040	1.250
			$\sum R = 1.676$

EXAMPLE 3.5

In a concrete-framed commercial building the external walling of brick has a U value of 1.8 W/m²K. The building has a gross perimeter of 180 m and is 3.6 m high. Thirty dense concrete pillars 180 mm wide penetrate the walling from inside to outside. The exposure is normal and the wall thickness is 300 mm.

   Find the overall U value.

Thermal conductivity of the concrete pillar $\lambda = 1.40$ W/mK.
For the concrete pillar:

$$U_2 = \frac{1}{R_{si} + (l/\lambda) + R_{so}}$$

$$= \frac{1}{0.12 + (0.30/1.40) + 0.06} \text{ W/m}^2\text{K}$$

$$= 2.54 \text{ W/m}^2\text{K}$$

$$\text{surface area of pillars} = 30 \times 0.180 \times 3.6 \text{ m}^2$$

$$= 19.44 \text{ m}^2$$

$$\text{gross wall area} = 180 \times 3.6 \text{ m}^2$$

$$= 648 \text{ m}^2$$

$$P_2 \text{ (pillars)} = \frac{19.44}{648} = 0.03$$

$$P_2 \text{ (walling)} = \frac{648 - 19.44}{648} = 0.97$$

Thus the overall value of U is given by:

$$U = (0.97 \times 1.8) + (0.03 \times 2.54) \text{ W/m}^2\text{K}$$

$$= 1.82 \text{ W/m}^2\text{K}$$

Where only one leaf of a structure containing a cavity is bridged, the resistance of each leaf is calculated separately using the proportional areas as appropriate and then the resistances can be added. The centre line of the cavity can be chosen as the dividing line between the two leaves and half the air space resistance added into each side of the structure. Heat bridges are thermal routes having a lower resistance than the surrounding material that cause distortions to the otherwise uniform temperature gradients. Precise calculation of overall thermal transmittance may require the use of a finite-element computer program that investigates the two- or three-dimensional heat conduction process taking place.

The thermal transmittance of a flat roof is calculated in the manner outlined for walls, but attention must be paid to tapered components. For a pitched roof:

$$U = \frac{1}{R_A \cos \beta + R_R + R_B}$$

where R_A is the combined resistance of the materials in the pitched part of the roof including the outside surface resistance ($m^2 K/W$), β is the pitch angle of the roof (degrees), R_R is the resistance of the roof void ($m^2 K/W$) and R_B is the combined resistance of the materials in the fiat part of the ceiling including the inside surface resistance ($m^2 K/W$).

Heat flow through solid ground floors in contact with the earth depends on the thermal resistance of the floor slab and the ground which, in turn, is largely determined by its moisture content. The thermal conductivity of the earth can vary from 0.70 to 2.10 W/mK depending upon the moisture content. Table 3.8 was evaluated for $\lambda = 1.40$ W/mK, which is about the same as for a concrete slab, and so the U values given can be used for floors of any thickness. Dense floor-finishing materials will not influence the quoted U values.

The thermal resistance of insulation placed under the screed of a solid floor, or on netting between the joists of a suspended timber floor, can be added to the reciprocal of the U value of the uninsulated floor and the new thermal transmittance can be calculated. An insulation material placed vertically around the edge of a concrete floor slab which has a thermal resistance of at least 0.25 $m^2 K/W$ and a depth of 1 m, for example, a 10 mm thickness of expanded polystyrene, will reduce the U value of the floor by the percentages shown in Table 3.9.

The thermal transmittance of windows depends on glazing and frame types and exposure. If a low-emissivity reflective metallic film is applied to the inside surface of the glass, then the internal

Table 3.8 U values for solid and suspended floors.

Length (m)	Breadth (m)	U (W/m²K)		
		Four exposed edges	Two perpendicular exposed edges	Suspended floor
100	100	0.10	0.05	0.11
40	40	0.21	0.12	0.22
20	20	0.36	0.21	0.37
10	10	0.62	0.36	0.59
10	4	0.90	0.54	0.83
4	4	1.22	0.73	0.96

Table 3.9 Corrections to U values of solid floors with edge insulation.

Length (m)	Breadth (m)	Reduction in U value (%)
100	100	16
40	40	18
20	20	19
10	10	22
10	4	25
4	4	28

Table 3.10 U values for typical windows.

Windows	U (W/m²K)		
	Sheltered	Normal	Severe
Single frame			
Wood frame	4.7	5.3	6.3
Aluminium frame	5.3	6.0	7.1
Aluminium frame with thermal break	5.1	5.7	6.7
Double glazing			
Wood frame	2.8	3.0	3.2
Aluminium frame	3.3	3.6	4.1
Aluminium frame with thermal break	3.1	3.3	3.7

surface resistance value can be significantly increased, resulting in a lower U value and reduced heat and light transmission from outside. Glass and metal window frames, in themselves, offer negligible resistance to heat flow, but when resistive materials are used the overall U value can be found using the proportional area method. Table 3.10 shows window U values assuming that the frame takes up 10% of the gross opening in the wall.

Heat loss from buildings

Heat loss occurs by convection and radiation from the outside of the building, and by infiltration of outdoor air. Heating equipment is sized on the basis of steady-state heat flows through the building fabric, with an estimation of the effect of non-steady influences relating to the thermal storage capacity of the structure, adventitious heat gains from people, lighting and machines, and the intermittency of heating system operation.

The steady-state heat loss Q_u through the building fabric is:

$$Q_u = \sum (AU)(t_{ei} - t_{ao}) \text{ W}$$

where $\sum(AU)$ is the sum of the products of the area and thermal transmittance of each room surface. Heat flows to adjacent rooms that are warmer than the outdoor air are found by using the appropriate temperature difference between them.

The ventilation heat Q_v required to warm the natural infiltration of outdoor air is:

$$Q_v = 0.33 \, NV \, (t_{ai} - t_{ao}) \text{ W}$$

The total heat requirement for each room is:

$$Q_p = Q_u + Q_v$$

The values of environmental and air temperature used in the calculations depend upon the type of heating system employed, and the following temperature ratios are used:

$$F_1 = \frac{t_{ei} - t_{ao}}{t_c - t_{ao}}$$

$$F_2 = \frac{t_{ai} - t_{ao}}{t_c - t_{ao}}$$

These two ratios are substituted into the equations for heat requirements Q_u and Q_v. The total heat requirement Q_p then becomes:

$$Q_p = \left[F_1 \sum (AU) + 0.33 F_2 NV \right] (t_c - t_{ao}) \text{ W}$$

For buildings with average external U values in the range 0.60–3.0 W/m^{2}K, including openings, which covers the majority of habitable structures, the temperature ratios have the following values (with an accuracy to 5.0%):

$$F_1 = 1.00 \quad F_2 = 1.10$$

for panel radiator heating systems:

$$F_1 = 0.92 \quad F_2 = 1.23$$

for forced warm-air heating systems. Further values are tabulated in the *CIBSE Guide*.

To check the comfort conditions produced by the heating system in a room we use:

$$t_{ai} = F_2 (t_c - t_{ao}) + t_{ao}$$

where t_c is the dry resultant temperature specified for the centre of the room from consideration of the application and t_{ao} has been specified for the location. The environmental temperature produced in the room is given by:

$$t_{ei} = F_1 (t_c - t_{ao}) + t_{ao}$$

EXAMPLE 3.6

An office building 20 m long by 10 m wide and 3 m high is to have a hot-water panel radiator heating system that will maintain a dry resultant temperature of 21°C at the centre of the room at an external air temperature of −4°C. There are 10 windows each of area 2 m^2 and two doors each of area 4 m^2. The roof can be taken as being flat. Infiltration of outside air amounts to 1.0 air change/h. Thermal transmittances are as follows: windows, 5.7 W/m^{2}K; walls, 0.5 W/m^{2}K; doors, 5.7 W/m^{2}K; roof, 0.3 W/m^{2}K; floor, 0.6 W/m^{2}K. Find the total rate of heat loss from the building under steady-state conditions, the room air temperature and the environmental temperature.

The calculation of $\sum(AU)$ is given in Table 3.11.
Using $F_1 = 1.0$, $F_2 = 1.10$, $N = 1.0$, $V = 600$ m^3, $t_c = 21°C$ and $t_{ao} = −4°C$

$$Q_p = (1 \times 415.6 + 0.33 \times 1.1 \times 1.0 \times 600)(21 + 4)$$

$$= 15\,835 \text{ W}$$

$$= 15.835 \text{ kW}$$

$$t_{ai} = 1.1[21 − (−4)] − 4°C$$

$$= 23.5°C$$

$$t_{ei} = 1[21 − (−4)] − 4°C$$

$$= 21°C$$

EXAMPLE 3.7

The air-conditioning system in a computer room breaks down, and it is thought that there would be a risk of condensation forming from the moisture in the air if the room air temperature were to fall below 10°C. Assess the likely room air temperature from the following information: room dimensions, 12 m × 9 m × 3.3 m high; dimensions of window in one long exterior wall, 2.5 m × 2.2 m; ventilation rate, 0.50 air changes/h. The adjacent rooms, those below and above, are all at $t_{ei} = 19°C$. The outside air temperature is −2°C. U values of the external wall, the window and the internal surfaces are 0.6, 2.8 and 1.6 W/m^{2}K respectively.

Table 3.11 Data for Example 3.6.

Surface	Area A (m^2)	U (W/m^{2}K)	AU (W/K)
Windows	20	5.7	114.0
Doors	8	5.7	45.6
Walls	152	0.5	76.0
Roof	200	0.3	60.0
Floor	200	0.6	120.0
			$\sum(AU) = 415.6$

The surrounding rooms will steadily transfer heat into the computer room and then this heat will escape through the one external wall and by natural ventilation. The air temperature of the computer room should stabilize at some value t_1 °C. A balance of heat flows into and out of the room can be made:

heat flow in = heat flow out

$$\sum (UA\ \Delta t) = \sum (UA\ \Delta t) + Q_v$$

We can assume that, initially, the computer room environmental temperature is the same as its air temperature. The internal partition, floor and ceiling surface area is 315 m², the window area is 5.5 m² and the external wall area is 34.1 m². Then:

$$\text{heat flow in} = 1.6\ \frac{W}{m^2 K} \times 315\ m^2 \times (19 - t_1)\ K$$

$$= 504(19 - t_1)\ W$$

$$\text{heat flow out} = 2.8\frac{W}{m^2 K} \times 5.5\ m^2 \times (t_1 + 2)\ K$$

$$+ 0.6\ \frac{W}{m^2 K} \times 34.1\ m^2 \times (t_1 + 2)\ K$$

$$+ 0.33\ \frac{W}{m^3 K} \times 0.5 \times 12 \times 9 \times 3.3\ m^3 \times (t_1 + 2)\ K$$

$$= 15.4(t_1 + 2) + 20.46(t_1 + 2) + 58.8(t_1 + 2)\ W$$

$$= 94.66(t_1 + 2)\ W$$

Therefore:

$$504(19 - t_1) = 94.66(t_1 + 2)$$

$$5.32(19 - t_1) = t_1 + 2$$

$$101 - 5.3t_1 = t_1 + 2$$

$$99 = 6.3t_1$$

Hence:

$$t_1 = 15.7°C$$

Therefore condensation is unlikely.

Where a building is occupied only occasionally, for example, a traditional heavyweight stone church or a brick-built assembly hall, the heating system is used intermittently and steady-state heat loss calculations are inappropriate. Admittance factors are used to evaluate the heat flow into the thermal storage of the structure, rather than through it. The heat output required for the heating system is

$$Q_p = \left[F_1 \sum (AY) + 0.33 F_2 NV \right] (t_c - t_{ao})\ W$$

The Y values given in Table 3.12 are for a 12 h on, 12 h off heating cycle. To obtain other cycle times, multiply the Y values by (12/cycle hours)$^{0.5}$. This gives higher heat input rates for shorter periods.

Table 3.12 Thermal transmittance and admittance factors for complete structural components with normal exposure.

Construction	U (W/m²K)	Y (W/m²K)
Walls		
220 mm brick, 13 mm light plaster	1.90	3.6
220 mm brick, 25 mm cavity, 10 mm plasterboard on dabs	1.50	2.5
220 mm brick, 25 mm cavity, 10 mm foil-backed plasterboard on dabs	1.20	1.8
220 mm brick, 20 mm glass fibre quilt, 10 mm plasterboard	1.00	1.4
220 mm brick, 25 mm polyurethane slab, 10 mm plasterboard	0.66	1.0
19 mm render, 40 mm expanded polystyrene slab, 200 mm lightweight concrete block, 13 mm light plaster	0.40	2.2
10 mm tile hanging, 25 mm air gap, 100 mm glass fibre quilt, 10 mm plasterboard	0.36	0.67
105 mm brick, 25 mm cavity, 105 mm brick, 13 mm dense plaster	1.50	4.4
105 mm brick, 50 mm UF foam, 105 mm brick, 13 mm light plaster	0.55	3.6
105 mm brick, 25 mm cavity, 100 mm lightweight concrete block, 13 mm light plaster	0.92	2.2
100 mm lightweight concrete block, 75 mm glass fibre, 100 mm lightweight concrete block, 13 mm lightweight plaster	0.29	2.4
Roof		
5 mm asbestos cement sheet	6.5	6.5
10 mm tile, loft space, 10 mm plasterboard	2.6	2.6
10 mm tile, loft space, 100 mm glass fibre quilt, 10 mm plasterboard	0.34	0.66
19 mm asphalt, 25 mm stone chippings, 150 mm heavyweight concrete block	2.3	5.2
19 mm asphalt, 13 mm fibreboard, 25 mm air gap, 75 mm glass fibre quilt, 10 mm plasterboard	0.40	0.69
Floor		
Concrete	—	6.0
Concrete, carpet or woodblock	—	3.0
Suspended timber and carpet	—	1.5
Partitions		
Heavyweight partition walls	—	3.0
Windows		
Single-glazed	—	5.6
Double-glazed	—	3.2

Boiler power

The boiler power required for a building is found from the sum of the following:

1. peak heat input rate $(Q_f + Q_v)$ W to the heating system;
2. heat loss from the distribution pipe or duct system, which can initially be taken as 10% of $(Q_f + Q_v)$ W and refined later when pipe sizes and lengths are known;
3. rate of energy supply Q_{HWS} W to the hot-water services system where this is supplied from the same boiler plant.

Then:

$$Q_{HWS} = \frac{\text{mass of stored water kg}}{\text{heating period}} \times 4.186 \, \frac{kJ}{kg \, K} \times (t_{HWS} - 10) \, K$$

The specific heat capacity of water is 4.186 kJ/kg K and the temperature of mains cold water is normally about 10°C. The mass of stored hot water at 65°C either will be 135 l for a small domestic residence or can be found from the number of occupants N and the expected daily hot-water usage per person, which will vary from 4 l/person for an office or shop to 70 l/person for a hotel. The time period to raise the storage cylinder contents to the desired temperature can be varied to suit the site conditions; 3 h is acceptable for housing.

EXAMPLE 3.8

Calculate the boiler power required for a commercial building having a peak heat loss of 900 kW, a low-pressure hot-water radiator heating system, 450 occupants and a 3 h heating period for the hot-water storage cylinder. Water is to be stored at 65°C and daily consumption is expected to be 65 l/person. One litre of water weighs 1 kg.

$$Q_{HWS} = \frac{450 \text{ people}}{3 \text{ h}} \times 65 \frac{l}{\text{person}} \times \frac{1 \text{ kg}}{1 \text{ l}} \times 4.186 \frac{kJ}{kgK} \times (65 - 10 \text{ K})$$
$$\times \frac{1 \text{ h}}{3600 \text{ s}} \times \frac{1 \text{ kWs}}{1 \text{ kJ}}$$
$$= 624 \text{ kW}$$

The boiler power is obtained as follows:

$$\text{building heat loss } Q_f + Q_v = 900 \text{ kW}$$
$$10\% \text{ distribution losses} = 90 \text{ kW}$$
$$Q_{HWS} = 624 \text{ kW}$$
$$\text{boiler power} = 1614 \text{ kW}$$

Thermal transmittance measurement

The current stock of buildings creates the need for the energy engineer to be able to discover the value of the thermal transmittance of existing structures. The building has design U values that were calculated in accordance with standard practice and regulations, but what is the reality of the designer's intentions? Does the design U value exist in the components that have been constructed? When the components of the building, such as walls, windows, corners of walls, and interfaces between walls and floors, are taken together as an integrated package, are the U values achieved? Has the process of construction destroyed the designer's work? Such possibilities have a lasting influence upon the energy consumption of the building.

A large proportion of the energy consumed in the UK is used to keep the inside of buildings warm. Monitoring the quality of in situ thermal insulation and for the retrofitting of additional insulation to existing structures is an important part of energy management. The built thermal transmittance (U value) of a structure can be calculated from measurements of air and surface temperatures from a thermocouple temperature instrument or multi-channel data logger (Figs 1.3, 1.11 and 1.12). It is not necessary to know the constructional details of the wall, roof, floor, door or window in order to discover its U value. The visiting surveyor will not wish to drill holes through brick, concrete and timber to measure the thickness of each material. Even if this is done, the quality of the materials remains largely unknown and assumptions about the water content and the integrity of each layer would have to be made. The constructional detail is unknown. There may be air spaces, vapour barriers and layers of thermal insulation in place, but these are hidden from view.

Figure 3.1 represents a cross-section through the unknown structure. It could be an external wall, internal wall, roof, floor, glazing or door. All that can be realistically assessed are the temperatures on either side, at nodes 1 and 4, and on the surfaces at nodes 2 and 3. A shielded surface-contact thermocouple probe can be used to measure each surface temperature. An exposed thermocouple junction or a sling psychrometer can be used to find the air temperatures. The values for the inside and outside surface film resistances, R_{si} and R_{so} m^{2}K/W, are assumed to be their normal, tabulated values for the appropriate applications. The heat transfer equations (Chapter 10) that describe the heat flow through the structure, Q W, are as follows.

For the whole structure:

$$Q = U \frac{W}{m^2 K} \times A \, m^2 (t_1 - t_4) \, K$$

For the interior film:

$$Q = \frac{1}{R_{si}} \frac{W}{m^2 K} \times A \, m^2 (t_1 - t_2) \, K$$

For the unknown structure:

$$Q = \frac{1}{R} \frac{W}{m^2 K} \times A \, m^2 (t_2 - t_3) \, K$$

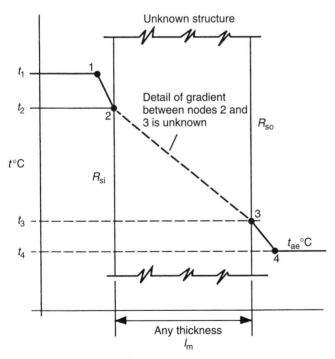

3.1 Temperature gradient through a structure.

where R m^{2}K/W is the resistance of the unknown parts of the construction. For the exterior film:

$$Q = \frac{1}{R_{so}} \frac{W}{m^2K} \times A \, m^2 (t_3 - t_4) \, K$$

The heat flow is considered to be under steady-state conditions: that is, it remains at a stable rate over several hours and certainly while the measurements are taken. This is the same assumption that is made for calculating thermal transmittances. It is also true that the daily cyclic variation in outdoor air temperature and solar heat gains, plus the intermittent cooling effects of wind and rain, cause unsteadiness in the flow of heat from the building. The analysis of such heat transfers requires dedicated software, weather data and a computer model of the whole building. An awareness of the overall problem is helpful, however.

There needs to be as large a temperature difference between indoors and outdoors as reasonably practical on the day of test. This is to minimize the effect of any errors in the measurement of the temperatures. When the indoor and external air temperatures are 20°C and 10°C, an error of 0.5°C in one of the temperatures will be $100 \times 0.5/(20 - 10)$%, 5% of the difference. If overall inaccuracies can be kept within 5%, a reasonably reliable outcome can be obtained. The heating system and weather should also be functioning under steady conditions during the test period. Take the values of R_{si} and R_{so} to be 0.12 and 0.06 m^{2}K/W, as they would be for walls with normal exposure. Use other values if necessary. If, on a test, the temperatures t_1, t_2, t_3 and t_4 are 21°C, 17°C, 0°C and −2°C the rate of heat flow Q, thermal transmittance U and resistance R of the structure can be calculated by using the R_{si} or R_{so} equations:

$$Q = \frac{1}{R_{si}} \frac{W}{m^2K} \times A \, m^2 (t_1 - t_2) \, K$$

The surface area A m^2 is taken as 1 m^2:

$$Q = \frac{1}{0.12} \times (21 - 17) \, W$$
$$= 33.33 \, W$$

The same answer results from the use of R_{so}:

$$Q = \frac{1}{R_{so}} \frac{W}{m^2K} \times A \, m^2 (t_3 - t_4) \, K$$
$$Q = \frac{1}{0.06} \times [0 - (-2)] \, W$$
$$= 33.33 \, W$$

Find the U value from:

$$Q = U \frac{W}{m^2K} \times A \, m^2 (t_1 - t_4) \, K$$
$$33.33 = U \frac{W}{m^2K} \times 1 \, m^2 \times [21 - (-2)] \, K$$

$$U = \frac{33.33}{[21 - (-2)]} \text{ W/m}^2\text{K}$$

$$= 1.45 \text{ W/m}^2\text{K}$$

This is an elderly wall, which has a higher thermal transmittance than for modern standards. Consideration can be given as to how much additional thermal insulation is possible. The thermal resistance of the existing structure, without the surface film resistances, can be found from:

$$Q = \frac{1}{R} \frac{\text{W}}{\text{m}^2\text{K}} \times A \text{ m}^2 (t_2 - t_3) \text{ K}$$

$$33.33 = \frac{1}{R} \frac{\text{W}}{\text{m}^2\text{K}} \times 1 \text{ m}^2 \times (17 - 0) \text{ K}$$

$$R = \frac{(17 - 0)}{33.33} \text{ m}^2\text{K/W}$$

$$= 0.51 \text{ m}^2\text{K/W}$$

When the thermal transmittance is known from design calculations or in situ measurements, the thickness of additional thermal insulation that is needed to reduce heat loss can be calculated. This may be desirable in order to align the building with current regulations and improve its energy-using efficiency. Outdated building designs will be less attractive to potential users than new or recently refurbished, low-energy consumption residential, commercial and industrial alternative sites.

The wall U value that was considered here could be lowered from 1.45 to, say, 0.4 W/m^{2}K by the addition of thermal insulation. If the insulation can be injected into the wall cavity no further constructional measures are needed. Where there is no cavity, or if rainwater penetration could result, then an additional internal or exterior layer of material is required. Thermal insulation may not be structurally rigid and it often does not provide a hard-wearing or weatherproof surface. Adding layers to either side of a wall necessitates architectural changes, particularly to fenestration and doorways. If polyurethane board and an internal surface finish of 10 mm plasterboard can be fitted to the interior surfaces, the necessary thickness of insulation can be calculated as follows.

From Table 3.1, the thermal conductivities are:

plasterboard: $\lambda = 0.16$ W/mK

polyurethane board: $\lambda = 0.025$ W/mK

new thermal resistance of whole structure $R_n = \dfrac{1}{U_n} \dfrac{\text{m}^2\text{K}}{\text{W}}$

$$= \frac{1}{0.4} \frac{\text{m}^2\text{K}}{\text{W}}$$

$$= 2.5 \text{ m}^2\text{K/W}$$

resistance of plasterboard $= \dfrac{0.01}{0.16} \dfrac{\text{mK}}{\text{W}}$

$$= 0.0625 \text{ m}^2\text{K/W}$$

resistance of existing wall $= \dfrac{1}{1.45} \dfrac{\text{m}^2\text{K}}{\text{W}}$

$$= 0.69 \text{ m}^2\text{K/W}$$

The additional thermal insulation that is needed is found by subtracting the existing thermal resistance, and that for the new surface finish, from the target thermal resistance:

$$\text{additional resistance needed} = (2.5 - 0.69 - 0.0625) \text{ m}^2\text{K/W}$$

$$= 1.748 \text{ m}^2\text{K/W}$$

$$\text{insulation resistance} = \frac{l \text{ mm}}{\lambda} \frac{\text{mK}}{\text{W}} \times \frac{1 \text{ m}}{10^3 \text{ mm}}$$

$$1.748 = \frac{l \text{ mm}}{0.025} \frac{\text{mK}}{\text{W}} \times \frac{1 \text{ m}}{10^3 \text{ mm}}$$

$$\text{insulation thickness } l \text{ mm} = 1.748 \frac{\text{m}^2\text{K}}{\text{W}} \times 0.025 \frac{\text{W}}{\text{mK}} \times \frac{10^3 \text{ mm}}{1 \text{ m}}$$

$$= 43.7 \text{ mm}$$

Materials are available in standard dimensions. The thickness to be used will be the next larger size, 50 mm. Check that the additional insulation calculations have been correctly made and find the real new U value:

$$R_n = \frac{1}{1.45} + \frac{0.01}{0.16} + \frac{0.05}{0.025} \frac{\text{m}^2\text{K}}{\text{W}}$$

$$= 0.69 + 0.0625 + 2.0 \text{ m}^2\text{K/W}$$

$$= 2.752 \text{ m}^2\text{K/W}$$

$$U_n = \frac{1}{R_n} \frac{\text{W}}{\text{m}^2\text{K}}$$

$$= \frac{1}{2.752} \frac{\text{W}}{\text{m}^2\text{K}}$$

$$= 0.36 \text{ W/m}^2\text{K}$$

The new thermal transmittance does not exceed the desired value of 0.4 W/m^{2}K and is suitable. If the wall has an air cavity between the inner and outer surfaces, it may be possible to inject urea formaldehyde or phenolic foam, or blown rock wool, and achieve the desired result without architectural effects. Another possibility is the addition of insulation and a protective layer to the exterior surface.

EXAMPLE 3.9

A 20-year-old 150 mm thick ribbed concrete flat roof over an office is supported on a structural steel frame. A typical cross-section through the roof is shown in Fig. 3.2. The concrete is waterproofed with 19 mm asphalt that is topped with 25 mm of white stone chippings. Beneath the concrete, there is a 400 mm deep unventilated air space for service cables and pipes. The ceiling riles are 12 mm thick fibreboard supported on a lightweight galvanized steel frame. The ceiling tile frame is suspended from the structural steel by galvanized wires and self-tapping screws. All the lighting, electrical and other services that are within the ceiling space are supported by hangers from the roof structural steel frame. The roof has normal exposure and its thermal transmittance is to be reduced to 0.25 W/m^{2}K. Thermocouple temperature sensors were used to assess the average thermal

3.2 Roof construction for Example 3.9.

transmittance of the roof structure. On the day of test, the temperatures t_1, t_2 and t_5 were 19°C, 17.6°C and 5°C. The temperature at node 4 could not be measured owing to the roughness of the surface. The temperature at node 3 could be measured but it is not needed. Describe the features of two methods that could be used to insulate the roof. Decide which materials would be suitable and find the correct thickness for the insulation.

From Tables 3.2, 3.3 and 3.5, $R_{si} = 0.1$ m²K/W, $R_{so} = 0.05$ m²K/W (low emissivity), $R_a = 0.17$ m²K/W (high emissivity).

From Table 3.1, the thermal conductivities of the materials are:

cast concrete, lightweight $\lambda = 0.38$ W/mK

asphalt $\lambda = 0.5$ W/mK

stone chippings $\lambda = 0.96$ W/mK

glass fibre quilt $\lambda = 0.06$ W/mK

polyurethane board $\lambda = 0.025$ W/mK

The options to be tried are as follows:

(a) Remove some ceiling tiles and lay a lightweight blanket, such as glass fibre, on top of the tiles. This depends on whether the fibreboard tiles, support wires and screws are able to hold the additional weight. Extra support rods may be needed. There will be considerable disturbance to the room usage. This may preclude installation work during normal working hours. Removing the tiles will disturb dust and debris from the void and necessitate a cleaning operation in the room. Indoor scaffolding will be needed. Care must be taken not to lay the insulation on top of luminaires and electrical cables, to avoid the overheating of lamps and wiring.

$$Q = \frac{1}{R_{si}} \frac{W}{m^2 K} \times A\ m^2 \times (t_1 - t_2)\ K$$

$$= \frac{1}{0.1} \times (19 - 17.6)\ W$$

$$= 14\ W$$

$$14\ W = U\ \frac{W}{m^2 K} \times 1\ m^2 \times (19 - 5)\ K$$

$$U = \frac{14}{19 - 5}\ W/m^2 K$$

$$= 1\ W/m^2 K$$

$$R_n = \frac{1}{0.25}\ \frac{m^2 K}{W}$$

$$= 4\ m^2 W/K$$

$$\text{resistance of the existing roof} = \frac{1}{1}\ \frac{m^2 K}{W}$$

$$= 1\ m^2 K/W$$

$$\text{insulation resistance} = (4 - 1)\ m^2 K/W$$

$$= 3\ m^2 K/W$$

$$\text{glass fibre thickness } l\ mm = 3\ \frac{m^2 K}{W} \times 0.04\ \frac{W}{mK} \times \frac{10^3\ mm}{1\ mm}$$

$$= 120\ mm$$

Glass fibre thickness to be used will be 150 mm.

$$R_n = \frac{1}{1} + \frac{0.15}{0.04}\ \frac{m^2 K}{W}$$

$$= 4.75\ m^2 K/W$$

$$U_n = \frac{1}{4.75}\ \frac{W}{m^2 K}$$

$$= 0.21\ W/m^2 K$$

The new thermal transmittance is below the desired value of 0.25 W/m²K and is suitable. Check that the new U value is correct:

$$R_n = 0.1 + \frac{0.012}{0.06} + \frac{0.15}{0.04} + 0.17 + \frac{0.015}{0.38} + \frac{0.019}{0.5} + \frac{0.025}{0.96} + 0.05\ \frac{m^2 K}{W}$$

$$= 4.73\ m^2 K/W$$

$$U_n = \frac{1}{4.73} \frac{W}{m^2 K}$$

$$= 0.21 \; W/m^2 K$$

(b) Remove the stone chippings from the roof and lay sheets of rigid polyurethane or phenolic foam. An adhesive can be used to hold the sheets in place. The stone chippings are then placed on top of the foam. The foam is water-repellent and rot-resistant. Installation on the outer surface of the roof will not cause disturbance indoors. The roof will become a warm-deck type (Chapter 10) and will gain the benefit of improved thermal storage capacity: that is, the building will remain warmed for longer periods. In summer, the concrete will be insulated from the solar heat gains and hot outdoor air, and will remain relatively cool:

$$\text{polyurethane thickness } l \text{ mm} = 3 \; \frac{m^2 K}{W} \times 0.025 \; \frac{W}{mK} \times \frac{10^3 \text{ mm}}{1 \text{ mm}}$$

$$= 75 \text{ mm}$$

Polyurethane thickness to be used will be 100 mm and the new U value will be 0.2 W/m^{2}K.
  The installation can be validated by measuring the three temperatures when steady-state conditions have been re-established. Calculate the ceiling surface temperature with the glass fibre insulation in place on a day when the indoor and outdoor air temperatures are 21°C and 3°C:

$$Q = 0.21 \; \frac{W}{m^2 K} \times 1 \; m^2 \times (21 - 3) \; K$$

$$= 3.78 \; W$$

$$Q = \frac{1}{R_{si}} \frac{W}{m^2 K} \times A \; m^2 \times (t_1 - t_2) \; K$$

$$3.78 = \frac{1}{0.1} \times (21 - t_2) \; K$$

$$t_2 = 21 - 3.78 \times 0.1°C$$

$$= 20.6°C$$

Questions

1. State what is meant by the following terms:

(a) thermal resistance;
(b) thermal conductivity;
(c) thermal resistivity;
(d) specific heat capacity;
(e) thermal transmittance;
(f) orientation and exposure;
(g) surface resistance;
(h) cavity resistance;
(i) emissivity;
(j) admittance factor;
(k) heavyweight and lightweight structures.

2. The following materials are being considered for the internal skin of a cavity wall:

    (a) 105 mm brickwork;
    (b) 200 mm heavyweight concrete block;
    (c) 150 mm lightweight concrete block;
    (d) 75 mm expanded polystyrene slab;
    (e) 100 mm mineral fibre slab and 15 mm plasterboard;
    (f) 40 mm glass fibre slab, 150 mm lightweight concrete block and 15 mm lightweight plaster.

    Compare their thermal resistances and comment upon their suitability for a residence.
3. Calculate the thermal transmittances of the following:

    (a) single-glazed window, severe exposure;
    (b) double-glazed window, sheltered exposure;
    (c) 220 mm brick wall and 13 mm lightweight plaster;
    (d) 220 mm brick wall, 50 mm glass fibre quilt and 10 mm plasterboard;
    (e) 105 mm brick wall, 10 mm air space, 40 mm glass fibre slab and 100 mm lightweight concrete block;
    (f) 40° pitched roof, 10 mm tile, roofing felt and 10 mm flat plaster ceiling with 100 mm glass fibre quilt laid between the joists;
    (g) 19 mm asphalt flat roof, 13 mm fibreboard, 25 mm air space, 100 mm mineral wool quilt and 10 mm plasterboard.

    All exposures are normal unless otherwise specified.
4. A lounge 7 m long × 4 m wide × 2.8 m high is maintained at a resultant temperature of 21°C and has 1.5 air changes/h of outside air at −2°C. There are two double-glazed wood-framed windows of dimensions 2 m × 1.5 m and an aluminium framed double-glazed door of dimensions 1 m × 2 m. Exposure is normal. One long and one short wall are external and constructed of 105 mm brick, 10 mm air space, 40 mm polyurethane board, 150 mm lightweight concrete block and 13 mm lightweight plaster. The internal walls are of 100 mm lightweight concrete block and are plastered. There is a solid ground floor with edge insulation. The roof has a thermal transmittance of 0.34 W/m²K. Adjacent rooms are at a resultant temperature of 18°C. Calculate the steady-state heat loss from the room for a convective heating system.
5. A single-storey community building of dimensions 20 m × 15 m × 3 m high has low-temperature hot-water radiant panel heaters. There are 10 windows of dimensions 2.5 m × 2 m. Natural infiltration amounts to 1 air change/h. Internal and external design temperatures are 20°C and −1°C. Thermal transmittances are as follows: walls, 0.6 W/m²K; windows, 5.3 W/m²K; floor, 0.5 W/m²K; roof, 0.4 W/m²K. Calculate the steady-state heat loss.
6. Calculate the environmental temperature produced in an unheated room within an occupied building, using the following information: room dimensions, 5 m × 4 m × 2.6 m high; a window of dimensions 2.5 m × 1.25 m in one long external wall; 0.5 air changes/h; external air temperature, 3°C; solid ground floor; surrounding rooms are all at an environmental temperature of 20°C. The thermal transmittances are as follows: external wall, 1 W/m²K; window, 5.3 W/m²K; floor, 0.7 W/m²K; internal walls, 1.2 W/m²K; ceiling, 1 W/m²K. Assume that the room environmental temperature is equal to the air temperature.
7. A single-storey factory is allowed to have 35% of its wall area as single glazing and 20% of its roof area as single-glazed roof-lights. Wall and roof U values are not to exceed 0.6 W/m²K. An architect proposes a building of dimensions 50 m × 30 m × 4 m high with a wall U value

of 0.4 W/m^{2}K, a roof U value of 0.32 W/m^{2}K, 20 double-glazed windows each of area 16 m^2 having a U value of 3.3 W/m^{2}K and 35 roof-lights each of area 10 m^2 having a U value of 5.3 W/m^{2}K. Does the proposal meet the allowed heat loss limit?

8. Calculate the boiler power required for a building with a heat loss of 50 kW and an indirect hot-water storage system for 20 people, each using 50 l of hot-water at 65°C per day. The cylinder is to be heated from 10°C in 2.5 h. Add 10% for pipe and cylinder heat losses.

9. A single-storey building has dimensions 40 m × 20 m × 4 m high with windows of area 80 m^2 and a door of area 9 m^2. It is to be maintained at a resultant temperature of 20°C when the outside is at −1°C and natural ventilation amounts to 1 air change/h. Thermal transmittances are as follows: walls, 1.8 W/m^{2}K; windows, 5.3 W/m^{2}K; door, 5 W/m^{2}K; floor, 0.6 W/m^{2}K; roof, 1.8 W/m^{2}K. A convective heating system is used. It is proposed to reduce the U values of the walls and roof to 0.4 and 0.3 W/m^{2}K respectively. Calculate the percentage reduction in heater power that would be produced.

10. List the ways in which existing residential, commercial and industrial buildings can have their thermal insulation improved. Discuss the practical measures that are needed to protect the insulation from deterioration.

11. Review the published journals and find examples of buildings where the existing thermal insulation has been upgraded. Prepare an illustrated presentation or article on a comparison of the outcomes from the cases found.

12. Write a technical report on the argument in favour of adding thermal insulation to existing buildings. Support your case by referring to government encouragement, global energy resources, atmospheric pollution, legislation, cost to the building user and the profitability of the user's company.

13. A flat roof over a bedroom causes intermittent condensation during sub-zero outdoor air temperatures. The roof has normal exposure. The owners want to eliminate the condensation and reduce the thermal transmittance to 0.15 W/m^{2}K. Thermocouple temperature sensors were used to assess the average thermal transmittance of the roof structure. On the day of test, the indoor air, ceiling surface and outdoor air temperatures were 16°C, 11°C and −2°C. Calculate the existing thermal transmittance of the roof and the thickness of expanded polystyrene slab that would be needed.

14. An external solid brick wall is to be insulated with phenolic foam slabs held on to the exterior brickwork with UPVC hangers. Expanded metal is to be fixed onto the outside of the foam and then cement rendered to a thickness of 12 mm. The wall has a sheltered exposure. The intention is to reduce the thermal transmittance to 0.3 W/m^{2}K. Thermocouple temperature sensors were used to assess the average thermal transmittance of the wall prior to the design work. On the day of test, the indoor air, interior wall surface and outdoor air temperatures were 15°C, 12.7°C and 6°C. Calculate the existing thermal transmittance of the wall and the thickness of phenolic foam that would be needed. If the foam is only available in multiple thicknesses of 10 mm, state the thermal transmittance that will be achieved for the wall. Calculate the internal surface temperature that should be found on the wall for a day when the indoor and outdoor air temperatures are 18°C and 0°C.

15. The roof over a car-manufacturing area consists of 4 mm profiled aluminium sheets on steel trusses. Wood wool slabs, 25 mm, are fitted below the roof sheets. The roof trusses remain uninsulated as they protrude through the wood wool. The trusses cause condensation to precipitate onto the vehicle bodies during cold weather. The roof is to be insulated with polyurethane board, which will be secured to the underside of the roof trusses. The roof has a normal exposure. The intention is to reduce the thermal transmittance to 0.25 W/m^{2}K. Thermocouple temperature sensors were used to assess the average thermal transmittance of the roof prior to the insulation. On the day of test, the indoor air under the roof was 13°C,

internal roof surface temperature was 11°C and the outdoor air temperature was 2°C. Calculate the existing thermal transmittance of the roof and the thickness of polyurethane that would be needed. The insulation is only available in multiple thicknesses of 10 mm. State the thermal transmittance that will be achieved for the roof. Calculate the internal surface temperature that should be found on the newly insulated roof for a day when the indoor and outdoor air temperatures are 16°C and −5°C.

16. Which of these buildings have a slow response, several hours, to variations in weather? More than one correct answer.

    1. Concrete- and steel-framed 20-storey offices.
    2. Traditional stone churches.
    3. London underground railway stations.
    4. Large volume single-storey industrial buildings having lightweight thermal insulation to corrugated sheet steel wall and roof cladding, for example, aircraft hanger, car factory.
    5. Small prefabricated building, transportable, temporary site accommodation, caravan, tent and marquee.

17. Which of these is correct?

    1. Thermal resistivity is the fire resistance property of a material.
    2. Thermal resistance is the total resistance to flow of water through a heating system circulation.
    3. Thermal conductivity is used in calculating the resistance of an electrical heating wiring system.
    4. Thermal resistance is a material component property and is measured in $m^2 K/W$.
    5. Thermal resistance is how many hours electrical cable insulation resists fire in the building.

18. Which of these is correct?

    1. The sheet of glass in a window provides a significant thermal resistance.
    2. Thermal conductivity of window glass is around 1 W/mK.
    3. Thermal conductivity of window glass is around 1 mK/W.
    4. Window glass is only used to keep wind out of the building.
    5. Windows create no thermal resistance to heat flow.

19. Which is the correct unit?

    1. Thermal conductivity $m^2 K/W$.
    2. Thermal transmittance $W/m^3 K$.
    3. Thermal conductivity mK/kJ.
    4. Thermal resistivity W/mK.
    5. Thermal resistivity mK/W.

20. What does $\sum (UA \, \Delta t)$ mean?

    1. Something in Greek language.
    2. Universal ASHRAE temperature difference used for building heat gain calculation.
    3. Integration of U values and areas during a time interval.
    4. Summation of thermal transmittance, surface area and indoor–outdoor air temperature difference of each external element of the building.
    5. All the U values, surface area and temperature differences added together for the whole building.

21. What does $0.33NV\Delta t$ mean?

    1. One-third of the volumetric air change rate multiplied by daily degree days above base temperature.
    2. 33% of normal building volume per degree temperature difference to calculate energy usage cost.
    3. A design guide to the plant room floor area likely to be required for air-handling units.
    4. A fraction of the nominal building volume multiplied by air temperature difference.
    5. Volumetric specific heat capacity of air, times number of air changes per hour, times room volume, times indoor–outdoor air temperature difference, calculates natural ventilation rate of heat loss for a heating system.

22. What are admittance values?

    1. Solar heat gain factors for windows and opaque structures.
    2. The opposite if resistance values.
    3. Number of people who can pass through the buildings' entry and transportation systems at peak flow periods.
    4. Always twice the thermal transmittance value.
    5. Thermal factors evaluating heat flows into thermal storage of the structure.

23. Which does not apply to admittance values?

    1. Y W/m^{2}K.
    2. Reciprocal of U value.
    3. Used instead of thermal transmittance in certain circumstances.
    4. Expresses heat flow inwards to a heavy mass structural component.
    5. Used for highly intermittently heated buildings.

24. Which explanation of thermal conductivity is correct?

    1. Ability of a material to conduct electricity.
    2. Property evaluating materials' ability to pass heat.
    3. Equal to resistivity multiplied by thickness.
    4. Units are W/m^{3}K.
    5. Units are W/m^{2}K.

25. Which is correct about thermal resistance?

    1. Calculated from data tables and computer programs.
    2. Calculated from material thickness divided by thermal conductivity.
    3. Calculated by dividing material thickness in metres by thermal resistivity in mK/W.
    4. Units are kJ/m^{2}K.
    5. Units are W/mK.

26. Which of these calculated values of thermal resistance is not correct?

    1. 110 mm of brickwork is 0.13.
    2. 150 mm of fibreglass roof insulation is 3.75.
    3. 100 mm concrete having a thermal conductivity of 2.0 W/mK is 0.05.
    4. A metal window frame 20 mm thickness has a thermal conductivity of 50.0 W/mK and has a thermal resistance of virtually zero.
    5. A low-energy building wall has 1.0 m thickness of phenolic foam having a thermal conductivity of 0.04 W/mK making a thermal resistance of 25.0 m^{2}K/W.

27. Which of these calculated values of thermal resistance is not correct?

    1. 110 mm of brickwork is 0.13.
    2. 150 mm of fibreglass roof insulation is 3.75.
    3. 100 mm concrete having a thermal conductivity of 2.0 W/mK is 0.05.
    4. A metal window frame 20 mm thickness has a thermal conductivity of 50.0 W/mK and has a thermal resistance of virtually zero.
    5. A low-energy building wall has 1.0 m thickness of phenolic foam having a thermal conductivity of 0.04 W/mK making a thermal resistance of 25.0 m²K/W.

28. Which of these is correct?

    1. U value is the sum of all thermal resistivities in a structure.
    2. R value is the sum of all thermal resistivities in a structure.
    3. Y value is the sum of all thermal resistivities in a structure.
    4. U value is the sum of all thermal resistances in a structure.
    5. R value is the sum of all thermal resistances in a structure.

29. Which is the correct description of a thermal transmittance exposure location?

    1. Sheltered is only for below-ground structures.
    2. Normal exposure means the surface faces the prevailing wind direction.
    3. Severe exposure applies to fifth floors anywhere.
    4. Severe exposure applies to floors above fifth floor in suburban districts.
    5. Sheltered exposure means the building is surrounded by trees.

30. Which is correct about an existing structure's thermal transmittance?

    1. Can only be calculated from design information.
    2. Cannot be measured in situ.
    3. Measurement requires a thermal imaging camera.
    4. Measure structural temperatures to calculate U value.
    5. Thermocouple temperature sensors have to be buried into drilled holes through the structure.

4 Heating

Learning objectives

Study of this chapter will enable the reader to:

1. state the applications for hot-water radiators, natural and fan convectors, embedded pipe radiant panel systems and overhead radiant panels;
2. discuss the use of centralized and decentralized forms of heating system;
3. state the applications for electrical heaters such as radiators, convectors and thermal storage radiators;
4. demonstrate the use of under floor and ceiling heating systems utilizing electrical energy;
5. apply appropriate heat emitters to the user's needs;
6. explain the use of warm-air heating methods;
7. understand the low-, medium- and high-pressure classifications and applications for water heating systems;
8. understand schematic pipe layouts and pump positioning;
9. design hot-water pipe systems;
10. understand the requirements of oil-firing equipment;
11. have an understanding of combustion;
12. describe flues for oil boilers;
13. calculate the room air temperature to be found during a commissioning test on a heating system;
14. understand the principles of electrical power generation and the use of combined heat and power plant;
15. describe the uses of district heating system;
16. understand the uses of a BEMS and its terminology;
17. differentiate between local control and overall supervision;
18. understand how a computer system enhances the engineer's work.

Key terms and concepts

absorption heat pumps 115; building energy management system (BEMS) 111; building management system (BMS) 112; ceiling heating 90; chemistry of combustion 103; chimney 102; class of oil 102; combined heat and power (CHP) 108; combustion air 103; computer-based control 111; dedicated 109; digital signal 112; district heating 109; duct 91; efflux velocity 105; electrical power generation 108; embedded pipe system 90; energy management system 111; equivalent length 98; evacuated tube radiant system 92; exothermic reaction 103; expansion vessel 95; fan convector 88; flue gas constituents 103; free-standing flue 102; geothermal heating system 114; grille 89; groundwater 114; heat exchanger 115; heating system performance testing low-, medium- and high-pressure hot water 106; microbore 95; microprocessor 111; modem 112; natural convector 88; neutral point 93; oil storage and handling 102; one-, two-, three- and four-pipe heating systems 93; outstation 111; panel and column radiators 87; Pascal 98; plant management system 112; plant status 112; pressure drop rate 99; programmable logic controller 111; pump head 98; pump performance curve 99; radiant panel 88; radiator 87; radiator temperature correction factor 97; Redwood oil viscosity 102; sealed system 95; skirting heater 89; storage heater 87; supervisor 112; thermal storage 88; turbulence 103; under floor heating 90; wall-flame, vaporizing and pressure jet burners 103; warm-air system 93; water flow rate 97; water velocity 97.

Introduction

Terminal heat emitters, such as radiators, convectors and warm-air methods, pipework layouts, and pipe and pump sizing are discussed. Oil-firing equipment is described and the combustion process is analysed. Basic flue arrangements are shown.

Heating systems only operate at their design heating duty when the outside air temperature coincides with that used for heat loss calculations; the commissioning engineer needs to relate heating performance on the day of test to the design figures. Such calculations are shown.

Electricity is generated at the expense of usable energy discharged to the atmosphere or the sea. The plant needed to convert this surplus into saleable heat for district heating is outlined. Interest in this subject will develop for various reasons, and the UK lags behind other European countries in the employment of combined heat and power stations.

The control and operational monitoring of heating, air conditioning and other building services has been enhanced by the use of computer-based techniques known as building energy management systems (BEMS). These are explained and clear links with other services are shown.

Heating equipment

A wide variety of heating equipment is available that can heat the occupied space either directly by combustion of a fuel or indirectly by utilizing air, water or steam as a heat transfer fluid. The cost of electricity reflects the complexity of its production and distribution, but from the user's point of view it is a refined source of energy, which can be converted with 100% efficiency. Electrical energy purchased at night can be used to heat water, concrete or cast iron in insulated containers. This stored heat is released when needed.

An economic balance is sought between capital and running costs for each application, bearing in mind the building's use. Automatic controls can monitor water and air temperatures, operational times and weather conditions to minimize fuel and electricity consumption. In order to take maximum advantage of a building's thermal storage capacity, optimum-start controllers are used to vary start and stop times for systems that are used intermittently. Computer control is employed in large buildings, where the capital cost can be offset by reduced energy consumption and personnel savings.

Heat emitters can be classified as follows.

Radiators

Heat emitters providing radiation come into this group. A steel single-panel radiator emits about 15% of its total heat output by radiation and the remainder by convection. Radiant output from multiple panel and column types may be a lower percentage of the total. Electric, gas and coal appliances produce large amounts of convection and are partly convectors.

Types of radiator are:

Hot water: single-, double- or triple-panel column radiators, skirting heaters, recessed panels, banks of pipes.
Electricity: off-peak storage heaters, radiant appliances, convectors, radiant ceiling systems.
Gas, coal and oil: radiant appliances; Fig. 4.1 shows a gas-fired domestic radiant and fan convective appliance often suitable for the elderly or infirm even when central heating provides air heating around the occupants. Air heating alone is not always enough warm for very sedentary residents.

The main characteristics of radiant, also providing convective heat output, appliances are as follows:

Steel single panels: Neat appearance, high heat output per square metre of surface area, easy to clean, narrow.
Steel double panels: Greater heat output per square metre of wall area used, difficult to clean, protrude into the room, more costly. Anti-corrosion chemicals needed in heating water for all steel materials.

4.1 Gas radiant and fan convector room heater.

4.2 Electrically heated oil-filled column steel radiator.

Cast iron panels: Heavy and more obtrusive, low heat output, very long service period.

Steel and cast iron columns: High heat output per square metre of wall area used, bulky, heavy, often mounted on feet, difficult to clean except the hospital pattern which are smooth finished; Fig. 4.2 shows an example.

Radiant panels: Flat cast iron or steel plates with water pipes bonded to their back. They are often mounted at high level in industrial workshops and require a large surface area.

Banks of pipes: Bare steel or copper pipes fitted at skirting level in rooms or storage areas to provide an inexpensive heating surface. Can be installed in floor trenches beneath a decorative floor grille allowing indoor foot traffic to use the floor space unrestricted. Traditional churches often have these and modern buildings have them at the foot of floor-to-ceiling glazed areas to counteract downdraughts.

Off-peak storage: Thermal storage heaters taking electricity at night during less expensive charging periods. The heat is stored at high temperature in cast iron or refractory bricks in an insulated casing. Heat is released continuously into the building unless the heater is fitted with a thermostatically controlled fan and a time switch that determines its operating period. The only other control is over the length of the charge period; this requires estimating the following day's weather pattern. Heaters are bulky and their weight requires attention to the floor structure to ensure sufficient strength.

Convectors

There are two types of convector, natural and fan.

Natural convectors

Natural convectors rely on gravity convection currents produced by the heater. Skirting heaters have a finned pipe inside a sheet metal casing as shown in Fig. 4.3. Their heat emission is about 480 W per metre run, they are light and easily handled and they are less obtrusive than taller equipment. Long lengths of unobstructed wall space are needed. Where they run behind furniture, the finned element is omitted and a plain pipe is installed to reduce heat output. They are always fitted onto two-pipe systems and the return pipe can be fitted inside the casing. Valves and air vents are enclosed in accessible boxes at the ends of continuous lengths. Natural convectors produce a uniformly rising current of warm air around the perimeter of the room and this is effective in producing a comfortable environment. There is negligible radiant heating.

Other natural convectors are either 1 m high or extend up to room height, as shown in Fig. 4.4. They create strong convection currents with little radiation and are particularly suitable for locations where elderly, very young or disabled people are being cared for as there are no hot surfaces that may cause skin burns or start fires.

Natural convectors have high heat outputs and can be built into walls, cupboards or adjacent rooms to improve their appearance. Electricity or low- or medium-temperature hot water can be used as the heating medium. The heating elements need periodic cleaning. Such heaters are used in locations where quiet operation and the lack of draughts or intense radiation are important design considerations, such as libraries, art galleries and antique furniture stores.

Fan convectors

Fan convectors have a similar construction to natural convectors with the addition of one or more centrifugal fans and an air filter. Heat output can be very high and fans may be operated at various fixed speeds or from variable-speed motors. Figure 4.5 shows a typical arrangement. Fan operation is controlled from built-in thermostats or remote temperature sensors.

Installation can be at low or high level and the heated air stream is directed away from sedentary occupants. Fan convectors can be usefully sited at doorways to oppose incoming cold air and rapidly reheat entrance areas.

4.3 Skirting convector heater.

4.4 Natural convector heater.

Sheet steel case
Removable pane
Finned tube

1.0 – 3.0 m high

Centrifugal fan and motor

4.5 Fan convector.

4.6 Balanced-flue gas-fired fan convector.

A two-pipe circuit must be used, and fan convectors are installed on separate circuits from hot-water radiators as their control characteristics are different. Constant-temperature hot-water is supplied to them, whereas radiators may have variable water temperatures to reduce heat output in mild weather.

Figure 4.6 is of a gas-fired balanced-flue fan convector that can be used when a central piped water circulation heating system is not practical. Figure 4.7 is of the outdoor balanced-flue terminal.

Figure 4.8 shows a typical electric off-peak storage heater, which may be a natural or fan convector.

Embedded pipes and cables

Low-temperature hot-water heating pipes or electric heating cables are buried in concrete walls, floors or ceilings to provide a large low-temperature surface that is maintained at a few degrees above room air temperature. Floor-to-ceiling air temperature gradients tend to be less than those

4.7 Balanced-flue terminal.

Insulated sheet steel case

Electric resistance heating elements

Concrete or cast iron storage blocks

4.8 Off-peak electric storage heater.

obtained with more concentrated forms of heat emission and a uniform distribution of comfort is produced. An example is shown in Fig. 4.9.

Soft copper pipes are laid in position on the concrete floor slab and held by clips, and the ends are connected to header pipes in service ducts. Joints are avoided for the under floor sections. Steel or plastic pipes may be used in some situations. Thermal expansion and contraction of the pipework must be accommodated and the floor surface temperature is limited to avoid damage to the structure, surface finishes or occupants. This is done by enclosing the pipe in a hard asbestos sleeve on water pipes operating at 85°C or by controlling water temperature to 45°C with a mixing valve system. Pipes are buried in the floor screed. Heating elements are evenly distributed to provide uniform radiation and convection to the occupants.

Electric ceiling heating can consist of buried cables or a flexible conducting mat fixed between the ceiling joists and plasterboard, as shown in Fig. 4.10. The mat is electrically insulated from the structure and connected to 240 V 50 Hz supplies to rise to a surface temperature of 40°C.

4.9 Embedded panel heating.

4.10 Radiant ceiling heating.

Radiant panels

Radiant panel systems employ either a high- or a low-temperature surface to transmit heat by radiation directly to the occupants, and to other unheated surfaces, producing an elevated mean radiant temperature. Comfort conditions can be maintained with lower air temperatures than with convective systems. This should result in economical running costs.

Convection heat output from the hot radiant source is minimized by placing thermal insulation over the reflector. High-temperature radiation is generated using gas combustion close to ceramic reflectors, which emit some heat in the visible part of the infrared region and consequently are seen to be contributing to a feeling of warmth. Domestic gas fires and industrial heaters are in this category. Covered pedestrian areas of shopping precincts can be warmed from recessed units in canopies.

The effect of using high-temperature panels is to produce a series of localized 'sun spots' over a small floor area. Careful siting is necessary to avoid overheating of people or objects.

Low-temperature systems utilize hot water, air or flue gas to heat a metal sheet or pipe, which emits long-wave infrared radiation outside the visible band. They can be installed in factory or office environments and produce a uniform overall warmth, assisted by re-radiation and convection from surfaces heated by the radiant source. Unlike convective systems, they are not adversely affected by room height. Complete systems can be suspended from the ceiling, leaving floors uncluttered.

An evacuated tube system is shown in Fig. 4.11. Flue products from the gas burners are drawn along steel pipes by a vacuum pump and discharged to the atmosphere.

4.11 Overhead industrial radiant tube heating system.

Warm air

Recirculated room air is heated either directly or indirectly by the energy source. Direct firing of combustion gases into the air is permissible only in large well-ventilated factory premises. All other applications require a fuel-to-air heat exchanger where the combustion products are enclosed in a sheet metal passageway. Room air is passed over the outside of this heating surface.

Heated air is passed through ducts to the occupied space. It is diffused into the room through a grille, which mixes it with room air convection currents and avoids draughts. Each grille has a damper to regulate the air flow. Extract grilles and ductwork return the air to the heater. Care is needed not to extract air directly from kitchens and bathrooms, as this would lead to odours and condensation in living areas.

The main advantage of warm-air systems is quick heating up and response to thermostatic control. A source of radiant heat is needed in the sitting room to complement the otherwise purely convective heating. A typical domestic installation is shown in Fig. 4.12, where the heater is fitted in a cupboard which is centrally located with respect to all the rooms. Stub ducts are used to connect the heater to the supply and recirculation grilles.

Hot-water heating

The basic arrangements of the various hot-water heating systems are shown in Figs. 4.13–4.16. Hot-water heating systems are classified by the temperature and pressure at which they operate (Table 4.1).

The pump position relative to the cold feed and vent pipe connections is important in systems with an open expansion tank. The water pressure rise across the pump can be considerable, and if the arrangement is incorrect water can be pumped up the vent pipe and discharged into the open tank. The connection of the cold-feed pipe to the circulation system is known as the neutral point, Fig. 4.18. It is here that the water pressure is always equal to the static height of water above it, with the pump exerting no additional pressure.

A satisfactory arrangement is shown in Fig. 4.18. The hydraulic gradient shows the variation of total water pressure throughout the circulation, that is, the sum of the static head and the pump head, some of which generates suction pressure between the neutral point and the pump inlet.

4.12 Ducted gas-fired warm-air heater installation.

4.13 Low-temperature hot-water one-pipe heating system.

The design of a pumped heating system is approached in the following way.

1. Calculate room heat losses.
2. Decide radiator and convector positions and then their sizes from manufacturers' literature.
3. Calculate the water flow rate for each heat emitter.
4. Design the pipework layout.
5. Mark the water flow rates on the pipework drawing and add them up all the way back to the boiler from the furthest heater, marking the drawing with each value.

4.14 Low-temperature hot-water two-pipe heating system.

4.15 Low- or medium-temperature hot-water microbore heating system.

4.16 Low- or medium-temperature hot-water heating system using a sealed expansion tank.

Table 4.1 Classification of hot-water heating systems.

Type	Flow temperature (°C)	Pressure
LTHW	80	Open to atmosphere, 1 bar
MTHW	120	7 m head above atmosphere, 2 bar
HTHW	Over 120	In excess of 7 m head, 3 bar

Note: LTHW, low-temperature hot-water; MTHW, medium-temperature hot-water (refer to Fig. 4.17); HTHW, high-temperature hot-water.

4.17 Pressurization equipment for a high-temperature hot-water heating system.

4.18 Neutral point in a heating system.

6. Choose pipe sizes from a chart or data, using an estimate of pressure loss rate and maximum allowable water velocity.
7. Calculate the pump head and water flow rate; compare with the pump manufacturer's stated performance curves and choose a suitable pump.

The temperature of water flowing through a heat emitter or along a pipe will drop from its flow temperature $t_f °C$ to its return temperature $t_r °C$ and lose heat Q W. If there is one heater in a room, Q will be equal to the room steady-state heat loss. Heat losses from distribution pipework may initially be assumed to be 10% of the room heat loss, and the radiator water flow rate should be increased by this amount.

The specific heat capacity (SHC) of water is 4.19 kJ/kgK. The heat balance equation is:

heat lost by water = radiator heat output Q + pipe heat loss

or

water flow rate $q \times SHC \times$ temperature drop $= Q + 10\% \times Q$

$$\text{water flow rate } q = \frac{1.1\,Q}{SHC(t_f - t_r)} \text{ kg/s}$$

The radiator manufacturer's heat output data will be for a fixed temperature difference between the mean water temperature and the room air temperature. Usually a figure of 55 K is used. Comparison of design conditions and literature data can be made using the following equations:

$$\text{radiator mean water temperature } MWT = \frac{t_f + t_r}{2}$$

$$\text{radiator heat output at 55 K difference} = \frac{\text{room heat loss}}{\text{temperature correction factor}}$$

where:

$$\text{temperature correction factor} = \left[\frac{((t_f + t_r)/2) - t_a}{55}\right]^{1.30}$$

and t_a is the room air temperature (°C).

EXAMPLE 4.1

A double-panel radiator is to be installed in a room where the air temperature is 22°C and the heat loss is 4250 W. Water flow and return temperatures are to be 82°C and 71°C respectively. An extract from a radiator manufacturer's catalogue for a temperature difference of 55 K is given in Table 4.2. Select a suitable radiator and find the water flow rate through it.

Table 4.2 Heat output from steel double-panel radiators of 500 and 700 mm height.

Radiator length (m)	Heat output (kW) for 55 K difference		
	500 mm	700 mm	900 mm
1.720	2.00	2.60	2.90
1.920	2.30	2.90	3.30
2.200	2.60	3.25	3.75
2.400	2.85	3.60	4.25
2.600	3.10	3.90	4.80

$$\text{temperature correction factor} = \left[\frac{(82 + 71)/2 - 22}{55} \right]^{1.30}$$

$$= 0.988$$

$$\text{radiator heat output at 55 K difference} = \frac{4250 \text{ W}}{0.988} = 4302 \text{ W}$$

The 2.6 m × 900 mm radiator is needed. Verify that the wall space is available and that there is not a clash with furniture and furnishings in the room. The water flow rate is:

$$q = \frac{1.1 \times 4.25}{4.19 \times (82 - 71)}$$

$$= 0.1014 \text{ kg/s}$$

A first estimate of pipe sizes can be made using an average pressure loss rate $\Delta p/EL$ for the whole system (EL is the equivalent length). Either an arbitrary figure of say 300 N/m^3 is chosen, or a figure is evaluated from the expected pump head. The index circuit length is found from the total flow and return pipe length from the boiler to the furthest radiator. Pipe fittings increase frictional resistance and an initial 25% increase in measured pipe length is made in order to find the equivalent length of the system:

$$\text{pump head } \Delta p \, \frac{N}{m^2} = EL \text{ m} \times \frac{\Delta p}{EL} \, \frac{N}{m^3}$$

This is often converted into head H m water gauge:

$$\text{pump head } H = \Delta p \, \frac{N}{m^2} \times \frac{m^3}{\rho \text{ kg}} \times \frac{kg}{g \text{ N}}$$

$$= \frac{\Delta p}{9.807 \times 1000} \text{ m}$$

$$= \frac{\Delta p}{9807} \text{ m}$$

Water flow rate through the pump is the sum of all the radiator water flow rates. Figure 4.19 shows typical pump performance curves. The allowed pressure loss rate can be assessed from

4.19 Pump performance curves.

the pump characteristic curves. For example, if pump B is to be used at a flow rate of 1 kg/s, the corresponding head developed is 3 m. This is equal to a pressure:

$$\Delta p = 9807\ H$$
$$= 9807 \times 3\ \text{m}$$
$$= 29\,421\ \text{N/m}^2$$

As,

$$29\,421\ \frac{\text{N}}{\text{m}^2} = EL\ \text{m} \times \frac{\Delta p}{EL}\ \frac{\text{N}}{\text{m}^3}$$

$$\text{allowed}\ \frac{\Delta p}{EL} = \frac{29\,421}{EL}\ \text{N/m}^3$$

If the measured index circuit is 50 m of pipework:

$$EL = 50\ \text{m} \times 1.25$$
$$= 62.5\ \text{m}$$

and,

$$\frac{\Delta p}{EL} = \frac{29\,421}{62.5} = 471\ \text{N/m}^3$$

This would be the maximum pressure loss rate, averaged over all the pipework, and pipe sizes would be read from the 460 N/m^3 line in Table 4.3 to ensure that the available pump head was not exceeded (Moss, 1996).

Table 4.3 Flow of water in copper pipes of various diameters.

$\Delta p/EL$ (N/m^3)	Water flow rate q (kg/s) for diameters of					
	6 mm	15 mm	22 mm	28 mm	35 mm	42 mm
200	0.013	0.065	0.174	0.381	0.656	1.060 $v = 1.0$
260	0.015	0.075	0.202	0.441	0.760	1.230
300	0.016	0.081	0.219	0.478	0.823	1.330
360	0.018	0.090	0.242	0.529	0.910	1.470
400	0.019	0.096	0.257	0.561	0.965	1.560
460	0.020	0.104	0.278	0.607	1.040	1.680 $v = 1.50$
500 $v = 0.50$	0.021	0.109	0.291	0.635	1.090	1.760
560	0.023	0.116	0.310	0.677	1.160	1.880
600	0.024	0.120	0.323	0.703	1.210	1.950
660	0.025	0.127	0.340	0.741	1.270	2.050
700	0.026	0.131	0.352	0.766	1.320	2.120
760	0.027	0.138	0.368	0.802	1.380	2.220
800	0.028	0.142	0.379	0.825	1.420	2.280

Note: v, water velocity (m/s); $\Delta p/EL$, rate of pressure loss due to friction.

Source: Reproduced from *CIBSE Guide* by permission of the Chartered Institution of Building Services Engineers.

EXAMPLE 4.2

Figure 4.20 shows the arrangement of a two-pipe low-temperature hot-water heating system serving three radiators. The pipe dimensions indicated apply to both flow and

4.20 Low-temperature hot-water heating system.

return pipes. The frictional resistance of the pipe fittings amounts to 25% of the measured pipe length. Flow and return water temperatures at the boiler are to be 85°C and 72°C respectively. Each radiator is situated in a room air temperature of 20°C. The heat outputs of radiators 1, 2 and 3 are 1, 2 and 3 kW respectively. Pump C of Fig. 4.20 is to be used. Find the pipe sizes for the system.

For radiator 1:

$$q_1 = \frac{1.1 \times 1}{4.19 \times (85 - 72)} = 0.02 \text{ kg/s}$$

Similarly, for radiators 2 and 3:

$$q_2 = 0.04 \text{ kg/s}$$
$$q_3 = 0.06 \text{ kg/s}$$

Water flow in the distribution pipework will be:

$$q_X = 0.12 \text{ kg/s}$$
$$q_Y = 0.10 \text{ kg/s}$$

Water flow rate through the pump is:

$$q_X = 0.12 \text{ kg/s}$$

Available pump head, from Fig. 4.19, is:

$$H = 6 \text{ m}$$

Pump pressure rise is:

$$\Delta p = 9807 \times 6 \text{ N/m}^2$$
$$= 58\,842 \text{ N/m}^2$$

The index circuit is from the boiler to radiator 3; thus the measured pipe length is:

$$2(10 + 3 + 10 + 3 + 10) \text{ m} = 72 \text{ m}$$

Therefore,

$$EL = 1.25 \times 72 \text{ m}$$
$$= 90 \text{ m}$$

and,

$$58\,842\ \frac{N}{m^2} = 90\ m \times \frac{\Delta p}{EL}\ \frac{N}{m^3}$$

$$\text{maximum available}\ \frac{\Delta p}{EL} = \frac{58\,842}{90}\ N/m^3$$

$$= 653.8\ N/m^3$$

Thus, from Table 4.2, using $\Delta p/EL = 600\ N/m^3$, the pipe sizes are:

$d_X = 15\ mm$

$d_Y = 15\ mm$

$d_Z = 15\ mm$

Note that the pressure loss rates in pipes Y and Z are 460 N/m^3 and less than 200 N/m^3 respectively. Not all the allowable pump head of 6 m will be absorbed in pipe friction losses. A gate valve beside the pump will be partially closed to increase the resistance of the system. Each radiator has two hand wheel valves, one for temporary adjustments by the occupant and the other for flow regulation by the commissioning engineer.

Oil-firing equipment

Fuel oil is graded in the Redwood no. 1 viscosity test according to its time of flow through a calibrated orifice at 38°C. Vaporizing and wall-flame burners in boilers of up to 35 kW heat output use 28 s oil, pressure jet burners use gas oil class D (35 s), and industrial boiler plant uses grade E (250 s), grade F (1000 s) and grade G (3500 s). Power stations may use 6000 s residual oil, heated to make it flow. This is the tar residue from crude oil distillation and can only be burnt economically on such a large scale.

Figure 4.21 shows a typical domestic oil storage and pipeline installation. In the UK domestic oils can be stored in outdoor tanks. Grades E, F and G require immersion heaters in the tank and pipeline heating to ensure flow.

4.21 Oil storage tank installation.

Wall-flame burners have a rotating nozzle, which sprays oil onto peripheral plates around the inside of a water-cooled vertical cylindrical combustion chamber. An electric spark ignites the oil impinging on the plates, establishing a ring of flame around the walls of the boiler. Correct oil flow rate from the reservoir is controlled by a ball valve.

Vaporizing burners consist of a vertical cylinder that is heated by the flame and evaporates further oil fed into its base from the reservoir flow control.

Pressure jet burners are usually confined to boilers in plant rooms as they produce more noise. Oil is pumped at high pressure through a fine nozzle, forming a conical spray in the furnace. Combustion air is blown into this oil mist from a centrifugal fan. The turbulent interaction of oil and air causes further atomization of the oil droplets, and the mixture is ignited by an electric spark.

Combustion

Combustion is an exothermic chemical reaction that liberates heat. Fuel must be intimately mixed with sufficient oxygen and raised to a temperature high enough for combustion to be maintained. All the carbon and hydrogen in the fuel are burnt into gaseous products that can be safely vented into the atmosphere. Hydrocarbon fuels are highly energy-intensive. They require little storage volume and their combustion is controllable.

The constituents of dry air are 21% oxygen, 79% nitrogen and less than 1% other chemicals such as carbon dioxide, carbon monoxide, nitrous oxides and rare gases, measured by volume. Nitrogen is inert and takes no part in the chemistry of combustion, but it is heated in its passage through the furnace. Typical chemical compositions of fuels are given in Table 4.4.

The quantity of air required for complete combustion and the composition of the products can be evaluated from the fuel chemistry. For methane (CH_4) the complete volumetric analysis would be

methane + air → carbon dioxide + water vapour + nitrogen

The chemical symbols for these are as follows: oxygen, O_2; nitrogen, N_2; carbon dioxide, CO_2; water vapour, H_2O. Therefore (after complete combustion) we have

$$CH_4 + 2O_2 + N_2 \rightarrow CO_2 + 2H_2O + N_2$$

Table 4.4 Fuel data.

	Anthracite	Gas oil	Natural gas
Moisture (%)	8.0	0.05	
Ash (%)	8.0	0.02	
Carbon (%)	78.0	86.0	
Hydrogen (%)	2.4	13.3	
Nitrogen (%)	0.9		2.7
Sulphur (%)	1.0	0.75	
Oxygen (%)	1.5		
Methane (%)			90.0
Hydrocarbons (%)			6.7
Carbon dioxide (%)			0.6

Source: Reproduced from *CIBSE Guide* by permission of the Chartered Institution of Building Services Engineers.

This means that one volume of CH_4, when reacted with two volumes of O_2 during complete combustion, will produce one volume of CO_2 and two volumes of water vapour. All measurements are at the same temperature and pressure. It is assumed that the water vapour is not condensed.

Some condensation is inevitable, however, and when sulphur (S) is present in the fuel, it combines with some of the O_2 to form sulphur dioxide (SO_2). If the gaseous SO_2 comes into contact with condensing water vapour and further O_2, weak sulphuric acid (H_2SO_4) may be formed in the flue. Coagulation of liquid H_2SO_4 and carbon particles from chimney surfaces leads to the discharge of acid smuts into the atmosphere, causing local damage to washing, cars and stonework. Acidic corrosion of the boiler and chimney greatly reduce their service period. The flue gas temperature is kept above the acid dew-point of about 50°C to avoid such problems.

It can be seen from the methane combustion equation that 2 m³ of O_2 are required to burn 1 m³ of CH_4 completely. This O_2 is contained in $2/0.21 = 9.52$ m³ of air, and this air contains, $9.52 - 2 = 7.52$ m³ N_2.

In order to ensure complete combustion under all operating conditions and to allow for deterioration of boiler efficiency between servicing, excess air is admitted. This ranges from 30% for a domestic pressure jet oil burner down to a few per cent in power station boilers where continuous monitoring and close control are essential. Excess O_2 from the excess air appears in the flue gas analyses. Measurement of O_2 and CO_2 levels reveals the quantity of excess air.

The presence of carbon monoxide (CO) in the flue gas indicates that some of the carbon in the fuel has not been completely burnt into CO_2 and that more combustion air is needed. The theoretically correct air-to-fuel ratio is the stoichiometric ratio. Figure 4.22 shows the variation of flue gas constituents with the air-to-fuel ratio.

The CO_2 content of oil-fired boiler plant flues will be about 12% at 30% excess air, the combustion air volume required per kilogram of fuel burnt will be about 14.6 m³ and the flue gas temperature leaving the boiler will be about 200°C. For domestic natural-draught gas-fired boilers, excess air may be 60%, the flue gas temperature will be 165°C and the CO_2 content will be around 7.5%.

Samples of flue gas taken during commissioning and routine servicing are tested for CO_2 and O_2 content by absorption into chemical solutions. The Orsat apparatus is typical. The smoke content is measured by drawing a sample through filter paper and comparing the discoloration with known values.

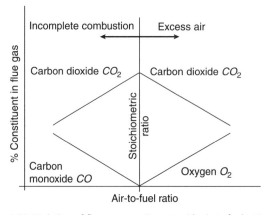

4.22 Variation of flue gas constituents with air-to-fuel ratio.

Flues

Flue systems for oil-fired boilers are of either the conventional brick chimney or free-standing pipe designs so as to discharge combustion products into the atmosphere and allow sufficient dilution so that fumes reaching ground level will not be noticeable. Chimneys from large boiler plant may be subject to minimum height specification by the local authority. Efflux velocity from the chimney can be increased by utilizing a venturi shape when fan assistance is used, effectively raising the chimney height.

The flue pipe diameter will be equal to the boiler outlet connection. Each boiler in a multiple installation should have its own flue. Flues must be kept warm to prevent acidic condensation, and are therefore constructed within the building. Some useful heat is reclaimed in this way. Figure 4.23 shows the necessary separation of flue pipe from combustible materials at an intersection with a floor. External free-standing pipes are constructed of double-walled asbestos or stainless steel to reduce heat loss. Figure 4.24 shows two suitable arrangements.

4.23 Separation of flue pipe from combustible materials in a floor.

4.24 Internal and external free-standing flue pipes.

Performance testing

A radiator heating system is subjected to a water pressure test at greater than its normal working pressure prior to thermal insulation. To check its thermal performance, an assessment is made of the expected room temperature for the external conditions prevailing during the test. Careful measurements of the external air temperatures at regular intervals, the amount of solar radiation, the wind speed, the occupancy, the use of internal lighting and the sources of heat gains are needed.

Under steady-state conditions:

Heat loss from room = heat output from radiator

$$UA\,(t_{ei} - t_{eo}) = \text{constant} \times \left[\frac{(t_f + t_r)/2 - t_{ai}}{55}\right]^{1.3}$$

The radiator constant is its heat emission for a 55 K temperature difference. The room heat loss is a series of constants multiplied by $(t_{ei} - t_{eo})$ K, including ventilation heat loss, and can be characterized by the air temperature difference $(t_{ai} - t_{ao})$ K. Thus $(t_f + t_r)/2$ can be replaced by the mean water temperature $t_m°C$. Therefore the heat balance equation simplifies to:

$$(t_{ai} - t_{ao})_1 = C(t_m - t_{ai})_1^{1.3}$$

and

$$(t_{ai} - t_{ao})_2 = C(t_m - t_{ai})_2^{1.3}$$

where subscripts 1 and 2 refer to the test day and design conditions respectively. These equations can be divided:

$$\frac{(t_{ai} - t_{ao})_1}{(t_{ai} - t_{ao})_2} = \left[\frac{(t_m - t_{ai})_1}{(t_m - t_{ai})_2}\right]^{1.3}$$

The left-hand side of the equation represents the variation of building heat loss with temperature, and the right-hand side represents the heating system performance at various room temperatures. A graph of each side against a base of room air temperature can be drawn to find the room air temperature that satisfies both sides of the equation.

EXAMPLE 4.4

A radiator heating system is designed to produce an internal air temperature of 23°C at an outside air temperature of −2°C with a mean water temperature of 85°C. It was tested on a calm cloudy day, before occupation, when the external air temperature remained stable at 5°C for 2 days. Water flow and return temperatures at the boiler were found to be 88°C and 76°C respectively during the test. Room internal air temperature remained stable at 26°C. State whether the heating system performance met its design conditions.

On the test day,

$$t_m = \frac{88 + 76}{2}\,°C = 82°C$$

and,

$$\frac{t_{ai} - 5}{23 - (-2)} = \left(\frac{82 - t_{ai}}{85 - 23}\right)^{1.3}$$

where t_{ai} is the expected internal air temperature on the test day. Then:

$$\frac{t_{ai} - 5}{25} = \left(\frac{82 - t_{ai}}{62}\right)^{1.3}$$

Assume a range of values for the internal air temperature of say 30°C, 25°C and 20°C. Evaluate each side of the equation (Table 4.5) and plot against t_{ai} (Fig. 4.25). The graphical solution reveals that an internal air temperature of 26.6°C satisfies both equations. This can be verified by substituting 26.6°C for t_{ai} in the heat balance equation. Thus the measured temperature on the test day is sufficiently close to the theoretically expected figure to say that the heating system meets its design specification.

Table 4.5 Radiator heat output test in Example 4.3.

t_{ai} °C	Heat loss	Radiator heat output		
	$(t_{ai} - 5)/25$	$(82 - t_{ai})/62$	$[(82 - t_{ai})/62]^{1.3}$	
30.0	1.0	0.839	0.80	
25.0	0.8	0.919	0.90	
20.0	0.6	1.000	1.000	

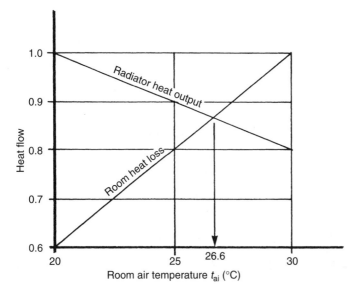

4.25 Variation of room heat loss and radiator heat output with room air temperature in Example 4.3.

Electrical power generation

Electricity is generated by alternators driven by steam turbines in power stations. The largest alternators produce 500 MW of electrical power at 33 kV. The steam is produced in a boiler heated by the combustion of coal or residual fuel oil, which could otherwise only be used for making tar. The oil is heated to make it flow through distribution pipework.

Nuclear power stations produce heat by a fission reaction and the active core is cooled by pressurized water (pressurized water reactor (PWR)), carbon dioxide gas (high-temperature gas-cooled reactor (HTGR)), liquid sodium (fast breeder reactor) or heavy water (Canadian deuterium (CANDU) system). This fluid then transfers its heat to water, boiling it into steam to drive conventional turbines.

Smaller alternators are driven by methane combustion in gas-turbine engines or by diesel engines. A large modern power station has four separate boiler-turbine-alternator sets, producing a total of 2000 MW at a maximum of 38% overall efficiency. Figure 4.26 shows the energy flows in a conventional power station.

Approximately half the input fuel's energy is dissipated in natural-draught cooling towers or sea water, depending on the plant location. Steam leaves the turbine at the lowest attainable sub-atmospheric pressure so that as much power as possible is extracted from it as it passes through the turbines. The temperature of the cooling water may be as low as 35°C, which is of little practical use unless a mechanical heat pump is employed to generate a fluid at 60°C–90°C. The heat could then be pumped to dwellings. Power stations are normally sited away from centres of population and heat transport costs are high.

During the next 25–100 years, the UK is going to have to make more efficient use of its indigenous hydrocarbon reserves, extend nuclear power generation capacity and develop alternative production methods such as tidal, wave, solar, wind, geothermal and hydroelectric plants.

Combined heat and power

Existing power stations generate electricity only, at as high an efficiency as possible. Combined heat and power (CHP) stations produce less electricity and more heat but improve overall fuel efficiency to about 50%, as indicated in Fig. 4.27.

Future CHP plants will be smaller than the present electricity-only plants and will be sited close to centres of industry and population. Coal-fired boilers will be used where practical. Fuel and ash will be mechanically handled and flue gases filtered to remove dust and impurities (Horlock, 1987).

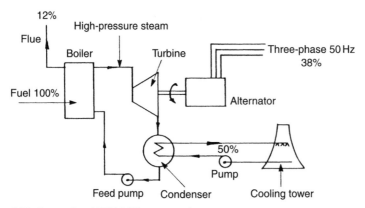

4.26 Conventional 2000 MW power station.

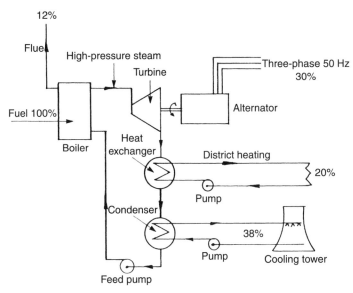

4.27 Combined heat and power plant.

District heating

District medium- or high-pressure hot-water heating, employing two-, three- or four-pipe underground distribution systems, will provide heat primarily to the largest and most consistent users, such as hospitals, factory estates and city centres. Further custom will be won from existing buildings by straight price competition. The street distribution layout is indicated in Fig. 4.28. Flow and return pipes will be well-insulated and may be installed inside one large-diameter pipe, which will form the structural duct and moisture barrier.

The CHP plant generates electricity for the locality and is connected into the national grid. It should also incinerate local refuse, utilizing the heat produced, and recycle materials such as metals and glass. It will provide hot water for sanitary appliances and air conditioning and, as these will be summer as well as winter heat loads, a method of separating them from the heating system will be used. This can be done with the three- and four-pipe arrangements shown in Figs 4.29 and 4.30 to economize on pump running costs and pipe heat losses during the summer.

The supply of heat to each dwelling will be controlled by an electric motorized valve, actuated by a temperature sensor in the heat exchanger, which will enable existing low-pressure hot-water systems to be connected. A heat meter, consisting of a water flow meter and flow and return temperature recorders, will continuously integrate the energy used, and quarterly bills could be issued through a directly linked computer.

Medium- and high-temperature hot-water heating systems are sealed from the atmosphere. Pressurization methods involve restraining thermal expansion, charging with air or nitrogen, or making use of the static head of tall buildings. As the boiling point of water increases with increasing pressure, high flow temperatures can be used. This permits a large drop in temperature from flow to return (50 K or more), and water flow rates can be reduced compared with low-pressure hot-water open systems. Pipe sizes are smaller and the system is more economical to install when used on a large scale.

District cooling from a central refrigeration plant serving air-conditioning units in commercial buildings can be developed alongside a CHP scheme. Underground chilled-water pipework will

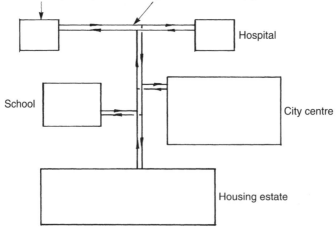

4.28 Medium- or high-pressure hot-water district heating system.

4.29 Three-pipe district heating system.

4.30 Four-pipe district heating system.

be separate from the heat network, and space, cost and acoustic advantages could be gained in comparison with individual systems. A higher standard of service should be available from centralized services, with fewer breakdowns and closer control of pollution.

Building energy management systems

Computer-based remote control and continuous monitoring of energy-using systems, such as heating, air conditioning, electrical power, lighting and transportation, provide a higher standard of service than can be achieved manually (Fig. 4.31). The following types of control can be used.

1. The programmable logic controller (PLC) is a dedicated microprocessor programmed to operate a particular plant item such as a boiler, refrigeration compressor or passenger lift.
2. The energy management system (EMS) is a dedicated microprocessor that is linked to all the energy- and power-using systems such as heating, air conditioning, electrical power, lighting, lifts, diesel generators and air compressors. It may appear as a metal box on a wall of the plant room, having numbered buttons and a single line of screen display for maintenance staff to use for carrying out a limited range of routine changes. Such a unit may serve as an outstation that is either intelligent, having its own microprocessor, or dumb, merely passing data elsewhere.

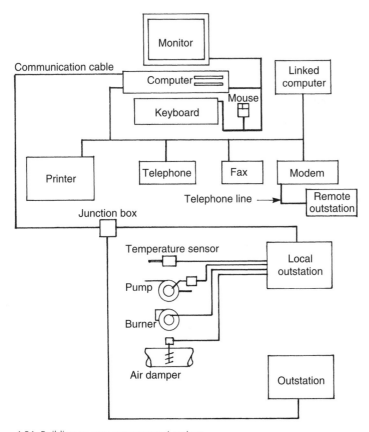

4.31 Building energy management system.

3. A building energy management system (BEMS) is a supervisory computer that is networked to microprocessor outstations, which control particular plant such as heating and refrigeration equipment. All the energy-using systems within one building are accessed from the supervisor computer, which has hard disk data storage, a display unit, a keyboard, a printer and mimic diagrams of all the services.

    Additional buildings on the same site can be wired into the same BEMS by means of a low-cost cable. Remote buildings or sites are linked to the supervisor through a telephone modem. A modem is a **mod**ulator-**dem**odulator box, which converts the digital signals used by the computer into telecommunication signals suitable for transmission by the telephone network to anywhere in the world.

4. The plant management system (PMS) is used to control a large plant room such as an electrical power generator or district heating station. A PMS can be anything from a small dedicated PLC on a water chiller to an extensive supervisory computer system.

5. The building management system (BMS) is used for all the functions carried out by the building including the energy services, security monitoring, fire and smoke detection, alarms, maintenance scheduling, status reporting and communications. Types of BMS range from systems serving one small office, shop or factory to systems serving government departments and international shopping chains, which carry out financial audits, stocktaking and ordering of supplies each night utilizing telecommunications. Suppliers of, say, refrigeration equipment maintain links with all their installations in clients' buildings and are informed of faults as they occur and often prior to clients being aware of the problem.

An outstation is a microprocessor located close to the plant that it is controlling and is a channel of communication with the supervisor. An intelligent outstation has a memory and processing capability that enables it to make decisions on control and to store status information. A dumb outstation is a convenient point for collecting local data such as the room air temperature and whether a boiler is running or not, which is then packaged into signals for transmission to the supervisor in digital code. Each outstation has its own numbered address so that the supervisor can read the data from that source only at a discrete time.

The supervisor is the main computer, which oversees all the outstations, PLCs and modems, contacting them through a dedicated wiring system using up to 10 V and handling only digital code. Such communication can be made every few seconds, and accessing the data can take seconds or minutes depending on the quantity of data and the complexity of the whole system, that is, the number of measurements and control signals transmitted.

The engineer in charge of the BEMS receives displayed and printed reports from the supervisor and has mimic diagrams (Fig. 4.32) of the plant, which enable identification of each pump, fan, valve and sensor together with the set points of the controllers that should be maintained. Alarm or warning status is indicated by means of flashing symbols and buzzers, indicating that corrective action by the engineer is required. Plant status, such as the percentage opening of a control valve and whether a fan is running, is recorded, but only the engineer can ascertain whether such information is correct, as some other component may have been manually switched off in the plant room by maintenance staff, or may have failed through fan belt breakage. Therefore a telephone line between the supervising engineer and the plant room staff is desirable to aid quick checking of facts.

The energy management engineer has seven main functions, which are accomplished by physical work, calculation and word processing:

1. supervision of the plant;
2. choosing the energy supply tariff;

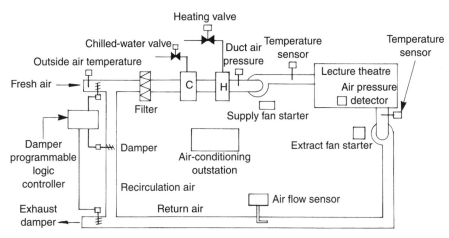

4.32 Mimic diagram on a BEMS screen.

3.   reading energy consumption meters and calculating consumption;
4.   organizing maintenance work and keeping records;
5.   liaising with all levels of staff and management;
6.   producing energy management reports;
7.   monitoring fire and security systems.

The BEMS can automate these functions as follows.

1.   It can supervise the plant continually by means of analogue sensors, analogue-to-digital interfaces and communication cables.
2.   It can perform remote reading of energy meters and of electricity, gas, oil and heat flow, integrate these with time to calculate total energy consumption, compare with desired values and control the plant to minimize consumption.
3.   It can compare electricity and gas consumption rates with published tariffs and advise which would be most economical. It may be possible to switch off electrical loads automatically to keep to the lower-cost tariff.
4.   It can continuously monitor fire detectors and security systems, such as door entry control and video camera operation, detect faults, inform about status and start alarm signals.
5.   It can access BEMS control and monitor information at outstations by means of passwords, which disseminate information to, and allow restricted use by, different types of employee.
6.   It can detect faults and alarm conditions as soon as they occur and display the warnings at the supervisory computer where they will be noticed without delay.
7.   It can carry out normal control functions for energy-using systems and supervise what is going on.

Data which are sent to an outstation include the following:

1.   measurement sensor data such as temperature, humidity, pressure, flow rate and boiler flue gas oxygen content;
2.   control signals to or from valve or damper actuators;
3.   plant operating status, which can be determined from the position of electrical switches which are open or closed, for example, to check whether a pump is operational.

Connections are made to an outstation by means of a pair of low-voltage cables, multi-core cable or fibre-optic cable (Haines and Hittle, 1983; Coffin, 1992; Levermore, 1992).

Geothermal heating

The Southampton geothermal heating system removes heat from salt water (brine), which is pumped up from 1.7 km depth by a unique down-hole pump. The well has been operational since 1981 and has had pumped circulation since 1988 (Southampton Geothermal Heating Company). The well-head water temperature is 74°C, which is sufficient to provide up to 2 MW of heating and hot-water services within the city centre when working in conjunction with an absorption heat pump. This renewable energy resource is the core of the district heating system.

The source of the warm water is a 16 m thick layer of porous Triassic Sherwood sandstone, which is maintained at 76°C by the natural circulation of groundwater and heat flow from greater depths. The water has an acidic pH of 6.0 and high suspended solid, chloride salt and ammonia concentrations. The well is lined with a 245 mm diameter steel pipe. Water is extracted at the rate of 12 l/s by a down-hole pump that is driven from a down-hole turbine and is at a depth of 650 m. The turbine is rotated by water from a ground-level pump, which passes 31.6 l/s and consumes 192 kW of electrical power. This pump has a 250 kW three-phase electric motor, which is operated at a variable speed through a variable-frequency inverter. Figure 4.33 shows the well-head and the 250 kW pump.

The well water is passed through a titanium flat-plate heat exchanger, shown in Fig. 4.34, before being run into the storm drain at a temperature down to 30°C. This discharge water temperature fluctuates owing to the variations in the district heating return water temperature and whether the heat pump is used. The use of diverting three-port flow control valves on the heating systems within the buildings served is discouraged, as these would increase the water temperature that is returned to the heat station. The use of the absorption heat pump (Chapter 5) also influences the discharge temperature to the storm drain that directs the groundwater into the River Test estuary.

The heat station contains two 400 kW gas-fired reciprocating combined heat and power units, CHP, each generating 550 kW of waste heat which is recovered into the thermal network,

4.33 Southampton geothermal well head, brine pump and filter (reproduced by courtesy of Southampton Geothermal Heating Company).

4.34 The titanium flat-plate heat exchanger transfers heat from the brine to the district water circulating through the Southampton geothermal heating system (reproduced by courtesy of Southampton Geothermal Heating Company).

together with a much larger 5.7 MW CHP unit. This larger unit is a Wartsila 18V32DF, an 18-cylinder high-efficiency dual fuel reciprocating engine capable of operating on natural gas or light fuel oil, 35 s Redwood viscosity, as used in water heating plant. Three additional water heaters rated at a total of 7 MW burn natural gas or oil as needed. The host station also contains district water treatment and pressurization to 4 bar, district heating pumps, generator control and switchgear panels, a computerized heat monitoring and charging system and an absorption heat pump.

Heat recovered from the exhaust gas heat exchanger fitted to the larger CHP unit is matched to that required by the absorption heat pump; these can work with the geothermal well or the large-scale district cooling system. The boilers also supply high-temperature hot-water for the absorption heat pump. The heat pump lowers the return water temperature from the district circuit so that more heat can be extracted from the flat-plate heat exchanger and the brine. The heat that is removed from the district circuit is put back into it after the water has passed through the flat-plate heat exchanger. This takes place at the higher-temperature part of the heat pump, in the condenser.

A two-pipe recirculatory heating pipe system distributes heat to the Civic Centre, Southampton Institute, television studios, shops, offices and hotels. Consumers pay charges from heat meters in their building. Each heat meter integrates the water flow rate passing through that consumer's building, the water specific heat capacity, along with flow and return temperatures, to evaluate remotely the MWh consumed. The consumer pays around £37/MWh (3.7 p/kWh) for the heat energy units and a further fixed monthly charge which includes a contribution towards the overhead costs and capital repayment to the supplier. Overall, these charges are set to offer around a 10% saving against the alternative cost of on-site heating and cooling. The consumer does not pay maintenance costs for the heat service up to the heat meter, as they do with their own plant and services. Connection into the geothermal circuit may be free of capital charge to the consumer with the initial connection costs rolled into the long-term energy charges. However, commonly the consumer pays a connection charge of much less than their expected heating and cooling plant installation cost. When the existing gas and oil-fired boilers in the Civic Centre and Institute are used, the heat supply system can have a capacity of 28.9 MW. This will be sufficient for a large part of the city centre within a 2 km radius.

A recently added district cooling-chilled water two-pipe distribution system has six rotary screw refrigeration compressors of 1.2 MW cooling capacity each to supply the larger buildings close to the heat station. An ice thermal storage bank is to be added so that as much off-peak electricity as possible can be used instead of during the higher priced peak daytime hours.

Figure 4.35 shows a simplification of the general arrangement of the heating system as it was originally installed. Recent additions of the CHP, exhaust gas heat exchanger, absorption and screw refrigeration chilled-water units have not been shown.

Figure 4.36 shows the construction of the underground cased pre-insulated pipe.

These descriptions of the heat station concept outline what can be done as alternatives to each building having its own heating and cooling plant with its attendant supervision, maintenance costs and owner responsibilities. Using a geothermal heat source greatly reduces greenhouse gas emissions as part of a sustainability initiative and is an example to us all of what can be done.

4.35 Simplified arrangement of the Southampton geothermal heating system.

Ductile iron pipe · End seals · Gasket · Polyurethane foam · Coating · Securing ring · Sealing ring · Packing · High-density polyethylene case

4.36 Underground district heating pipe (reproduced by courtesy of Southampton Geothermal Heating Company).

Questions

1. Sketch and describe two different types of heating system for each of the following applications: house, office, commercial garage, shop, warehouse and heavy engineering factory.
2. Why may the water in large heating systems be pressurized? Explain how pressurization systems work.
3. How do heating systems alter the mean radiant temperature of a room? Give examples.
4. What factors are included in the decision on the siting of a heat emitter? Give examples and illustrate your answer. What safety precautions are taken in buildings occupied by very young, elderly, infirm or disabled people?
5. How can radiant heating minimize fuel costs while providing comfortable conditions? Give examples.
6. Sketch the installation of a ducted warm-air heating system in a house and describe its operation.
7. List the characteristics of electrical heating systems and compare them with other fuel-based systems.
8. Outline the parameters considered when deciding whether to use a one- or two-pipe distribution arrangement for a radiator and convector-low pressure hot-water heating system.
9. Three rooms have heat losses of 2, 4 and 5 kW respectively. Double-panel steel radiators are to be used on a two-pipe low-pressure hot-water system having flow and return temperatures of 85°C and 72°C respectively. Room air temperatures are to be 20°C. Choose suitable radiators from Table 4.2 and calculate the water flow rate for each.
10. Sketch and describe a microbore heating installation serving hot-water radiators. State its advantages over alternative pipework systems.
11. A medium-pressure hot-water heating system is designed to provide a heat output of 100 kW with flow and return temperatures of 110°C and 85°C respectively. Calculate the pump water flow rate required in litres per second.
12. Find the dimensions of a double-panel steel radiator suitable for a room having an air temperature of 15°C when the water flow and return temperatures are 86°C and 72°C respectively, and the room heat loss is 4.25 kW.

13. The two-pipe heating system shown in Fig. 4.20 is to be installed in an office block where radiators 1, 2 and 3 represent areas with heat losses of 12, 20 and 24 kW respectively. Water flow and return temperatures are to be 90°C and 75°C respectively. The pipe lengths shown are to be multiplied by 1.5. Pump A (Fig. 4.19) is to be used. Pipe heat losses amount to 10% of room heat losses. The friction loss in the pipes is equivalent to 25% of the measured length. Find the pipe sizes.

14. A hot-water radiator central heating system is commissioned and tested while the average outdoor air is 3°C and there is intermittent sunshine and a moderate wind. The building is sparsely occupied. Water flow and return temperatures at the boiler are 90°C and 80°C respectively, and the room average temperature is 27°C. The heating system was designed to maintain the internal air at 22°C at an external air temperature of −1°C with flow and return temperatures of 85°C and 73°C respectively. State whether the heating system met its design specification and what factors influenced the test results.

15. List and discuss the merits of the methods used to generate electrical power.

16. Discuss the application of CHP systems in relation to density of heat usage, local and national government policy, possible plant sites, complexity of existing underground services, ground conditions, costs of competing fuels, type and age of buildings, traffic disruption during installation and better control of pollution. (The term 'density of heat usage' refers to the actual use of heat in megajoules per unit ground plan area m^2, including all floors of buildings and appropriate industrial processes requiring the sort of heat to be sold.)

17. Where will a computer-based building management system usually not be found?

    1. Public hospital.
    2. Prison.
    3. Car-manufacturing plant.
    4. 500-person office building.
    5. 100-room hotel.

18. What does the term BMS mean?

    1. Building access system.
    2. Building maintenance system.
    3. Building management system.
    4. Building monitored security.
    5. Business manual security.

19. How often does the building management system communicate data with sensors and actuators?

    1. Continuously.
    2. Once per hour.
    3. Daily.
    4. Every few seconds.
    5. Annual reports.

20. What forms does building management system data not take when passing through the communications cabling?

    1. Alternating current of over 1.0 A.
    2. Light pulses through fibre-optic cables.
    3. Internet protocol data packets.
    4. Electrical direct current below 0.10 A.
    5. Voltage of 10 V maximum.

21. What is a 'point' in a building management system?

    1. No such thing.
    2. Water chiller, flow control valve and temperature sensors.
    3. Water system.
    4. Air-handling unit.
    5. Printed circuit board in an outstation.

22. What opens and closes a water flow control valve or an air damper in a computer-based building management system?

    1. Electric or pneumatic motor.
    2. Manually operated and wheel valves.
    3. Hydraulic actuator.
    4. Geared drive.
    5. Winding gear.

23. Which of these are 'controllers' in a building management system?

    1. An engineer sat at a personal computer.
    2. The personal computer.
    3. Printed circuit boards hidden away in metal or plastic boxes in plant rooms.
    4. Water and air temperature sensors.
    5. Hand-held devices.

24. Which of these may be found in a computer-based building management security system? Choose as many as appropriate.

    1. Armed guards.
    2. Intruder protection bars at windows.
    3. Digital or video camera recording.
    4. Guard dogs.
    5. Record of personnel movements.
    6. Identity badging and door swipe cards.
    7. Fibre-optic cable network communications.
    8. Telephones.
    9. Asset tracking e-tags.

25. Which of these comments are factually correct about a building management system and are not just an opinion?

    1. Physical security protection is now out of date.
    2. Allows one person to control and monitor a large facility.
    3. Digital recording cameras stop illegal break-ins and escapes.
    4. Turn off the power source and it is useless.
    5. RS232 and RS484 are types of automatic control system.

26. What does the mechanical services switchboard, MSSB, do?

    1. Router for all telephone calls between Property Services staff.
    2. Automatically controls all air conditioning and transportation systems on the campus.
    3. It is the manually operated switchboard for all mechanical services systems within the building.
    4. Switches all the electrical sub-circuits for the whole building.
    5. Only needed in buildings that do not have a computer-based building management system.

27. Which is not correct about the heating hot water, HHW, source:

    1. Often called a boiler.
    2. A boiler is a gas-fired or electrically heated device which generates steam.
    3. HHW circulating at 75°C is steam at atmospheric pressure when it leaks out of the pipes.
    4. Is usually a gas-fired water heater at 75°C.
    5. Is a sealed water system having a thermal expansion tank.

28. Outstation and main control boxes of a building management system are normally found in:

    1. Terminal air-conditioning units.
    2. Plant rooms.
    3. Computer rooms.
    4. Services shafts.
    5. The office of the building management personnel.

29. Programming and commissioning of a building management system is carried out:

    1. With a screwdriver.
    2. At the server computer.
    3. Remotely through the internet.
    4. By calibrating room air temperature sensors with a thermometer.
    5. With a laptop computer communicating directly with each control box.

30. What types of cable system are used for building management system communications cables?

    1. 240 V alternating current.
    2. Screened TV Arial cable.
    3. RS485 and RS232 copper 10 V twisted pair.
    4. Mineral-insulated copper conduit.
    5. Any earthed cable.

31. What is a room-sealed gas appliance? More than one correct answer.

    1. Gas-fired heater installed outside a room in the outside air.
    2. Gas-fired warm-air ducted heater installed within a sealed cupboard.
    3. Any balanced-flue gas appliance.
    4. No such thing.
    5. Gas-fired heater where combustion air enters the unit directly from outdoors through the same wall terminal where flue gas is discharged into the outdoor air.

32. What does an air curtain do?

    1. Provides a security screen at an international airport flight passenger entrance.
    2. A draught to overcome prevailing wind direction at a doorway.
    3. Creates a downward or horizontal air stream across an open doorway.
    4. Makes entering a building very draughty.
    5. Wastes energy from fan power.

33. Which reports do the energy auditor want from the BMS?

    1. Daily schedules of zone temperatures.
    2. Lift and security camera usage data.
    3. Zone temperatures, mechanical and electrical systems schematic drawings, detailed energy consumption, weather data, gas and electrical hourly peak demands each month and trend graphs.
    4. All fault reports.
    5. Energy auditor is too remote from day to day use of the building and does not see any reports.

34. What is the name of the generic data communication system used in BMS?

    1. Open system.
    2. BACNet.
    3. LONtalk.
    4. Ethernet.
    5. GSM telephone.

35. Which communication protocol (language) passes through BMS data systems?

    1. Binary 32 data bit streams.
    2. TCP/IP.
    3. HTTP.
    4. Wi-Fi.
    5. Token ring.

36. What does TCP/IP stand for?

    1. Television control programming, internet post.
    2. Transmission control protocol, internet protocol.
    3. Telephone control program, internet protocol.
    4. Transmission control program, internet packages.
    5. Telephone communication package, internal protocol.

37. Which of these acronyms is not related to data communications within building management computer-based systems?

    1. BACNet.
    2. LONtalk
    3. MODbus.
    4. GSMnet.
    5. ARCNet.

38. What does this formula calculate?

$$\frac{Q}{SHC(t_f - t_r)}$$

    1. Heat emission from a radiator.
    2. Supply air volume flow rate for an air-conditioned room.
    3. Average temperature of hot water in a storage calorifier.
    4. Required hot water flow rate to a heat emitter when Q represents the heat emission.
    5. Thermal transmittance of a structure from heat flow Q.

39. Which of these is the equivalent length for a pumped heating water system circulation?

    1. Total length of the longest flow and return circuit.
    2. Total length of all pipes in the heating system.
    3. That flow and return circulation which creates the pump pressure rise.
    4. Length of the longest circuit to the furthest heat emitter.
    5. Total length of all pipes plus 25%.

40. Which units specify pump head?

    1. N/m^2.
    2. kN/m^2.
    3. N/m^3.
    4. bar.
    5. metre water gauge.

41. Which is true about combustion?

    1. All fuels contain nitrogen.
    2. Air is mainly nitrogen.
    3. Nitrogen is combustible.
    4. Nitrogen has inertia.
    5. Nitrogen in air accelerates the combustion process.

42. Which of these chemical symbols is correct?

    1. CO_3, carbon trioxide.
    2. CO, calcium oxide.
    3. HO, hydrogen oxide.
    4. CO_2, carbon oxide.
    5. H_2O, water vapour or liquid.

43. Which statement relating to combustion is correct?

    1. CO_2 means two molecules of carbon monoxide.
    2. $2O_2$ means two molecules of oxygen.
    3. $2H_2O$ means two atoms of hydrogen plus two atoms of ozone.
    4. CO is carbon oxide.
    5. SO_2 means sodium dioxide.

44. Which statement relating to combustion is correct?

    1. Nitrogen in air assists combustion.
    2. Air contains 23.2% oxygen by volume.
    3. Air contains 21% oxygen by volume.
    4. Outdoor air normally contains 12% CO_2.
    5. Internal combustion engine traffic cleans up free hydrocarbons in the atmosphere.

45. Which of these combustion equations is correct?

    1. Fuel + air + heat = nitrogen + water vapour + heat.
    2. Hydrogen + oxygen = water vapour + heat.
    3. Carbon + hydrogen + oxygen + nitrogen = carbon dioxide + nitrogen.
    4. Hydrocarbon fuel + air = carbon dioxide + water vapour + nitrogen.
    5. Carbon + oxygen + nitrogen = carbon dioxide + heat .

46. What does stoichiometric ratio mean?

    1. Optimum efficiency.
    2. Maximum oxygen in flue gas.
    3. Maximum carbon dioxide in flue gas.
    4. Poor combustion.
    5. 100% excess air provided.

47. Which is not correct about combined heat and power plant?

    1. Reject steam from turbine at higher pressure and temperature than in conventional plant.
    2. About 30% efficient at converting fuel energy into electricity.
    3. Must always be gas-turbine-driven generation.
    4. Can be gas-turbine-driven generation.
    5. Plant overall thermal efficiency around 50%.

48. Which is correct about conventional power generation?

    1. Steam boilers work at the lowest practical pressure.
    2. Boilers generate steam at the highest practical pressure.
    3. Wet steam leaves the boilers.
    4. Modern power plant does not use steam.
    5. Steam is an out of date heating method.

49. Which is correct about nuclear-sourced conventional power generation?

    1. Nuclear power stations never create any greenhouse gases.
    2. They will become the sole means of generating electricity.
    3. They are too dangerous to build.
    4. Spent nuclear fuel rods are safe to handle.
    5. Spent nuclear fuel rods remain radioactive for thousands of years.

50. Which is correct about nuclear-sourced conventional power generation?

    1. Uranium is combusted to produce steam.
    2. Uranium fusion releases heat.
    3. Fission of Uranium releases heat.
    4. Uranium corrodes into lead in releasing heat.
    5. Radiation from Uranium releases heat.

51. Which is correct about nuclear-sourced conventional power generation?

    1. Pressurized water nuclear reactor uses water to drive the electricity generator.
    2. Pressurized water nuclear reactor generates steam to drive turbines.
    3. Sodium gas-cooled nuclear reactors are the most popular type.
    4. Carbon dioxide gas-cooled nuclear reactors are the most popular type.
    5. Oil-cooled nuclear reactors are the most popular type.

52. Which is correct for district heating?

    1. Heat supplied is charged to the consumer on a flat rate such as £/m^2 floor area.
    2. Heat supplied is charged at an annual fixed price to consumer.
    3. Heat supplied is charged at an annual fixed price irrespective of how much is consumed.

4. Heat supplied is charged to consumer through heat metering.
5. Impossible to measure amount of heat energy actually supplied to any consumer.

53. Which is correct about the use of carbon dioxide sensors?

1. Detect ingress of pollution from road traffic.
2. Used to vary the supply of outdoor air into rooms having VAV systems.
3. Control the intake of outdoor air into an air-handling unit.
4. Warn the fire and smoke detection systems of a fire source.
5. Used to control underground car park mechanical ventilation systems.

54. What limits the thermal efficiency of a solar water heating panel, tube or concentrator?

1. Weak solar irradiance.
2. Surface area of collector.
3. Lack of thermal insulation on the back of a flat panel.
4. Excessive water flow rate.
5. Heat loss from heated water back to the environment.

55. Which component of a building management system controller takes input and output direct current voltages from sensors and actuators, changing them into computer data?

1. Multiplexer.
2. The Ethernet.
3. Analogue-to-digital converter.
4. EPROM and RAM chips.
5. Arithmetic logic unit, ALU.

56. Which of these statements about HTML is correct?

1. Another name for binary code.
2. Internationally accepted standard of control system protocol.
3. Proprietary name of open access control protocol.
4. A mathematical program language.
5. Language of TCP/IP.

57. What does N/m^3 stand for?

1. Nanometres per m^2 pressure drop per metre run of pipe.
2. Neurons per cubic metre of room volume.
3. Newton's per square metre pressure drop per metre run of pipe or duct.
4. Newton's per cubic metre is a density.
5. Nano-particles of radon gas per cubic metre of air in a building.

5 Ventilation and air conditioning

Learning objectives

Study of this chapter will enable the reader to:

1. recognize the physiological reasons for fresh air ventilation of buildings;
2. calculate fresh air requirement;
3. understand the basic design criteria for air movement control;
4. describe the four combinations of natural and mechanical ventilation;
5. describe the working principles of air-conditioning systems;
6. calculate ventilation air quantities;
7. understand psychrometric cycles for humid air;
8. calculate air-conditioning heating and cooling plant loads;
9. describe the various forms of air-conditioning system;
10. state where reciprocating piston, screw and centrifugal compressors are suits;
11. understand the coefficient of performance;
12. explain the states of refrigerant occurring within a vapour-compression refrigeration cycle;
13. explain the operation of refrigeration equipment serving an air-conditioning system;
14. comprehend the absorption refrigeration cycle;
15. explain how ventilation rates are measured;
16. choose suitable materials for air-conditioning ductwork;
17. understand the relationship between CFCs and the environment;
18. know the uses of CFCs, and good practice and handling procedures;
19. be able to discuss the problem of SBS;
20. know the symptoms, causes and possible cures for SBS;
21. relate the daily cyclic variation of air temperatures to the need for air conditioning.

Key terms and concepts

absorption 154; air change 127; air-conditioning systems 147; air flow rate 127; air temperature variation 161; air velocity 127; anemometer 156; biocide treatment 160;

Introduction

The reasons for ventilation lead into an understanding of the necessary combinations of natural and mechanical systems and air conditioning, which means full mechanical control of air movement through the building (Kut, 1993). The calculation of air changes and air flow rates from basic human requirements, and then for removal of heat gains, is fundamental. The sizing of air ducts and the calculation of heater and cooler loads utilizing the psychrometric chart of humid air properties is shown.

Systems of air conditioning ranging from small self-contained units to large commercial applications are described. Appropriate refrigeration elements are developed, from thermodynamic principles, into complete installations in a form that is easily understood. Both vapour-compression and vapour-absorption cycles are explained.

Sick building syndrome (SBS) has been attributed to air conditioning, but has been found to be due to an amalgam of possible causes, none of which is singly identifiable as the culprit. The possible causes and solutions relating to user complaints are discussed.

Chlorofluorocarbons (CFCs) were widely used in thermal insulation and refrigeration until their potential for environmental damage resulted in the Montréal Protocol agreement. The effects of this agreement on the building services industry are discussed.

Ventilation requirements

An attempt to calculate the quantity of ventilation air needed for habitation can be made by considering the CO_2 production P m³/s of a sedentary adult, the concentration C_s% of CO_2% in outdoor air and the maximum threshold CO_2 concentration C_r% for the occupied space. Normally, $P = 4.7 \times 10^{-6}$ m³/s per person, $C_s = 0.03$% and $C_r = 0.5$%. If Q m³/s is the rate of fresh air flowing through a room to provide acceptable conditions, a CO_2 balance can be made:

CO_2 increase in ventilation air $= CO_2$ produced by occupant

Now,

flow rate of CO_2 entering room $= QC_s$ m³/s

flow rate of CO_2 leaving room $= QC_r$ m³/s

Therefore,

$$QC_r - QC_s = P \text{ m}^3/\text{s}$$

Hence,

$$Q(C_r - C_s) = P \text{ m}^3/\text{s}$$

and,

$$Q = \frac{P}{C_r - C_s} \text{ m}^3/\text{s}$$

$$= \frac{4.7 \times 10^{-6}}{(0.5 - 0.03) \times 10^{-2}} \text{ m}^3/\text{s}$$

$$= 0.001 \frac{\text{m}^3}{\text{s}} \times \frac{10^3 \text{ l}}{1 \text{ m}^3}$$

$$= 1 \text{ l/s}$$

Other factors have a stronger influence on ventilation requirement:

1. bodily heat production, about 100 W per person during sedentary occupation;
2. moisture exhaled and evaporated from the skin, about 40 W per person for sedentary occupation;
3. body odour;
4. fumes from smoking.

These factors greatly outweigh the CO_2 requirement. The recommended fresh air supply per person is 10 l/s, which is increased by up to 50% in the event of expected heavy tobacco smoke. Building Research Establishment Digest 206: 1977 gives design curves for open-plan and small offices of room height 2.7 m and floor space per person 4.5 m^2. A small office requires 2.25 air changes/h of outdoor air, and this will normally be provided by natural ventilation. A single-person workstation occupies around 10 m^2 floor area when half of the walkway is included, so an outdoor air supply of 10 l/s per person equates to 1 l/s m^2 floor area.

The ventilation system should not produce monotonous draughts but preferably variable air speed and direction. Facilities for manual control of ventilation terminals allow sedentary workers some freedom of choice over their environment. Careful location of ventilation grilles and control of both the temperature and velocity of moving air in mechanical systems can ensure that neither cool nor hot draughts are caused. The maximum air velocity that can be perceived at neck level is related to its temperature. If the values given in Table 5.1 are not exceeded, then annoyance should be avoided.

Table 5.1 DIN criteria for air movement at the neck.

Moving-air temperature (t_{ai} °C)	20	22	24	26
Maximum air speed (v m/s)	0.10	0.20	0.32	0.48

Source: Reproduced from the *IHVE Guide* (CIBSE, 1986 [IHVE, 1970]) by permission of the Chartered Institution of Building Services Engineers.

Grille manufacturers' data reveal the length of the jet of air entering the room for particular air flow rates. Additionally, the air jet may rise or fall depending on whether it is warmer or cooler than the room air. Moving air currents tend to attach themselves to a stationary surface and follow its contours, the Coanda effect. When this boundary layer flow either comes to the end of, say, a wall or hits a bluff body, such as a beam or luminaire, it may be suddenly detached and cause turbulent flow; this appears as a draught, which is an uncomfortable air movement.

An air jet entering a room should be allowed to mix with the room air before entering the occupied space. This can be done either above head height with cooling systems or in circulation spaces at low level with heating systems. The design criteria for ventilation systems are as follows:

1. correct fresh air quantity, minimum normally 10 l/s per person;
2. avoidance of hot or cold draughts by design of the air inlet system;
3. some manual control over air movement;
4. mechanical ventilation to provide a minimum of 4 air changes/h to ensure adequate flushing of all parts of rooms;
5. air change rates that can be increased to remove solar and other heat gains;
6. air cleanliness achieved by filtration of fresh air intake and recirculated room air.

Natural and mechanical systems

The four possible combinations of natural and mechanical ventilation are as follows.

1. Natural inlet and outlet: utilizing openable windows, air bricks, louvres, doorways and chimneys. Up to about 3 air changes/h may be provided, but these depend upon prevailing wind direction and strength, the stack effect of rising warm air currents, and adventitious openings around doors and windows.
2. Natural inlet, mechanical outlet: mechanical extract fans in windows or roofs and ducted systems where the air is to be discharged away from the occupied space owing to its contamination with heat, fumes, smoke, water vapour or odour. This system can be used in dwellings, offices, factories or public buildings. A slight reduction in air static pressure is caused within the building, and external air flows inwards. This inflow is facilitated by air inlet grilles, sometimes situated behind radiators or convector heaters. There is no filtration of incoming impurities. This system is used particularly for toilet or kitchen extraction, smoke removal from public rooms and heat or fume removal from industrial premises. Heating of the incoming air is essential for winter use.
3. Mechanical inlet, natural outlet: air is blown into the building through a fan convector or ducted system to pressurize the internal atmosphere slightly with a heated air supply. The air leaks out of the building through adventitious openings and permanent air bricks or louvres. This system can be used for offices, factories, large public halls or underground boiler plant rooms.
4. Mechanical inlet and outlet: where natural ventilation openings would become unable to cope with large air flow rates without disturbing the architecture or causing uncontrollable draughts, full mechanical control of air movement is assumed. This may augment natural ventilation at times of peak occupancy or solar heat gain. When a building is to be sealed from the external environment, then a full air-conditioning system is used. Figure 5.1 shows the basic arrangement of a single-duct system.

5.1 Single-duct air-conditioning system.

The designers of new buildings bear a heavy responsibility for the future environmental condition of the planet. A building that is a large user of primary energy, that is, fossil fuel, to power its mechanical systems, usually air conditioning, is a charge on society for 50 or more years. Such buildings are unlikely to have their mechanical heating, cooling, lifts, lighting and computer power supply networks removed for reasons of economy by future tenants. Once the pattern of energy use is set for a building by the architect and original client, it remains that way until it is demolished.

The increasing need for comfortably habitable buildings in the temperate climate of the British Isles since the 1960s led to demands for air conditioning. Warm weather data for the UK shows that an external air temperature of 25°C d.b. is normally only exceeded during 1% of the summer period June to September (CIBSE, 1986, figure A2.11). (Compare this to those areas of the world within 45°C of latitude from the equator, where the outside air temperature exceeds 30°C d.b.; people living and working in such areas would probably consider the use of air conditioning in the UK a waste of energy resources.) These 4 months total 120 days and include weekends, a public holiday and many people's annual holidays from their places of work. Therefore, only one day per summer is likely to have an outdoor air temperature that exceeds 25°C d.b. When a long, hot summer is experienced in southern England, there can be several days which exceed 25°C d.b., but this may not be repeated for a few years. In the rest of the UK, lower outdoor air temperatures are experienced most of the time. The UK has a basically maritime climate, with the Atlantic Ocean, English Channel and North Sea providing humid air flows at air temperatures that are modified by the evaporative cooling effect of the seas. Long, hot summers are produced by strong winds from eastern Europe. These winds have travelled across the hot and dry land mass of northern Europe and then been reduced in temperature due to the evaporation of water from the North Sea. Such evaporation of sea water is aided by high wind speed. Sea water is evaporated into the wind and a change of phase occurs in the water. In order to evaporate water into moisture that is carried by the air stream, the water is boiled into steam, a change of phase. This phase change can only take place with the removal of sensible heat from the air and the water, as the latent heat of evaporation has to be provided, just as in a steam boiler.

Removal of sensible heat from the air lowers its dry-bulb temperature. This is the principle of evaporative cooling that is employed in cooling towers and in the evaporative cooling heat exchangers that are used for indoor comfort cooling in hot, dry climates. The occasional long, hot summers in the UK coincide with the increase of outside air wet-bulb temperature that is associated with the evaporative cooling effect from the sea. The result is warm and humid weather in south-east England. It is the combined effect of the higher dry- and wet-bulb outdoor air temperatures which produces the humid air that many people find uncomfortable. The building designer has to decide how to address periods of discomfort due to warm outdoor weather conditions and explain the likely outcome to the client.

The principal source of indoor thermal discomfort in the UK is the manner in which modern buildings are designed and then filled with people and heat-producing equipment. Buildings that were constructed of masonry and brick, with small, openable windows, overhanging eaves from pitched tiled roofing, hinged wooden external or internal slatted shutters; permanent ventilation openings from air bricks, chimneys and roof vents, and north light windows for industrial use, have a large thermal storage mass to smooth out the fluctuations of the outdoor air temperature. Such buildings from the designs of the late nineteenth and early twentieth centuries are better able to provide shaded and cool habitable zones than the exposed glazing high-rise buildings that have been popular since the 1960s. Deep core floor plan buildings that are unreachable by outdoor air to provide the required ventilation and exterior façades which are deliberately exposed to solar heat gain with no provision for external shading screens, trees or shadows cast from surrounding buildings are designed to create interior discomfort. Current buildings are also packed to capacity with productive workers who each need a personal computer and permanent artificial illumination in order to carry out their tasks. The other mistake is to employ people to work in these glass boxes only during daylight hours that coincide with the peak cooling-load times. The Mediterranean countries close many of their businesses during the afternoons, close the wooden window shutters and sleep during the hottest part of the day, coming out in the late afternoon to start work again.

Energy conservation engineers can assist the building maintenance manager to make the best use of what has been built. There will be many technical methods that are available to reduce the use of energy by the mechanical and electrical systems within the building, but at best, these are likely to reduce energy use by a maximum of 25%, often much less can be achieved. Retrofitting a building with energy-saving systems may accompany a change of use or an overall refurbishment. There has been a strong move in the direction of low-energy-use buildings within the UK during recent years. Greenfield sites in the mild climate of the British Isles have provided building designers with more options than are available in more extreme climates. The outdoor air temperatures that are used for the design of the heating and cooling systems in the UK are within the range of −3°C d.b to 30°C d.b. The minimum overnight outdoor air temperature may drop to −17°C d.b. once per year during a severe winter. The extreme low and high outdoor air temperatures that can be experienced are of little importance to the designers of commercial and most industrial buildings, unless indoor air temperatures are of critical importance for the condition of products, human or animal safety, or industrial processes. The users of most buildings are expected to withstand colder or warmer indoor temperatures for a few days per year and not complain too much. In extreme cases, workplace agreements are made to allow work to be stopped for short periods so that the occupants of the building can move away from the workstation for recuperation. An example is a production factory where 10 min breaks in an 8 h shift are agreed for production-line workers to go outside the building when the indoor air temperature exceeds 30°C d.b., which it does frequently during long, hot summers, due to the minimalist thermal insulation value of the building's walls and roofing.

The design team of a new building need to develop a method of solving the complex issues surrounding the task of providing a building which satisfies the:

- basic needs of the owner;
- architectural and local planning design philosophy;
- requirements of legislation;
- access, spatial, visual, aural and thermal comfort needs of the occupants;
- use of energy during the 50-year service period of the building;
- sources of energy that are available for the building and the maintainability of the whole complex.

Those matters which relate to the need, or otherwise, for air conditioning can be summarized as follows:

1. the local design weather conditions;
2. the indoor design set points for zone air temperature and relative humidity;
3. the allowable variation in the indoor design air conditions;
4. the number of occasions during each year that divergence from the specified indoor design conditions are allowable;
5. the time periods when the building will be occupied by the main users and the service personnel;
6. the sources of primary energy available for the building, their long-term reliability, storage requirements and safety considerations;
7. whether renewable energy can be used on the site;
8. the building usage;
9. the means by which the building can be heated;
10. the outdoor air ventilation quantity requirements;
11. the peak energy requirements for heating and cooling;
12. the location and sizes of plant and service shaft spaces available;
13. whether natural ventilation or assisted natural ventilation can provide the required air flow through the building;
14. whether mechanical cooling systems are needed to maintain the specified peak design conditions;
15. the need to provide accurate indoor environmental conditions for equipment, material storage and handling, or an industrial process;
16. whether low-cost cooling systems can be used;
17. whether there is a real need to provide a mechanical means of air conditioning;
18. if there is a process requirement for closely conditioned air within the building;
19. energy recovery strategies available;
20. the maintainability of the mechanical services and how replacement plant can be provided without major structural works becoming necessary.

These considerations all have an impact on the building design team's decision-making process. The local weather conditions that create the maximum heating and cooling loads during the occupied part of the day, or night, will determine the size of the heating and cooling plant that are required. Occupancy times can sometimes be varied in order to minimize the plant capacity that is needed, for example, by using the Mediterranean region off-peak working principle, however unpalatable this may seem to northern European practices. The architectural and engineering designers have the option to experiment with the thermal insulation value of the exterior envelope of the building in varying the area of glazing, walling and roofing and the thermal transmittance of each component. Life-cycle costing of each alternative thermal design will reveal the total cost of constructing and using the building for 50 years, with reasonable assumptions on price changes each year and the cost of refitting the building every 20 years because of improvements in technology.

Selection of the indoor design air set points for temperature and relative humidity will be determined by legislative and comfort standards. Short-term variances within an allowable range can minimize the use of energy at peak times. It may be possible to avoid the installation of mechanical cooling systems where the users of the building agree to accept regular, short-term divergence from the standard design conditions. The provision of ceiling-mounted or portable fans with comfort breaks and refrigerated drinks dispensers might allow a building to avoid the installation of mechanical cooling, depending upon the number and frequency of divergences from the normal standards. Allowing the indoor air temperature to rise to 25°C d.b. in summer before switching on the air-conditioning chiller during the afternoon, and whether the building is occupied, the time is before 3.30 p.m. and the outdoor air temperature is above 25°C d.b.; all these conditions can be accessed through the computer-based building management system, if one is provided. Automatic control programming can be set to minimize the use of electrical energy by the mechanical cooling system as a deliberate strategy by the designers and operators of the building. Dynamic thermal analysis software is used to model buildings and their services systems to assess the indoor air conditions that are expected to occur.

The provision of outdoor air ventilation is a legislative requirement based on the activities of and the numbers of people within the enclosure. The minimum quantity of outdoor air must be maintained for each person to comply with the standards. Outdoor air ventilation does more than provide air for breathing; it also flushes the building with outdoor air to remove heat and to control the concentrations of odours and atmospheric pollutants that are produced within the building. Toilet exhaust air flows often partly match the inflow rate of outdoor air in commercial buildings. Any balance of flow between the exhaust quantity and inflow of outdoor air may be allowed to leak through doors, permanent ventilators and other openings in the structure, such as gaps around window frames. The inclusion of carbon dioxide sensors in the return or exhaust air ducts allows the building management system to minimize the opening of the outside air intake motorized dampers and consequently reduce the heating or cooling load of the air-conditioning plant and save energy. Carbon dioxide in air that is removed from the occupied space is a direct assessment of the number of occupants and their activity level, allowing energy use to match the instantaneous load on the building. Other means of assessing room occupancy are available from infrared or ultrasonic sensors. If the designers of the building are able to know the divergence in the patterns of occupancy of rooms or spaces, then decisions can sometimes be taken to diversify the provision of lighting, heating and cooling zones to minimize the use of mechanical and electrical plant and systems, with benefits in the initial plant capacity and in the use of energy in the long term.

Where incoming outdoor unconditioned air can be preconditioned, an air-to-air flat-plate heat exchanger may be used. This is a compact box around the size of a suitcase having counter flow passages for incoming and outgoing air separated by aluminium foil plates. Incoming winter air is warmed by the room air being exhausted; in summer, room air at the correct temperature cools the outdoor ventilation air stream.

Natural ventilation can be used in atria and industrial buildings where the stack effect of height within the building is used to create air movement. Low-level outside air intakes and roof-level air extractors or openable ventilation units can be mechanically controlled to match the heating and cooling load on the building to the flow of air through the spaces. The avoidance of draughts around sedentary occupants will always be the main challenge. Low-level radiant heating from warmed floors or overhead radiant panels can offset high rates of air movement. Anyone who has sat within cathedral ceiling spaces will recognize the potential discomfort problems in cold weather within intermittently used high thermal-mass buildings, for example, stone churches and sports halls. The use of natural ventilation in the UK is accompanied by the problem of allowing the uncontrolled ingress of the moisture that is present in the outside air. Any surface within the building that is below the dew-point temperature of the space will accumulate damp and create long-term mould growth. This becomes a comfort, health and maintenance cost if damage to the building is to be avoided. In climates where the outdoor air temperature rises above 20°C d.b. on most days of the year, the use of outdoor air natural ventilation is often out of the question. This is due to the combination of intense solar radiation heat gains in these regions and the lack of a natural cooling function by the outdoor air.

Hot dry climates within the Middle East and southern continents make use of evaporative low-cost cooling. Figure 5.2 shows an Australian type where outdoor air of above 25°C d.b. and around 10–30% saturation is drawn inwards through wetted vertical panels and blown directly into occupied rooms as quickly as possible. The only operational cost being that for a 400 W axial fan and a small water consumption in this case. Larger units serve commercial premises and may have an indirect gas-fired heat exchanger to provide winter heating. The supply air has a high humidity but is at reduced dry-bulb air temperature. All cooling is created from evaporation of water, latent heat transfer, but provides sensible cooling of the conditioned space through high air flow. Air is released from the building through window and screened door openings.

Within the tropics, the constant high moisture content of the outdoor air makes it unsuitable for natural ventilation practice in commercial buildings. Buildings in the UK that have an internal atrium have restored a historical precedent in turning the building inside-out. The exterior surfaces need less glazing as the occupants' view is directed inwards towards a planted open space that has natural daylight. The atrium can be used to return conditioned air back to the air-handling plant room without the need for a return air fan, collect exhaust air and expel it through roof openable vents and facilitate the removal of smoke during an emergency. Heat produced from the occupants, fluorescent lighting, computers and electrical equipment assists the upward flow of air away from the occupied zones. Atria are used in all climates from sub-zero through 40°C d.b. environments and fully provide indoor office, retail shopping malls, casino, hotel and entertainment spaces throughout the world.

5.2 Ducted evaporative cooler used in hot climates.

The options that are available to the designers in the selection of the method of controlling the environmental conditions within the building include the following.

1. Natural ventilation – applicable when the external climate, the use and design of the building permit it. Mild climate localities usually close to the coast where the sea is warm; internal air conditions are allowed to vary widely and are directly related to the external weather conditions; central heating system; cooling may be provided from packaged direct expansion refrigeration units within each zone; manually operated internal or external shading blinds; passive solar architecture that may include thermal storage walls; chilled-water beams or flat panels may be installed at high level within offices to provide limited cooling (the chilled water in the beams and panels is maintained at a temperature that is above the room air dew-point so that there is no condensation on the exposed surfaces or within the ceilings, avoiding dehumidification and control of the zone relative humidity); chilled beams and panels provide no ventilation air.

2. Assisted natural ventilation – a development of natural ventilation, as earlier, where applicable. Mechanically operated ventilation louvers and exhaust air fans improve the control of air flow through the occupied spaces; the incoming outside air may be cooled with a water spray evaporative cooler in hot, dry climates; evaporative cooling is a low-cost means of cooling the incoming outside air, which is exhausted by either natural openings of doors and windows, or by exhaust air fans in confined zones; the incoming outdoor air may be cooled through a specific temperature range, say 10 K, to provide limited cooling by means of direct expansion or chilled-water refrigeration plant.

3. Mechanical ventilation which only passes outside air through the zone. This usually applies to moderate climates such as the UK where minimal cooling is required; in climates where the outside air temperature exceeds 30°C d.b. the flow rate of outside air is likely to be insufficient to provide enough cooling for zone temperature control in an air-conditioning system; where it is possible to locate the exhaust air duct alongside the incoming outside air duct, within the ceiling space or a plant room, an air-to-air fiat-plate heat exchanger is used to transfer heat between the incoming and outgoing air streams; the outgoing exhaust air is already at the correct zone temperature, of around 22°C d.b. (it cannot be recycled as it is vitiated with carbon dioxide, odours and atmospheric pollutants); heat transfer works throughout the year (in winter, the incoming outdoor air is preheated by up to 10 K from the outgoing exhaust air; in summer, the incoming outdoor air is precooled by up to 10 K from the outgoing exhaust air); heat transfer efficiency is around 55%; a similar heat transfer can be obtained from a recuperative heat wheel that transfers heat from the outgoing exhaust air to the incoming outside air, generating up to a 10 K temperature change in the outside air stream; the heat transfer medium of the wheel may be strips of mylar film.

4. Mechanical ventilation with recirculated room air. The maximum quantity of conditioned room air is recirculated to save energy use at the heating and refrigeration plant; the outside air motorized dampers are modulated from closed to fully open to control the zone air temperature without the use of the mechanical refrigeration plant for as long a time as possible (this provides low cost cooling to the building); when the refrigeration plant has to be used, the outside and exhaust air motorized dampers are moved to their minimum outside air positions, often around 10% open, allowing the maximum use of recirculated room air; a range of ducted air-conditioning systems are in use, including single-duct, dual-duct, induction units, fan coil units and variable air volume systems.

The single-duct system works in the following way. Some of the air extracted from the room is exhausted to the atmosphere and as much as possible is recirculated to reduce running costs of heating and cooling plants. Incoming fresh air is filtered and mixed with that recirculated; it is then heated by a low-, medium- or high-pressure hot-water or steam finned pipe heat exchanger or an electric resistance element. The heated air is supplied through ducts to the room. The hot-water flow rate is controlled by a duct-mounted temperature detector in the extract air, which samples room conditions. The electrical signal from the temperature detector is received by the automatic control box and corrective action is taken to increase or reduce water flow rate at the electrically driven motorized valve at the heating coil.

During summer operation, chilled water from the refrigeration plant is circulated through the cooling coil and room temperature is controlled similarly.

A temperature detector in the fresh air duct will vary the set value of the extract duct air temperature – higher in summer, lower in winter – to minimize energy costs. A low-limit temperature detector will override the other controls, if necessary, to avoid injection of cold air to the room.

The building is slightly pressurized by extracting only about 95% of the supply air volume, allowing some conditioned air to leak outwards or exfiltrate.

Energy savings are maximized by recirculating as much of the conditioned room air as possible. Room air recirculation with economy-cycle motorized dampers can, sometimes, be retrofitted to existing systems as an energy conservation measure. In mild climates, such as in the UK, full outside air systems are also used. These have no recirculation air ducts; either a flat-plate heat exchanger or run-around pipe coils can be installed to preheat and precool the incoming outside air to save energy. Such heat exchangers are around 55% efficient, which is not as good as recirculation.

EXAMPLE 5.1

A room 15 m × 7 m × 2.8 m high is to have a ventilation rate of 11 air changes/h. Air enters from a duct at a velocity of 8.5 m/s. Find the air volume flow rate to the room and the dimensions of the square duct.

The air flow rate is given by:

$$Q = \frac{N \text{ air changes}}{\text{hour}} \times \frac{V \text{ m}^3}{\text{air change}} \times \frac{1 \text{ h}}{3600 \text{ s}}$$

where room volume V m^3 = 1 air change. Hence,

$$Q = \frac{NV}{3600} \text{ m/s}$$
$$= \frac{11 \times 15 \times 7 \times 2.8}{3600} \text{ m}^3/\text{s}$$
$$= 0.9 \text{ m}^3/\text{s}$$

Also,

$$Q \text{ m}^3/\text{s} = \text{duct cross-sectional area } A \text{ m}^2 \times \text{air velocity } v \text{ m/s}$$

Therefore:

$$A = \frac{Q}{v} = \frac{0.9}{8.5} \text{ m}^2 = 0.106 \text{ m}^2$$

If the duct side is l m, then $A = l^2$ m^2. Therefore:

$$l = \sqrt{A} \text{ m}$$

$$= \sqrt{0.106} \text{ m}$$

$$= 0.325 \text{ m}$$

EXAMPLE 5.2

A lecture theatre has dimensions 25 m $\times$ 22 m $\times$ 6 m high and has 100 occupants; 8 l/s of fresh air and 25 l/s of recirculated air are supplied to the theatre for each person. A single-duct ventilation system is used. If 10% of the supply volume leaks out of the theatre, calculate the room air change rate and the air volume flow rate in each duct.

$$\text{supply air quantity} = (8 + 25)\frac{l}{s} \times \frac{1 \text{ m}^3}{10^3 \text{ l}}$$

$$= 0.033 \text{ m}^3/\text{s}$$

Hence,

$$Q = 0.033 \times 110 \text{ m}^3/\text{s}$$

$$= 3.63 \text{ m}^3/\text{s}$$

Now,

$$Q = \frac{NV}{3600}$$

and hence,

$$N = \frac{3600 \, Q}{V}$$

$$= \frac{3600 \times 3.63}{25 \times 22 \times 6} \text{ air changes/h}$$

$$= 3.96 \text{ air changes/h}$$

Leakage from the theatre is:

$$Q_1 = 10\% \times Q$$

$$= 0.1 \times 3.63 \text{ m}^3/\text{s}$$

$$= 0.36 \text{ m}^3/\text{s}$$

Quantity of air extracted from the theatre is:

$$Q_e = 3.63 - 0.36 \text{ m}^3/\text{s}$$

$$= 3.27 \text{ m}^3/\text{s}$$

Quantity of fresh air entering the ductwork is:

$$Q_f = \frac{8 \times 110}{10^3} \ m^3/s$$

$$= 0.88 \ m^3/s$$

Quantity of recirculated air is:

$$Q_f = Q - Q_r$$

$$= 3.63 - 0.88 \ m^3/s$$

$$= 2.75 \ m^3/s$$

Exhaust air quantity is:

$$Q_{ex} = Q_e - Q_r$$

$$= 3.27 - 2.75 \ m^3/s$$

$$= 0.52 \ m^3/s$$

EXAMPLE 5.3

There are 35 people in a gymnasium, each producing CO_2 at a rate of $10 \times 10^{-6} \ m^3/s$. If the maximum CO_2 level is not to exceed 0.4%, find the air supply rate necessary. The outdoor air CO_2 concentration is 0.03%. Explain what this means to the ventilation designer.

$$Q = \frac{P}{C_r - C_s} \ m^3/s$$

$$= \frac{35 \times 10 \times 10^{-6}}{(0.4 - 0.03) \times 10^{-2}} \ m^3/s$$

$$= 0.095 \ m^3/s$$

This is a very small air flow of outdoor air through a gymnasium where strenuous activity by 35 athletic people is to take place. The outdoor air of 95 l/s may be able to be provided by natural ventilation through controlled low- and high-level vents to the outside. Additional recirculated air flow to control the air temperature with heating or cooling systems is provided if the climate required mechanical air treatment.

Removal of heat gains

Ventilation air is used to remove excess heat gains from buildings. Two types of heat gain are involved: sensible and latent.

Sensible heat gains result from solar radiation, conduction from outside to inside during hot weather, warm ventilation air, lighting, electrical machinery and equipment, people and industrial processes. Such heat gains affect the temperature of the air and the building construction.

Latent heat gains result from exhaled and evaporated moisture from people, moisture given out from industrial processes and humidifiers. These heat gains do not directly affect the temperature of the surroundings but take the form of transfers of moisture. They can be measured in weight of water vapour transferred or its latent heat equivalent in watts.

The latent heat of evaporation of water into air at a temperature of 20°C and a barometric pressure of 1013.25 mb is 2453.61 kJ/kg. Thus the latent heat (*LH*) required to evaporate 60 g of water in this air is:

$$LH = 60 \text{ g} \times \frac{1 \text{ kg}}{10^3 \text{ g}} \times 2453.61 \frac{\text{kJ}}{\text{kg}}$$

$$= 147.22 \text{ kJ}$$

If this evaporation takes place over, say, 1 h, the rate of latent heat transfer will be:

$$LH = 147.22 \frac{\text{kJ}}{\text{h}} \times \frac{1 \text{ h}}{3600 \text{ s}} \times \frac{10^3 \text{ J}}{\text{kJ}} \times \frac{\text{W s}}{\text{J}}$$

$$= 40.9 \text{ W}$$

This is the moisture output from a sedentary adult.

Removal of sensible heat gains to control room air temperature is carried out by cooling the ventilation supply air and increasing the air change rate to perhaps 20 changes/h. Figure 5.3 shows this scheme. The temperature and moisture content of the supply air increase as it absorbs the sensible and latent heat gains until it reaches the desired room condition. The net sensible heat flow will be into the room in summer and in the outward direction in winter.

Rooms that are isolated from exterior building surfaces have internal heat gains from people and electrical equipment, producing a net heat gain throughout the year. The heat balance is as follows:

net sensible heat flow into room = sensible heat absorbed by ventilation air

Therefore:

SH = air mass flow rate × specific heat capacity × temperature rise

5.3 Schematic representation of heating, cooling and humidity control of an air-conditioned room.

where *SH* is the sensible heat. The specific heat capacity of humid air is 1.012 kJ/kgK. The volume flow rate of air is normally used for duct design.

$$Q \frac{m^3}{s} = \text{air mass flow rate} \frac{kg}{s} \times \frac{m^3}{\rho\, kg}$$

The density of air at 20°C d.b. and 1013.25 mb is $\rho = 1.205$ kg/m³.

The supply air temperature t_s can have any value and as density is inversely proportional to the absolute temperature, from the general gas laws,

$$SH\, kW = Q \frac{m^3}{s} \times 1.205 \frac{kg}{m^3} \times \frac{(273+20)}{(273+t_s)} \times (t_r - t_s)\, K \times 1.012 \frac{kJ}{kgK} \times \frac{kW\,s}{1\, kJ}$$

For summer cooling, t_r is greater than t_s during a net heat gain. For winter heating, t_r is less than t_s as there is a net heat loss, and so the temperature difference $(t_s - t_r)$ must be used in the equation. It is more convenient to rewrite the equation in the form,

$$Q = \frac{SH\, kW}{t_r - t_s} \times \frac{273 + t_s}{357}\; m^3/s$$

EXAMPLE 5.4

An office 20 m × 15 m × 3.2 m high has 30 occupants, 35 m² of windows, 25 (2 × 30) W fluorescent tube light fittings, a photocopier with a power consumption of 1500 W and conduction heat gains during summer amounting to 2 kW. Solar heat gains are 600 W/m² of window area. The sensible heat output from each person is 110 W. The room air temperature is not to exceed 24°C when the supply air is at 13°C. Calculate the supply air flow rate required and the room air change rate. State whether the answers are likely to be acceptable.

The sensible heat gains are summarized in Table 5.2. Then:

$$Q = \frac{29.3}{24 - 13} \times \frac{273 + 13}{357}\; m^3/s$$

$$= 2.13\; m^3/s$$

Table 5.2 Summary of sensible heat gains in Example 5.4.

Source	Quantity (W/m² × m²)	SH (W)
Windows	600 × 35	21 000
Occupants	110 × 30	3300
Lights	25 × 2 × 30	1500
Photocopier		1500
Conduction		2000
	SH gain =	29 300
		= 29.3 kW

Some engineers prefer to calculate the mass flow rate of air flow through the air-conditioned space. This is easily found by multiplying the volume flow rate by the density of air at that location, thus,

$$\text{supply air mass flow rate} = 2.13 \, \frac{m^3}{s} \times 1.205 \, \frac{kg}{m^3} = 2.57 \, kg/s$$

Air mass flow rate does not change with temperature, volume flow does.

$$Q = \frac{NV}{3600}$$

$$N = \frac{3600 \, Q}{V}$$

$$= \frac{3600 \times 2.13}{20 \times 15 \times 3.2} \, \text{air changes/h}$$

$$= 8 \, \text{air changes/h}$$

Between 4 and 20 air changes/h are likely to create fresh air circulation through an office without causing draughts and should be suitable for the application.

EXAMPLE 5.5

A room has a heat loss in winter of 32 kW and a supply air flow rate of 3.5 m^3/s. The room air temperature is to be maintained at 22°C. Calculate the supply air temperature to be used.

For winter, the equation is:

$$Q = \frac{SH}{t_s - t_r} \times \frac{273 + t_s}{357}$$

This can be rearranged to find the supply air temperature t_s required:

$$Q \times 357(t_s - t_r) = SH(273 + t_s)$$

$$Q \times 357t_s - Q \times 357t_r = 273SH + SH \times t_s$$

$$Q \times 357t_s - SH \times t_s = 273SH + Q \times 357t_s$$

Thus, for winter:

$$t_s(Q \times 357 - SH) = 273SH + Q \times 357t_r$$

$$t_s = \frac{273SH + 357Qt_r}{357Q - SH} \, °C$$

In this example:

$$t_s = \frac{273 \times 32 + 357 \times 3.5 \times 22}{357 \times 3.5 - 32}$$

$$= 29.75°C$$

The heated or cooled air supply may also have its humidity modified by an air-conditioning plant so that the percentage saturation of the room air is controlled. A water spray or steam injector can be used for humidification. The refrigeration cooling coil lowers the air temperature below its dew-point to condense the moisture out of the air.

Figure 5.4 shows how the properties of humid air are determined from a psychrometric chart. If g_r and g_s are the moisture contents of the room and the supply air respectively, then a heat balance can be written for the latent heat gain absorbed by the previously calculated supply air flow rate Q:

$$\text{latent heat gain to room} = \text{latent heat equivalent of moistures}$$
$$\text{removed by the conditioned air}$$

$$LH \text{ kW} = \text{moisture mass flow rate } \frac{\text{kg}}{\text{s}} \times \text{latent heat of evaporation } \frac{\text{kJ}}{\text{kg}}$$

$$= \text{air mass flow rate } \frac{\text{kg}}{\text{s}} \times \text{moisture absorbed } \frac{\text{kg } H_2O}{\text{kg dry air}} \times 2453.61 \frac{\text{kJ}}{\text{kg } H_2O}$$

$$LH \text{ kW} = Q \frac{\text{m}^3}{\text{s}} \times 1.205 \frac{\text{kg}}{\text{m}^3} \times \frac{273 + 20}{273 + t_s} \times (g_r - g_s) \frac{\text{kg } H_2O}{\text{kg air}} \times 2453.61 \frac{\text{kJ}}{\text{kg } H_2O}$$

$$Q = \frac{LH}{g_r - g_s} \times \frac{273 + t_s}{866\,284} \text{ m}^3/\text{s}$$

The denominator can be rounded to 860 000 with an error of less than 1.0%.

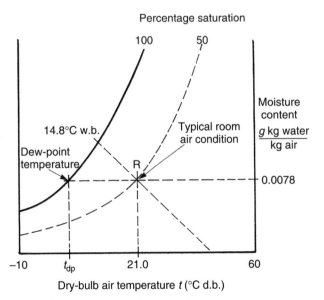

5.4 Sketch of the CIBSE psychrometric chart.

EXAMPLE 5.6

The people in the office in Example 5.4 each produce 40 W of latent heat. Find the supply air moisture content to be maintained, given that the room air is to be at 50% saturation and the corresponding moisture content g_r is 0.008905 kg H_2O/kg air.

$$LH = 30 \text{ people} \times 40 \, \frac{W}{\text{person}} \times \frac{1 \text{ kW}}{10^3 \text{ W}} = 1.2 \text{ kW}$$

$$t_s = 13°C \quad Q = 2.13 \text{ m}^3/s$$

air mass flow rate $= 2.13 \text{ m}^3/s \times 1.205 \text{ kg/m}^3 = 2.57 \text{ kg/s}$

$$Q = \frac{LH}{g_r - g_s} \times \frac{273 + t_s}{860\,000} \text{ m}^3/s$$

We obtain,

$$g_r - g_s = \frac{1.2 \times (273 + 13)}{2.13 \times 860\,000} \text{ kg } H_2O/\text{kg air}$$

Hence,

$$g_s = 0.008905 - 0.000187 \text{ kg } H_2O/\text{kg air}$$

$$= 0.008718 \text{ kg } H_2O/\text{kg air}$$

Psychrometric cycles

Heating and cooling processes in air-conditioning equipment can be represented on the psychrometric chart in the following manner.

Heating

Heating is performed with a low- or high-pressure hot-water, possibly steam, finned pipe heating coil, electric resistance heater or fuel-fired heat exchanger as shown in Fig. 5.5. *SE* is the specific enthalpy, total heat content, of the air, as read from the chart.

Cooling

Cooling is performed by passing chilled water, brine solution or refrigerant through a finned pipe coil. When the coolant temperature is below the air dew-point, condensation occurs and the air will be dehumidified. Figure 5.6 shows the two possible cycles.

Mixing

Mixing of two airstreams occurs when the fresh air intake joins the recirculated room air. The quantity of each air stream is regulated by multi-leaf dampers operated by electric or pneumatic motors under the direction of an automatic control system. Varying the intake of fresh air between the minimum amount during peak summer and winter conditions and 100% when free atmospheric cooling can be achieved during mild weather and summer evenings can result in minimizing the energy costs of the heating and refrigeration plants. Figure 5.7 shows the operational process.

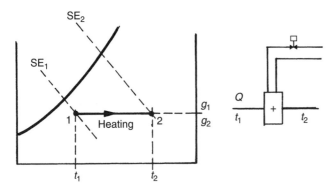

5.5 Air heating depicted on a psychrometric chart.

5.6 Cooling and dehumidification.

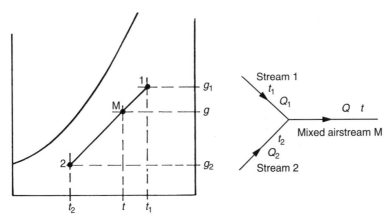

5.7 Psychrometric cycle for the mixing of two airstreams.

The mass flow balance for the junction is:

mass flow of stream 1 + mass flow of stream 2 = mixed mass flow

or,

$$Q_1\rho_1 + Q_2 = Q\rho$$

The enthalpy balance, taking the specific heat capacity as constant, is:

$$Q_1 t_1 + Q_2 t_2 = Qt$$

Dividing through by Q gives,

$$\frac{Q_1}{Q} t_1 + \frac{Q_2}{Q} t_2 = t$$

The mixed air temperature and moisture content lie on the straight line connecting the two entry conditions and can be found by the volume flow rate proportions as indicated by the equation.

Humidification

In winter, incoming fresh air with a low moisture content can be humidified by steam injection, banks of water sprays, evaporation from a heated water tank or a spinning disc atomizer. A preheater low-pressure hot-water coil usually precedes the humidifier to increase the water-holding capacity of the air. This also offsets the reduction in temperature of the air owing to transference of some of its sensible heat into latent energy, which is needed for the evaporation process. Figure 5.8 shows such an arrangement.

A temperature sensor in the humidified air is used as a dew-point control by modulating the preheater power to produce air at a consistent moisture content throughout the winter. For comfort air conditioning, the room percentage saturation will be 50% ± 10%. This permits a wide range of humidifier performance characteristics.

The humidification process often follows a line of constant wet-bulb temperature. The water spray temperature is varied to alter the slope of the line on the psychrometric chart. A complete psychrometric cycle for a single-duct system during winter operation is shown in Fig. 5.9. The preheating and humidification stages have been omitted, as close humidity control is deemed not to be needed in this case. A typical summer cycle is shown in Fig. 5.10.

Some reheating of the cooled and dehumidified air will be necessary because of practical limitations of cooling coil design. Part of the boiler plant remains operational during the summer. Reheating can be avoided by using a cooling coil bypass which mixes air M and air C to produce the correct supply condition. Heating and refrigeration plant capacities are found from the enthalpy changes and specific volume, read from the chart, and the air volume flow rate:

$$\text{heat transfer rate} = Q\,\frac{m^3}{s} \times \frac{kg}{v_s\,m^3} \times (SE_1 - SE_2)\,\frac{kJ}{kg} \times \frac{kWs}{kJ}$$

5.8 Preheating and humidification.

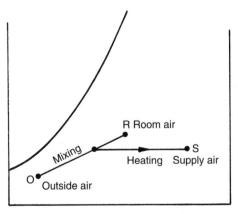

5.9 Winter psychrometric cycle for a single-duct system.

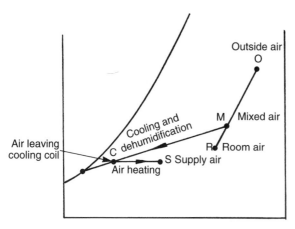

5.10 Summer psychrometric cycle for a single-duct system.

EXAMPLE 5.7

Outside air at −5°C d.b., 80% saturation enters a preheater coil and leaves at 24°C d.b. The air volume flow rate is 6.5 m³/s. Find (a) the outdoor air wet-bulb temperature and specific volume, (b) the heated air moisture content and percentage saturation, and (c) the heating coil power.

From the CIBSE psychrometric chart:

(a) −5.9°C and 0.7615 m³/kg;
(b) 0.00198 kg H_2O/kg air and 10%;
(c) SE_2 at 24°C d.b. and 0.00198 kg H_2O/kg is 29.0 kJ/kg
     SE_1 at −5°C d.b. and 0.00198 kg H_2O/kg is −0.073 kJ/kg

$$\text{Heater duty} = 6.5\ \frac{m^3}{s} \times \frac{kg}{0.7615\ m^3} \times [29 - (-0.073)]\ \frac{kJ}{kg} \times \frac{kWs}{kJ}$$

$$= 248.2\ kW$$

EXAMPLE 5.8

A cooling coil has water passing through it at a mean temperature of 10°C, an air flow of 10.25 m³/s air enters the coil at 28°C d.b., 23°C w.b. and leaves at 15°C d.b. Find the leaving air wet-bulb temperature and percentage saturation. Calculate the refrigeration capacity of the coil in equivalent tonnes refrigeration capacity, ton(r), when 1 ton(r) is equal to 3.527 kW(r).

Plot the cooling and dehumidification line on the psychrometric chart in the manner shown in Fig. 5.5 with a target point of 10°C on the saturation curve. The air leaves the cooling coil at 15°C d.b., 14.2°C w.b. and 91% saturation.

Specific enthalpy of the air entering the coil is 68 kJ/kg and is 40 kJ/kg on leaving. The specific volume of the air entering the coil is 0.874 m³/kg.

$$\text{Refrigeration capacity} = 10.25\ \frac{m^3}{s} \times \frac{kg}{0.874\ m^3} \times (68 - 40)\ \frac{kJ}{kg} \times \frac{kWs}{kJ}$$

$$= 328.4\ kW(r)$$

$$= 328.4\ kW(r) \times \frac{1\ ton(r)}{3.517\ kW(r)}$$

$$= 93.4\ ton(r)$$

Refrigeration capacity is commonly rated in ton(r) as this is the energy needed to freeze a US tonne of ice over a period of 24 h,

$$1\,TR = \frac{2000\ lb}{24\ h} \times \frac{144\ Btu}{lb} \times \frac{1\ kWh}{3412\ Btu}$$

$$= 3.517\ kW(r)$$

EXAMPLE 5.9

A 6 m³/s recirculated room air at 22°C d.b., 50% saturation is mixed with 1.5 m³/s of incoming fresh air at 10°C d.b., 6°C w.b. Calculate the mixed air dry-bulb temperature. Plot the process on a psychrometric chart and find the mixed air moisture content.

$Q_1 = 6\ m^3/s$, $Q_2 = 1.5\ m^3/s$, $Q = $ supply air flow 7.5 m³/s

Using the equation:

$$t = \frac{Q_1}{Q}\, t_1 + \frac{Q_2}{Q}\, t_2$$

$$t = \frac{6}{7.5} \times 22 + \frac{1.5}{7.5} \times 10$$

$$= 19.6°C$$

From the chart,

$$g = 0.0076\ kg\ H_2O/kg\ air$$

Air-conditioning systems

Single duct

The single-duct system (Fig. 5.11) is used for a large room such as an atrium, a banking hall, a swimming pool, or a lecture, entertainment or operating theatre. It can be applied to groups of rooms with a similar demand for air conditioning, such as offices facing the same side of the

5.11 Single-duct all-air installation in a false ceiling.

building. A terminal heater coil under the control of a temperature sensor within the room can be employed to provide individual room conditions.

A variable air volume (VAV) system has either an air volume control damper or a centrifugal fan in the terminal unit to control the quantity of air flowing into the room in response to signals from a room air temperature sensor. Air is sent to the terminal units at a constant temperature by the single-duct central plant, according to external weather conditions. A reducing demand for heating or cooling detected by the room sensor causes the damper to throttle the air supply or the fan to reduce speed until either the room temperature stabilizes or the minimum air flow setting is reached.

Air flow from the diffuser is often blown across the ceiling to avoid directing jets at the occupants. As a result of the Coanda effect the air stream forms a boundary layer along the ceiling and entrains room air to produce thorough mixing and temperature stabilization before it reaches the occupied part of the room. When the VAV unit reduces air flow, there may be insufficient velocity to maintain the boundary layer, and in summer cool air can dump or drop from the ceiling onto the occupants, resulting in complaints of cool draughts.

Dual duct

In order to provide for wide-ranging demands for heating and cooling in multi-room buildings, the dual-duct system, as shown in Fig. 5.12, is used. Air flow in the two supply ducts may, of necessity, be at a high velocity (10–20 m/s) to fit into service ducts of limited size. Air turbulence and fan noise are prevented from entering the conditioned room by an acoustic silencer.

In summer, the hot duct will be for mixed fresh and recirculated air, while the cold duct is for cooled and dehumidified air. The two streams are mixed in variable proportions by dampers controlled from a room air temperature detector. During winter, the cold duct will contain the untreated mixed air and the air in the hot duct will be raised in temperature in the plant room. The system is used for comfort air conditioning as it does not provide close humidity control. It reacts quickly to changes in demand for heating or cooling when, for example, there is a large influx of people or a rapid increase in solar gain.

Induction

Induction is a less costly alternative to the all-air single- and dual-duct systems for multi-room applications. The central air-conditioning plant handles only fresh air, perhaps only 25% of the supply air quantity for an equivalent single-duct system. All the humidity control, and also some

5.12 Dual-duct installation in a false ceiling and detail of the mixing box.

of the heating and cooling for the building, is achieved by conditioning the fresh air intake in the plant room.

Primary fresh air is injected through nozzles into the induction unit in each room. These units may be in the floor, in the ceiling void or under the windowsill. Because of the high-velocity jets, the local atmospheric pressure within the unit is lowered and air is induced into it from the room. The induced air may enter at three or four times the volume flow rate of primary air, and it flows through a finned pipe bank and dust filter before mixing with primary air and being supplied to the room.

The secondary air flow rate can be manually adjusted using a damper. Either hot or chilled water is passed through the room coil depending upon demand. A two-, three- or four-pipe distribution system will be used. The two-pipe system requires a change-over date from heating to cooling plant operation, but a three-way valve can blend hot and chilled water from the three-pipe arrangement. The third alternative has separate hot- and chilled-water pipe coils and pipework.

The extract ductwork and fan removes 90% of the primary air supply and exhausts it to the atmosphere. All recirculation is kept within the room and this greatly reduces duct costs and service duct space requirements. Figure 5.13 shows a typical installation in an office.

Fan coil units

Heating and cooling loads that prove to be too great for induction units can be dealt with by separate fan and coil units fitted into the false ceiling of each room or building module. Better air filtration can be achieved than with the induction unit. A removable access hatch below the unit is required to facilitate motor and filter maintenance.

Care is taken to match the fan-generated noise to the required acoustic environment. As with the other systems, the extracted air can be taken through ventilated luminaires to remove the lighting heat output at source and avoid overheating the room. The supply and extract

5.13 Induction unit installation in a multi-storey building.

5.14 Fan coil unit installation in a false ceiling.

ducts only carry the fresh air. All recirculation is confined to the room. A typical layout is shown in Fig. 5.14.

Packaged unit

A packaged unit is a self-contained air-conditioning unit comprising a hermetically sealed refrigeration compressor, a refrigerant evaporator coil to cool room air, a hot-water or electric resistance heater battery, a filter, a water- or air-cooled refrigerant condenser and automatic controls. Packaged units can either be completely self-contained, needing only a supply of electricity, or piped to central heating and condenser cooling-water plant. Small units are fitted into an external wall

Supply air

Electronic heater

Refrigeration coil

Drip tray

Fan

Filter

Recirculated room air

Condenser coil and propeller fan in outside air duct

Refrigeration compressor

5.15 Packaged air-conditioning unit.

and have a change-over valve to reverse the refrigerant flow direction. This enables the unit to cool the internal air in summer and the external air in winter.

Heat rejected from the condenser is used to heat the internal environment in winter. In this mode of operation it is called a heat pump. A separate ventilation system may be needed. Compressor and fan noise levels are compared with the acceptable background acoustic environment. Maintenance requirements are filter cleaning, bearing lubrication and replacement of the compressor when it becomes too noisy or breaks down.

Split system units have a separate condenser installed outside the building. Two refrigerant pipes of small diameter connect the internal and external equipment boxes. This allows greater flexibility in siting the noise-producing compressor. Ducted models provide conditioning and ventilation and are often sited on flat roofs. Figure 5.15 shows a typical through-the-wall installation.

Vapour-compression refrigeration

The electrically driven vapour-compression refrigeration system is the principal type used. Its rival, the absorption cycle, burns gas to produce cooling but has a coefficient of performance of around 1, whereas vapour compression has a coefficient of performance in the range 2–5, and so it is cheaper to operate. Compressor types are as follows.

1. Single- or multi-cylinder reciprocating piston compressor with spring-loaded valves: domestic refrigerators and small air conditioners have hermetically sealed motor–compressor units which are sealed for their service period, that is, about 10 years.

    A condensing unit comprises a sealed compressor, a refrigerant condenser, a liquid receiver, pipework and controls. Refrigerant pipework is installed on site from this unit to a finned pipe forced-draught air-cooling coil in the air-conditioning system.

Large air-cooling plant comprises a multi-cylinder in-line or V-formation compressor, a shell and tube refrigerant to a water evaporator producing chilled water, and a shell and tube refrigerant to a water condenser where the refrigerant vapour is condensed into liquid and the heat given out is carried away by a water circuit to a cooling tower on the roof.

2.   The centrifugal compressor is used in large chilled-water plants where the noise and vibration produced by the reciprocating type would be unacceptable. A centrifugal impeller of small diameter is driven through a step-up gearbox from a three-phase electric motor. The lack of vibration and compactness of the very high-speed compressor makes siting the plant easier.

3.   The screw compressor has two meshed gears, which compress the refrigerant in the spaces between the helical screws. One gear is driven by an electric motor through a step-up gearbox. The compressor operates at high speed and has very low noise and vibration levels.

The operation of a vapour-compression refrigeration plant is shown in Fig. 5.16.

Refrigerants commonly used are non-toxic fluids with high latent heat. Refrigeration plant with a capacity of up to about 175 kW, and motor cars, uses refrigerant HFA134A (replaced R12, which is CCl_2F_2), which boils at $-29.8°C$ in the atmosphere. In a typical system it will be evaporated at 5°C under a pressure of 3.6 bar and condensed at 40°C at 9.6 bar. Larger plant uses fluorinated hydrocarbon R22 ($CHClF_2$), which has a greater refrigerating effect per kilogram but is more expensive.

The coefficient of performance (COP) is an expression of cycle efficiency and is found from

$$COP = \frac{\text{heat absorbed by refrigerant in the evaporator W}}{\text{power consumption by the compressor W}}$$

The vapour-compression cycle can be represented on a pressure–enthalpy diagram for the refrigerant as shown in Fig. 5.17. Referring to Figs 5.16 and 5.17, compression 1–2 raises the temperature of the refrigerant dry superheated vapour from about 20°C to 60°C, where it can then be cooled and condensed at a sufficiently high temperature to reject the excess heat from the building to the hot external environment.

It condenses at 40°C and collects in the liquid receiver. This warm high-pressure liquid passes through an uninsulated pipe so that it is subcooled to below its saturation temperature (about 20°C) at the expansion valve located alongside the evaporator.

5.16 Vapour-compression refrigeration system.

5.17 Pressure–enthalpy diagram for a refrigerant showing the vapour-compression cycle.

The pressure rise produced through the compressor is dissipated in friction through the fine orifice in the valve. Such a sudden pressure drop is almost an adiabatic thermodynamic process. This is represented by the vertical line 3–4 on the pressure–enthalpy diagram. Some heat loss from the valve body takes place so that 3–4 will be slightly curved. Condition 4 is at the lower pressure of the evaporation process, where the refrigerant temperature has dropped to 5°C. Some of the refrigerant liquid has flashed into vapour.

The liquid and flash vapour mixture flows through the evaporator, where it is completely boiled into vapour and is then given a small degree of superheat (path 4–1). It then enters the compressor as dry low-pressure superheated vapour at 20°C. A suction temperature-sensing phial controls the refrigerant flow rate by means of a liquid-filled bellows on the thermostatic expansion valve. This matches refrigerant flow to the refrigerating effect required by the air-conditioning system and ensures that liquid droplets are not carried into the compressor, where they could cause damage.

Large plant has refrigerant pressure controllers, which reduce compressor performance by unloading some cylinders. Lubricating oil contaminates the refrigerant leaving the compressor. This is separated gravitationally and returned to the crankcase.

The pressure–enthalpy diagram depicts the reverse Carnot cycle. In the other direction, the cycle is for an internal combustion engine. The theoretical ideal coefficient of performance is given by,

$$\text{Ideal COP} = \frac{T_1}{T_2 - T_1}$$

where T_1 is the evaporation absolute temperature (K) and T_2 is the condensation absolute temperature (K). With the temperatures previously used,

$$\text{Ideal COP} = \frac{273 + 5}{(273 + 40) - (273 + 5)}$$
$$= 7.94$$

Friction losses from fluid turbulence, heat transfers to the surroundings, and mechanical and electrical losses in the compressor all reduce this value to 2–4 in commercial equipment. The electricity consumption of fans and pumps adds to the running costs.

5.18 Refrigeration plant serving an air-conditioning system, showing typical fluid temperatures.

Figure 5.18 shows the installation of a chilled-water refrigeration plant serving one of the cooling coils in an air-conditioning system in a large building. Each air-handling system has filters, heaters, humidifiers and recirculation ducts appropriate to the design. There are several separate air-handling plants, each serving its own zone. Zones are decided by the similarity of demand for conditioning. The south-facing orientation has a cyclic requirement for cooling that is distinct from that of the other sides of a building. Internal areas require cooling throughout the year. These differing needs are often met by having separate zones for each area.

Absorption refrigeration cycle

An example of a two-drum absorption refrigeration cycle is shown in Fig. 5.19. The input heat source may be gas, steam or hot water from a district heating scheme, or rejected heat from a gas-turbine electricity-generating set.

The generator (1) contains a concentrated solution of lithium bromide salt in water. Pure water is boiled off this solution and condenses on the cooling-water pipes, which are connected to an external cooling tower. The generator drum pressure is sub-atmospheric at 0.07 bar, with boiling and condensation taking place at 38°C. Water leaves the condenser (2), and then passes through an expansion valve (3), where its pressure and temperature are lowered to 0.01 bar and 7°C. It then completely evaporates while being sprayed over water pipes in the evaporator. These pipes are the chilled-water circuit at 6–10°C, which supplies the refrigeration for the air-conditioning cooling coils. Water vapour in the evaporator drum (4) is sucked into a weak lithium bromide solution by the salt's affinity for water. Latent heat given up by the water vapour as it condenses into the solution is removed from the cooling tower by cooling-water pipes. The weak solution (5)

5.19 Two-drum absorption refrigeration cycle.

is pumped back into the higher-pressure generator drum to complete the cycle. This pump is the only moving part of some systems. Concentrated salt solution (6) is passed down to the absorber via a heat exchanger and pressure-reducing valve to replace the salt removed. The production of chilled water is equivalent to about half the heat input to the generator.

The gas domestic refrigerator works on an absorption system using liquid evaporation to provide the cooling effect. Figure 5.20 shows the features of modern equipment. A solution of ammonia in water is heated in the boiler (A) by a small gas flame. This is the only energy input. Ammonia gas is driven off the solution and then condensed to a liquid in the air-cooled condenser (B) outside the refrigerator cabinet. The liquid ammonia then passes, with some hydrogen, into the evaporator (C) inside the refrigerator cabinet. This is the ice box. The ammonia completely evaporates while absorbing the heat from the cabinet. The two gases are then led into the absorber (D), where the ammonia is absorbed by a weak solution trickling down the absorber. The strong ammonia solution produced is then driven back into the boiler, while the hydrogen gas, which is not absorbed, passes to the evaporator. The weak solution trickling down the absorber is provided from the boiler. Both types of absorption refrigerator provide cooling from a source of heat, they have few or no moving parts and they only require low-power pumps. This makes the cycle suitable for solar power input. The equipment is vibrationless and very quiet in operation.

Air-cooled condenser

B

Evaporator C

Refrigerator cabinet

Boiler A

Gas burner   Absorber D

5.20 Gas-fired domestic absorption refrigeration system.

Ventilation rate measurement

Measurement of room ventilation rate may be required for research into the energy consumption of heated buildings or to carry out commissioning tests on warm-air heating, ventilation or air-conditioning systems. Three basic methods are available.

Smoke

Provided that smoke detectors and alarms are deactivated and suitable warning is given, the room can be filled with smoke and the ventilation system switched on. The time to clear the smoke is used to calculate the air change rate and volume flow rate. Smoke candles or an oil-burning generator are used.

Anemometer

The air velocity through each ventilation grille and any obvious gaps around doors and windows is measured using a suitable anemometer: a rotating vane for large grilles, and a thermistor, mini-vane or pitot-static tube for small airways. The air flow rates into and out of the room are calculated from the airway areas and the average air velocities through them.

Tracer gas

A non-toxic tracer gas (nitrous oxide or helium) is released into the room and thoroughly mixed with portable fans to fill the complete volume. Samples of room air are taken at intervals and passed through an analyser, which measures the concentration of tracer gas. The room air change rate is calculated from two known concentrations and the time interval between them. This technique can be used for naturally ventilated buildings and produces accurate results.

The katharometer measures air electrical conductivity and gives an output of percentage concentration of tracer in air. An infrared analyser uses a source of infrared radiation and passes it down two tubes to receiving photocells. One tube contains a reference gas and the other the sample of room air. The different gases absorb different amounts of radiation, and the variation in the signals from the photocells is calibrated as the percentage of tracer gas in the air.

Consider the injection of tracer gas into a room such that its concentration is C_r%. A stirring fan is used. Ventilation of the room is at the rate of N air changes per hour. The tracer gas concentration falls to C_τ% during time interval τ. The room air change rate can be found as follows:

$$N = \frac{1}{\tau} \ln\left(\frac{C_r}{C_\tau}\right)$$

so that,

$$N\tau = \ln\left(\frac{C_r}{C_\tau}\right)$$

and hence,

$$e^{N\tau} = C_r/C_\tau$$

Thus the concentration C_τ at time τ is given by:

$$C_\tau = C_r \times e^{-N\tau}$$

EXAMPLE 5.10

Nitrous oxide tracer gas is admitted into a building and mixed with the internal air to achieve a 8% concentration. After 45 min, the tracer gas concentration has fallen to 1.5%. Calculate the air change rate per hour.

$$N\, \frac{\text{air changes}}{\text{h}} = \frac{1}{\tau\, \text{h}} \times \ln\left(\frac{C_r}{C_\tau}\right)$$

Now,

$$\tau = 45\, \text{min} = 0.75\, \text{h}$$
$$C_r = 8\%$$

and,

$$C_\tau = 1.5\%$$

Hence,

$$N = \frac{1}{0.75} \ln\left(\frac{8}{1.5}\right)$$
$$= 1.33\, \ln\, 5.33$$
$$= 2.23\, \text{air changes/h}$$

Materials for ventilation ductwork

The materials used for ventilation ductwork are listed in Table 5.3. Thin-gauge galvanized mild steel sheet ducts are the most popular because of their low cost. Prefabricated ducts and fittings allow rapid site erection. Circular, rectangular, flat or spirally wound circular ducts are generally used. Joints are pop-riveted and sealed with waterproof adhesive tape, hard-setting butyl bandage or heat-shrunk plastic sleeves. Large ducts have bolted angle-iron flanges, which also act as support brackets. Stiffening steel strips or tented sheets are used to reduce the drumming effect on flat duct sides caused by air turbulence. Bare metal is painted with metal oxide or zinc chromate paint. Ducts are thermally insulated with resin-bonded glass fibre boards or expanded polystyrene.

An air pressure consisting of the design operational pressure plus 250 pascal (Pa) ($1\,Pa = N/m^2$) is applied after installation for test purposes. The maximum allowed leakage rate is 1% of the system design air flow rate and leaks must not be audible.

Chlorofluorocarbons

Chlorofluorocarbons (CFC) are numbered to represent the chemical combination. Those in common use are listed in Table 5.4.

When CFCs are released into the atmosphere as a result of leakage from refrigeration systems, the production of expanded foam, venting to the atmosphere during maintenance of refrigeration compressors, the use of aerosols, chemical cleaning or the destruction of refrigerators, they find their way to the upper atmosphere, where they are broken down by the action of ultraviolet solar radiation and chlorine is released. This degradation will continue for many years. Atmospheric ozone is destroyed by the chlorine, and it is reported that the resulting increase in the levels of ultraviolet radiation reaching the earth's surface will cause ecological damage as well as an increase in skin cancer.

Table 5.3 Ventilation ductwork materials.

Material	Application	Jointing technique
Galvanized mild steel sheet	All ductwork	Riveted slip joints, machine-formed snap-lock, flanged, butyl cement bandage, heat-shrunk sleeve
UPVC and polypropylene	Prefabricated systems for housing and toilet extract	Flanged, socket and spigot
Resin-bonded glass fibre	Low-velocity and domestic warm-air heating	Butt, sleeved, socket and spigot
Asbestos cement	Prefabricated circular and rectangular for flues and chemical exhausts	Socket and spigot
Flexible glass fibre, proofed fabric reinforced with galvanized spring wire helix	Short connections from a duct to a terminal unit	Jubilee clip, waterproof tape
Aluminium, copper, wired glass, stainless steel	Kitchen extract hoods, ornamental use	Flanges
Brick, concrete, timber, fibre- or plasterboard	Recirculation airways within suspended ceilings and floors, surfaces sealed against dust release	As appropriate

Table 5.4 Chlorinated fluorocarbons in common use.

Number	Use	Ozone depletion potential
R11	Foam insulation and furniture	1.0
R12	Refrigeration systems of all size	1.0
R22	Larger refrigeration plant	0.05
R113	Solvent cleaner	1.0
R502	Refrigeration	0.33
Halon	Fire extinguishers	1.0

In 1986, 100 000 tonnes of CFCs were manufactured, and the 1987 Montréal Protocol agreement was signed by most countries with the target of reducing production to zero by the year 2000. In the immediate time-scale, R22 has a sufficiently low ozone depletion potential (ODP) to be used until a suitable replacement is found. Current use of R12, which is the most common refrigerant fluid, particularly in small refrigeration plants and domestic refrigerators and freezers, is replaced by hydrogen fluorine alkaline HFA134A. CFC refrigerants are miscible with the mineral lubricating oil used in the compressor, but HFA134A is not and requires synthetic polyglycol alkaline (PGA) lubricant.

Good practice in the use of CFCs involves the following:

1. avoid leakage by correct use of pipe materials, engineering design, testing and maintenance procedures;
2. recover fluid from the system by using a vacuum pump;
3. return CFCs to the manufacturer;
4. employ reliable contractors;
5. use alternative chemicals.

R12 returned to the manufacturer is either cleaned and recycled with new R12 by bulk mixing or pyrolysed at 2500 K in a furnace, where it is completely destroyed and the flue gas is filtered as necessary.

Any replacement for commonly used CFCs must have the following properties:

1. no chlorine content;
2. ODP = 0;
3. non-flammable;
4. low toxicity;
5. similar boiling temperature and vapour pressure to R12;
6. miscible with compressor lubricant and easily separated;
7. thermodynamic and fluid flow properties compatible with currently installed refrigeration systems;
8. cost-effective;
9. no new environmental risks.

Sick building syndrome

Indoor environments may be made artificially close to the warm spring day that most people would like to inhabit, but it is not the genuine atmosphere and will be polluted with synthetic particles and vapours plus other contributory factors:

- tobacco smoke
- body odour
- deodorants
- vapours from cleaning fluids, photocopiers, paints and furnishings
- dust
- bacteria
- noise
- flickering lamps
- glare from artificial illumination and the sun
- carpets
- polyvinyl chloride (PVC)
- paper
- formaldehyde
- volatile chemicals
- bacteria grown in stagnant water in humidifiers
- treated water aerosols distributed from showers, washing facilities or fountains
- open-plan office
- too many people.

The air temperature, humidity and air movement, which will seem either stagnant or too draughty, can rarely please more than 95% of the occupants and frequently please a lot fewer. The total environmental loading upon the occupants may rise to an unacceptable level, which can be low for those who are hypersensitive, that is, physically and psychologically unable to fight off such a bombardment of additional foreign agents to the body.

Sick building syndrome, SBS, is epitomized by the occupants' exhibiting a pattern of lethargy, headaches, dry eyes, eye strain, aching muscles, upper respiratory infections, catarrh and aggravated breathing problems such as asthma, upon returning to their workplace after the weekend. Apparent causes are sealed windows, air conditioning, recirculated air, recirculating water humidifiers, high-density occupation, low negative ion content, smoking, air-ductwork corrosion, airborne micro-organisms, dust, and excrement from dust mites in carpets. SBS can be defined as a combination of health malfunctions that noticeably affect more than 5% of the building's population.

This means that there should be sick house syndrome as well. Perhaps there is, or perhaps we are more tolerant at home. Cases of formaldehyde vapour irritation after cavity insulation have been noted. The pattern of house occupation is different, and variation of climatic controls is easily achieved.

Relief can be gained by operating windows, temperature control, sun blinds, air grilles, by a brisk walk or by going outdoors to stimulate the body to sweat toxins out.

It has been easy to blame air conditioning for SBS, but the cause is more complex and has much to do with the standards we demand of our buildings, the psychological influence of having to go into the workplace at all and the total internal environment created. Naturally ventilated buildings often have a higher bacteria and dust count than air-conditioned buildings, which use filters and have sealed windows.

Outbreaks of Legionella diseases have been attributed to the growth of bacteria in stagnant water in wet cooling towers. These bacteria are distributed on air currents and breathed in by those susceptible to infection, sometimes with fatal results. Dry heat exchangers are preferred for discharging surplus building heat gains back into the external atmosphere but they are rather large. Adequate cleanliness and biocide dosing of recirculated cooling-tower water is mandatory.

The cure for SBS requires the following actions:

1. measure pollutants to identify causes;
2. remove recirculating water humidifiers from air-conditioning plant and replace only with direct steam or water injection;
3. allow individuals to have control over local air movement, direction and temperature;
4. clean recirculating water systems such as wet cooling towers and remaining humidifiers and treat them with biocides;
5. ensure that fresh air ventilation ductwork, filters, heating and cooling coils and grilles are internally and externally clean and fully functional;
6. inspect and clean air-conditioning systems and potential dust-traps regularly;
7. appraise the lighting system to maximize natural illumination and reduce glare.

Air temperature profile

The recommended upper limit for the room environmental temperature for normally occupied buildings is 27°C (CIBSE, 1986, Section A8). The external design air temperature for comfort in offices in London (CIBSE, 1986, Table A2.22) may be chosen as 29°C d.b., 20°C w.b. Higher outdoor air temperatures occur. The indoor limit of 27°C will be exceeded in naturally ventilated buildings in the UK and in warmer locations (Chadderton, 1997a, chapter 3). The elevation of indoor temperature above that of the outdoor air is caused by a combination of the infiltration of external air, solar radiation and indoor heat gains. In parts of the world where high solar radiation intensity and continuously higher external temperatures are common, for example, Sydney with 35°C d.b. and 24°C w.b., the necessity for controlled air circulation and refrigeration can be recognized.

Environmental temperature is a combination of mean radiant and air temperatures (Chapter 1). Intense solar radiation through glazing during the summer can lead to the mean radiant temperature being higher than the air temperature. The temperature of the air in an office, factory or residence may need to be kept to an upper limit of, say, 26°C d.b. in order to limit the environmental temperature to 27°C. Such conditions are tolerable, but not comfortable, for sedentary work. Considerable discomfort is experienced when strenuous physical activity is conducted.

The indoor air temperature fluctuates through each 24-h period owing to the position of the sun relative to the building. South-facing rooms that have a large area of glazing are likely to be exposed to the greatest indoor air temperatures. Figure 5.21 shows the variation of outdoor air, indoor air, window glass and predicted outdoor air temperatures for a south-facing office in Southampton on Sunday 27 June 1993. The windows and doors remained closed throughout the weekend. The office had only natural ventilation, no mechanical cooling, open light grey slatted venetian blinds, and had been used normally for the preceding week. The exterior wall had a 70% glazed area.

The general profile of the outdoor air temperature t_{ao} that is expected can be calculated from a sine wave (Jones, 1985, p. 113):

$$t_{ao} = t_{max} - \frac{t_{max} - t_{min}}{2} \times \left[1 - \sin \frac{\theta \pi - 9\pi}{12} \right]$$

The 24-h clock time is θ hours: that is, a time between 0 and 24 h. The predicted outdoor air temperature curve has been calculated for maximum and minimum values, t_{max} and t_{min}, of 28°C d.b. and 13°C d.b. on an hourly basis for 24 h. This corresponds to the conditions after a week of warm sunny weather in June 1993 in Southampton. This has been a common occurrence since 1990. A thermocouple temperature logger, similar to that in Figure 1.3, measured the

5.21 Temperature profiles: south-facing office, Sunday 27 June 1993.

outdoor air, indoor air and the internal surface temperature of the outer pane of the double-glazed window, hourly. Figure 5.21 shows that the measured outdoor air temperature follows the general shape of the predicted sine wave. The thermocouple that was adhering to the glass showed a combination of two factors: first, that the air temperature in the cavity between the panes of double glazing rose to 35°C; second, that the glass absorbed some of the incident solar radiation and was raised in temperature. The internal room air temperature was measured in a shaded location at just above desk height. The room air temperature remained between 25°C d.b. and 31°C d.b. During normal use, the same office produced an internal air temperature of 25°C d.b. when the outdoor air temperature peaked at 27°C d.b.

While such an example does not prove conclusively that all south-facing rooms in the UK need to be air conditioned for human thermal comfort, it does give some evidence to strengthen the argument in favour of mechanical cooling. Working in air temperatures that move above 24°C d.b. in naturally ventilated spaces that have significant solar radiation can be noticeably uncomfortable. Whether the performance of human productivity or effectiveness becomes impaired is arguable. Low-cost cooling systems can be designed that make use of cool parts of a building to lower the temperature of the areas that are exposed to solar radiation. Heat pump systems, mechanical ventilation and evaporative water-cooling towers can be used to limit room air temperatures, without the need to involve high-cost refrigeration equipment.

EXAMPLE 5.11

A south-facing office in Basingstoke has natural ventilation and a large area of glazing. The maximum and minimum outdoor air temperatures are expected to be 26°C d.b. and 12°C d.b. on a summer day. Calculate the outdoor air temperature that is expected at 1600 h. Find the indoor air temperature that will be generated if solar radiation heat gains raise the indoor air by 1.5°C from that of the outdoor air. Comment upon the thermal comfort conditions that are provided for the office and make recommendations.

$\theta = 16$ h for the time of 1600 h

$$t_{ao} = t_{max} - \frac{t_{max} - t_{min}}{2} \times \left(1 - \sin \frac{\theta\pi - 9\pi}{12}\right)$$

$$= 26 - \frac{26 - 12}{2} \times \left(1 - \sin \frac{16\pi - 9\pi}{12}\right)$$

$$= 26 - 7 \times (1 - \sin 1.833)$$

The 1.833 is in radians and not degrees. Switch on the radians mode of the calculator then press the SIN X key to find the answer of sin $1.833 = 0.966$. Alternatively, in degree mode, multiply 1.833 by 360 and then divide the answer by (2π). This produces an angle of 105°. Press the SIN key and find:

sin $105 = 0.966$.

There are 2π radians in 360°.

$$t_{ao} = 26 - 7 \times (1 - 0.966)$$

$$= 25.8°C \text{ d.b.}$$

indoor air temperature $t_{ai} = 25.8 + 1.5°C$ d.b.

$$= 27.3°C \text{ d.b.}$$

The calculated indoor air temperature exceeds that recommended. While this may prove be acceptable for reasonable thermal comfort, there is solar radiation and possible glare during the afternoon in summer. External solar shading or interior blinds are recommended if seating cannot be relocated away from direct solar glare. Sedentary office workers would benefit from measures to increase the throughput of cooler air from the north side of the building or by means of a mechanical cooling system.

Questions

1. A banking ball is cooled in summer by an air-conditioning system that provides an air flow rate of 5 m^3/s to remove sensible heat gains of 50 kW. Room air temperature is maintained at 23°C. Derive the formula for calculating the supply air temperature and find its value.
2. A room has a sensible heat gain of 10 kW and a supply air temperature of 10°C d.b. Find the supply air rate required to keep the room air down to 20°C d.b.
3. Figure 5.22 shows a west-facing window in a warm climate at a latitude of 35° south around midday. Explain how solar control is being achieved here and how and if it may be applied to commercial buildings in other latitudes.
4. Ten people occupy an office and each produces 50 W of latent heat. The supply air flow rate is 0.5 m^3/s and its temperature is 12°C d.b. If the room is to be maintained at 21°C d.b. and 50% saturation, calculate the supply air moisture content.
5. The cooling coil of a packaged air conditioner in a hotel bedroom has refrigerant in it at a temperature of 16°C. Room air enters the coil at 31°C d.b. and 40% saturation and leaves at 20°C d.b. at a rate of 0.5 m^3/s.

   (a) Is the room air dehumidified by the conditioner?
   (b) Find the room air wet-bulb temperature and specific volume.
   (c) Calculate the total cooling load in the room.

5.22 Shaded window.

6. A department store has 340 people in an area of 35 m × 25 m that is 4 m high. Smoking is permitted.

   (a) Calculate the fresh air quantity required to provide 12.5 l/s per person.
       If the air change rate is not to be less than 5 changes/h, find the following.
   (b) supply air quantity;
   (c) percentage fresh air in the supply duct;
   (d) extract air quantity if 85% of the supply air is to be mechanically withdrawn;
   (e) recirculated air quantity;
   (f) ducted exhaust air quantity.

7. Air enters an office through a 250 mm × 200 mm duct at a velocity of 5 m/s. The room dimensions are 5 m × 3 m × 3 m. Calculate the room air change rate.

8. Show two methods of allowing fresh air to enter a room where extract ventilation is by mechanical means and the incoming air is not to cause any draughts.

9. Discuss the relative merits of centrifugal and axial flow fans used in ventilation systems for occupied buildings.

10. Sketch and describe the arrangements for natural and mechanical ventilation of buildings. State two applications for each system.

11. Describe the operating principles of four different systems of air conditioning. State a suitable application for each.

12. State, with reasons, the appropriate combinations of natural and mechanical ventilation for the following: residence, city office block, basement boiler room, industrial kitchen, internal toilet accommodation, hospital operating theatre, entertainment theatre.

13. Explain, with the aid of sketches, how the external wind environment affects the internal thermal environment of a building.

14. A four-storey commercial building is to be mechanically ventilated. Air-handling plant is to be sited on the roof. Each floor has dimensions 20 m × 10 m × 3 m and is to have 6 air changes/h. Of the air supplied, 10% is allowed to exfiltrate naturally and the remainder is extracted to roof level. The supply and extract air ducts run vertically within a concrete service shaft and the limiting air velocity is 10 m/s. Estimate the dimensions required for the service shaft. Square ducts are to be used and there is to be at least 150 mm between the duct and any other surface.

15. List the procedure for the design of an air-conditioning system for an office block.

16. A lecture theatre has dimensions 15 m × 15 m × 4 m and at peak occupancy in summer has sensible heat gains of 30 kW and latent heat gains of 3 kW. Room and supply air temperatures are to be 23°C d.b. and 14°C d.b. respectively. Room air moisture content is to be maintained at 0.008 kg H_2O/kg air. Calculate the supply air volume flow rate, the room air change rate and the supply air moisture content.

17. To avoid draughts, a minimum supply air temperature of 30°C d.b. is needed for the heating and ventilation system serving a public room. The room has an air temperature of 21°C d.b. and a sensible heat loss of 18 kW. It is proposed to supply 2 m^3/s of air to the room. Calculate the supply air temperature that is required. If it is not suitable, recommend an alteration to meet the requirements.

18. Describe the operation of the vapour-compression refrigeration cycle and sketch a complete system employing chilled-water distribution to cooling coils in an air-conditioning system.

19. Discuss the uses of the absorption refrigeration cycle for refrigerators and air-conditioning systems.

20. Show how refrigeration systems can be used to pump heat from low-temperature sources, such as waste water, outdoor air arid solar collectors, to produce a usable heat transfer medium for heating or air-conditioning systems.

21. Measurements in a mechanically ventilated computer room showed that tracer gas concentration fell from 10% to 3% in 5 min. Calculate the air change rate.

22. A gymnasium of dimensions 20 m × 12 m × 4 m is to be mechanically ventilated. The maximum occupancy will be 100 people. The supply air for each person is to comprise 20 l/s of fresh air and 20 l/s of recirculated air. Allowing 10% natural exfiltration, calculate the room air change rate, the air flow rate in each duct and the dimensions of the square supply duct if the limiting air velocity is 8 m/s.

23. Where does sick building syndrome apply?

    1. Architectural design failures.
    2. Perception that exterior design of a building does not fit in successfully with existing local architecture.
    3. Interior of a building that looks to be designed by a sick mind.
    4. Polluted interior atmosphere.
    5. Poor quality external environment makes users of the building susceptible to airborne upper respiratory ailments and overall sickness.

24. Which is the reason to use ice thermal storage in a HVAC refrigeration system?

    1. Reduce water chiller plant room space requirement.
    2. Reduce number of water chillers needed.
    3. Reduce water chiller run time.
    4. Install smaller capacity refrigeration compressors.
    5. Reduce energy cost.

25. How can an off-peak ice-making chiller be more efficient to operate than a daytime water chiller?

    1. Greater temperature difference between evaporation and condensing temperatures.
    2. Lower outdoor night-time dry- and wet-bulb air temperatures.
    3. Reduced electrical tariff.
    4. Reduces peak hours electrical demand kW.
    5. It is not more energy-efficient.

26. Which of these can affect asthma sufferers?

    1. Excess of outside air ventilation.
    2. House dust mites and mould spores.
    3. Warm indoor air.
    4. Humid and warm indoor air.
    5. Matters other than those related to ventilation.

27. Which is the most efficient way of recovering energy from room air?

    1. Recirculation.
    2. Sensible heat recovery thermal wheel.
    3. Total heat recovery thermal wheel.
    4. Plate heat exchanger.
    5. Run-around pipe coils.

28. How does a run-around pipe coil system function?

    1. Closed pipe loop passes water by gravity circulation between heat source and sink locations.
    2. Outgoing air-duct water-cooled coil; closed cycle water pipework; warmed water pumped through incoming outdoor air-duct coil, recovering useful heat.
    3. Refrigerated evaporator coil in outgoing waste heat duct passes useful heat to condenser coil in the incoming air duct.
    4. Refrigeration system recovers waste heat in an outgoing air with a chilled-water coil in exhaust air duct.
    5. Cold-water feed pipe to domestic hot-water system preheated in an outdoor air coil in warm weather.

29. What does a chilled beam mean?

    1. Steel structural beam exposed in the occupied room and cooled by a supply air stream from a directional grille.
    2. Refrigerated pipe within a room.
    3. Chilled-water pipes alongside structural floor beams.
    4. Exposed steel beam at high level in a room having chilled-water pipes attached.
    5. Natural convector chilled-water finned pipe.

30. Which water temperature flows through a chilled beam?

    1. 6–12°C.
    2. 4–18°C.
    3. Below room air dew-point.
    4. Minimum of room air temperature minus 10°C.
    5. Above room air dew-point.

31. Why might an under floor air distribution system, UFAD, have benefits?

    1. Keeps feet cool.
    2. Quieter than ducts within ceiling.
    3. Supply air within the floor void cools concrete floor slab thermal mass.
    4. Keeps under floor power and communications cables cool.
    5. There are no benefits as it costs more.

32. How could a chilled-water cooling coil distribute bacteria into occupied air-conditioned rooms?

    1. It cannot, as air temperature remains too cool.
    2. It will not under normal operation.
    3. Condensate water trap between drain tray and sewer always maintains a water seal.
    4. Water seal in P-trap between drain tray and sewer may become dehydrated and allow sewer gases to pass into the air-handling unit and supply duct.
    5. It will not when adequately maintained in accordance with codes and standards.

33. How could a cooling-water tower become a health hazard?

    1. It cannot while adequately maintained.
    2. Very easily, bird droppings may create bacterial growth in cooling water immediately after monthly servicing work and inspection.
    3. Chemical dosing with biocide does not allow any cooling-tower water contamination.
    4. Cooling tower is outdoors and so is no more a health hazard than an ornamental fountain.
    5. Cooling-tower water is always too cool to support growth of bacteria, mould or algae.

34. What can be done to maintain the health and safety of the internal surfaces of ducted ventilation and air-conditioning systems?

    1. Replace aged air ducts.
    2. Increase air velocity to blow deposits out of ducts and terminal units outside of occupied hours.
    3. Change air filters regularly.
    4. Internal visual inspection, compressed air brushing, scraping and vacuum cleaning.
    5. Nothing more than maintaining air filters to keep air and ducts clean.

35. Where can sinusitis, asthma, pneumonia and skin dermatitis originate?

    1. Mould spores in warm humid uncleaned air and water building services systems.
    2. Contaminated outdoor air.
    3. Low air humidity.
    4. Contacting people with breathing infections.
    5. Warm humid air in crowded buildings or transportation.

36. When carbon dioxide level in occupied rooms is sensed for control of ventilation airflow, what is the maximum set point used, approximately, in parts per million, ppm.

    1. 100
    2. 700
    3. 250
    4. 1000
    5. 5000

37. When the coefficient of performance during heating, COP_H, of a vapour-compression refrigeration cycles is 3, which of these is the correct compressor power input to generate 750 kW of heating?

    1. 750 kW.
    2. 2250 kW.
    3. 100 kW.
    4. 250 kW.
    5. 75 kW.

38. When the coefficient of performance during cooling, COP_R, of a vapour-compression refrigeration cycles is 2.25, which of these is the correct compressor power input to generate 225 kW of cooling?

    1. 225 kW.
    2. 22.5 kW.
    3. 506 kW.
    4. 100 kW.
    5. 2.25 kW.

39. What does ventilating a building mean?

    1. Maintaining indoor air circulation.
    2. Exhausting room air to outdoors.
    3. Minimum provision of 4 air changes/h.
    4. Removal of moisture, odours and carbon dioxide from rooms.
    5. Provision of adequate outdoor air into each room.

40. How should conditioned supply air enter a room?

    1. Directed at the ceiling.
    2. Directed at walls.
    3. Should mix with room air outside the occupied room volume.
    4. Strike the floor in walkways and mix with room air.
    5. Diffuse imperceptibly with room air.

41. Which is a suitable design criterion for air conditioning?

    1. Some manual control over air movement preferred.
    2. Windows should always be openable.
    3. Manual adjustment of air-conditioning system must be avoided.
    4. Automatic comfort control essential for every occupant.
    5. Sealed windows avoided as they always generate discomfort.

42. Which is correct about air filters?

    1. Air filter air pressure drop reduces as dust load increases.
    2. Increasing dust load in a filter causes no reduction in air flow to the rooms.
    3. Air filtration at the air-handling unit completely stops atmospheric dust and dirt from entering the building.
    4. Increasing dust load in a filter reduces supply air flow rate.
    5. A dirty air filter has an air pressure drop of around ten times that of a clean filter.

43. Which is not correct for natural ventilation in the UK?

    1. Must generate at least 4 air changes/h.
    2. Often associated with low-energy buildings.
    3. Original means of ventilating pre-1950's traditional brick buildings.
    4. Always justifiable in this climate.
    5. Often only provides up to 3 air changes/h.

44. What is meant by hybrid ventilation?

    1. System has latest technology in computer control.
    2. Only works in tall buildings.
    3. Combines active control of natural ventilation with mechanical air movement systems.
    4. Unnatural combination of mechanical systems with manually operated ventilators.
    5. Combination of systems invented by Hyme Bridowski in 1935.

45. What drives the fan in a large air-handling unit?

    1. Diesel engine prime mover.
    2. Three-phase electric motor.
    3. Single-phase synchronous alternating current motor.
    4. 1000 V AC motor.
    5. 240 V AC motor.

46. Which is a small fan drive motor, such as in a FCU?

    1. DC variable-speed motor as they are lowest cost.
    2. 415 V AC motor if they require around 850 W power output.
    3. Three-phase for any size as this is the most energy-efficient type.
    4. 240 V, three-phase synchronous electric motor.
    5. 240 V, single-phase direct drive motor.

47. AHU chilled-water flow control is by:

    1. Modulating damper.
    2. Electronic control system.
    3. Modulating water flow valve.
    4. Manually set only once by the commissioning engineer.
    5. All valves remain fully open to maximize available cooling during hot weather.

48. An office 15.0 m × 7.0 m × 2.8 m has 11.0 air changes/h from air supplied through a duct where it flows at a velocity of 8.5 m/s. Which two answers are correct?

    1. Supply air flow rate is 1.20 m^3/s.
    2. Supply air flow rate is 750.0 l/s.
    3. Supply air flow rate is 0.9 m^3/s.
    4. Duct dimensions are 325 mm × 325 mm.
    5. Duct dimensions are 650 mm × 325 mm.

49. A retail shop 22.0 m × 6.5 m × 3.5 m has 7.5 air changes/h from air supplied through a duct where it flows at a velocity of 9.5 m/s. Which two answers are correct?

    1. Supply air flow rate is 0.085 m^3/s.
    2. Supply air flow rate is 1043 l/s.

3. Supply air flow rate is 10.4 m³/s.
4. Duct dimensions are 335 mm × 335 mm.
5. Duct dimensions are 990 mm × 990 mm.

50. A conference hall 55.0 m × 27.0 m × 3.6 m has 15.0 air changes/h from air supplied through a duct where it flows at a velocity of 4.5 m/s. Which two answers are correct?

1. Supply air flow rate is 22.275 m³/s.
2. Supply air flow rate is 5940 l/s.
3. Supply air flow rate is 2.228 m³/s.
4. Duct dimensions are 1000 mm × 5000 mm.
5. Duct dimensions are 2000 mm × 2475 mm.

51. Which of these statements on air-conditioning systems is correct? More than one answer is correct.

1. Air-handling units are usually the largest item of plant.
2. Air-handling units are always manufactured off-site and delivered in one piece to any building.
3. Some air-handling units are large enough for a person to walk inside.
4. Air-handling units are where the room supply air is conditioned.
5. Air-handling units do not contain any moving parts.

52. Why are motorized dampers fitted into the outside air and return air intakes to the air-handling unit in a large air-conditioning system?

1. Close of the air supply during a storm.
2. Stop sucking dust into the building.
3. Vary the winter and summer intake of outdoor air.
4. Shut the air conditioning down at night.
5. Fully open up during fire mode.

53. Identify which statement correctly describes the operation of the vapour-compression refrigeration cycle:

1. A compressor pump drives liquid refrigerant around the system.
2. Refrigerant condenses at 20°C to reject heat from the building.
3. Refrigerant gas vaporizes at 30°C and at high pressure to absorb heat from the building.
4. An expansion valve raises refrigerant gas pressure.
5. Heat is absorbed from the building by vaporizing refrigerant at low pressure at around 5°C.

54. Which statement is correct?

1. A cooling tower cannot be a source of infectious bacteria.
2. The cooling tower rejects heat from the building to the outdoor atmosphere.
3. Cooling towers are only operational during the summer.
4. Cooling towers sprays mains water into the air.
5. Cooling-tower water systems are occasionally dosed with biocide.

55. What does VRV mean?

1. Variable refrigerant volume.
2. Volume refrigerated valve.

3. Vacuum recycled vanadium.
4. Variable refrigeration value.
5. Valid refrigerant valence.

56. How is the coefficient of performance, COP, of a refrigeration system maximized?

1. Discharging waste heat to the atmosphere at the highest temperature.
2. Discharging waste heat to the atmosphere at the lowest temperature.
3. Evaporating refrigerant at the lowest possible temperature.
4. Evaporating refrigerant at the lowest possible temperature.
5. Using the smallest possible temperature increase between evaporation and condensation of the refrigerant.

57. Which is correct about air pressurization of buildings?

1. Supply and exhaust air quantities must be equal.
2. Outdoor wind environment creates internal air pressurization.
3. When exhaust air volume exceeds supply air quantity, building is pressurized.
4. When exhaust air volume exceeds supply air quantity, building is depressurized.
5. Supply air fans do not create building air pressurization.

58. Which are the two types of heat transfer taking place during ventilation of a building?

1. Latent and radiant.
2. Sensible and convection.
3. Latent and conduction.
4. Sensible and radiant.
5. Sensible and latent.

59. Which does not correctly describe heat transfer?

1. Sensible heat is removed from air when water droplets spray into warm air and vaporize.
2. Latent heat transfer occurs when water is evaporated into steam vapour.
3. Evaporative coolers and cooling towers rely on latent heat transfer to remove sensible heat from the water passing through.
4. Evaporative coolers work less efficiently in warm humid climates.
5. Cooling towers already have saturated air, so there is no latent heat transfer with the circulating water.

60. Why is a psychrometric chart used?

1. Shows temperature profile through a wall.
2. Calculates latent heat demand.
3. Calculates sensible heat load on the building.
4. Shows physical properties of humid air.
5. Plots air dry-bulb temperature against atmospheric pressure.

61. Which is correct for sensible heating processes on a psychrometric chart?

1. Curved line between two dry-bulb temperatures.
2. Vertical straight line.
3. Any line at 45° to the horizontal.
4. Horizontal line.
5. A line concentric with the dew-point curve.

62. Which of these is a sensible heating process line on a psychrometric chart?

    1.   Straight line between two dry-bulb temperatures at constant specific enthalpy.
    2.   Line between two dry-bulb temperatures at constant percentage saturation.
    3.   Straight line between two dry-bulb temperatures at constant moisture content from right to left.
    4.   Angled straight line between two dry-bulb temperatures from left to right.
    5.   Straight line between two dry-bulb air temperatures at constant moisture content from left to right.

63. Which of these does not correctly describe a cooling process line on a psychrometric chart?

    1.   Cannot be precisely drawn on the chart due to variation of air percentage saturation within the air spaces around a cooling and dehumidification coil.
    2.   Only the end points of the line are known precisely.
    3.   Line drawn represents overall picture of cooling process through the coil.
    4.   Curved line downwards from right to left between two moisture contents.
    5.   Straight line angled downwards between two pairs of coordinates from air dry-bulb temperature and moisture content.

64. Which of these describes the leaving air condition when warm humid air enters a chilled-water cooling coil?

    1.   Higher moisture content.
    2.   Higher specific enthalpy.
    3.   Same moisture content.
    4.   Lower dry-bulb air temperature and around 90% saturation.
    5.   100% saturated air at same moisture content.

65. Which correctly describes cooling processes on a psychrometric chart?

    1.   Reduces percentage saturation.
    2.   Reduces air wet-bulb temperature.
    3.   Maintains air at constant specific enthalpy.
    4.   Maintains constant air wet-bulb temperature.
    5.   Does not change air specific volume.

66. Which does not correctly describe humidification processes on a psychrometric chart?

    1.   Water sprays onto a chilled-water cooling coil.
    2.   Steam injection provides better air cleanliness.
    3.   Straight line moving away from 100% saturation curve.
    4.   Straight line moving towards the 100% saturation curve.
    5.   Adiabatic saturation line.

67. Which is correct about a VAV air-conditioning system?

    1.   Setting a minimum supply air flow rate at each terminal unit avoids dumping cool air onto occupants.
    2.   Always requires terminal reheat coil.
    3.   Stands for vortices-activated valve.
    4.   Stands for volume-activated variable flow system.
    5.   One terminal unit serves two zone orientations.

68. Which is a correct description of the dual-duct air-conditioning system?

    1. Duplicated supply and return air ducts.
    2. A reduced cost design.
    3. Simultaneous heating and cooling to adjacent rooms.
    4. Not used in commercial office buildings.
    5. Appropriate for low-energy new buildings.

69. Which is a correct description of the induction unit air-conditioning system?

    1. Each room terminal unit has a secondary air circulation fan.
    2. Recirculation air is induced away from the air discharged to atmosphere by suction from the supply air fan.
    3. Recirculated room air is neither heated nor cooled.
    4. Each terminal induction unit has a 2-, 3- or 4-pipe heating and chilled-water distribution.
    5. Recirculated room air does not need filtration.

70. What does FCU air-conditioning system stand for?

    1. Full conditioning unit.
    2. Face console unit.
    3. Full compressor unit.
    4. Fan coil unit.
    5. Failed compressor unit.

71. Which is a correct description of the fan coil unit air-conditioning system?

    1. Terminal unit used in an induction system.
    2. An FCU is a small AHU.
    3. Anything having a fan.
    4. Room air conditioner with a fan.
    5. Electric heating coil with a supply air fan.

72. Which is appropriate for an FCU?

    1. Can be around the size of a suitcase.
    2. Does not always contain a fan.
    3. The unit that takes heated and cooled streams of air from different ducts, mixes them and supplies conditioned air into the zone.
    4. Potential direct replacement for an induction unit.
    5. Always large enough to walk around inside it.

73. Which of these applies to packaged room air-conditioning units?

    1. Always connected to a ducted air system.
    2. Always connected to a central chilled-water plant system.
    3. Each unit has a refrigeration compressor.
    4. Always very quiet operation.
    5. Power demand not exceeding 250 W.

74. Which is not correct for packaged room air-conditioning units?

    1. Small applications such as home single office and motel room.
    2. Stand-alone unit used for large computer server rooms.
    3. Silent and have no servicing requirement.

   4.   May be connected to the BMS.
   5.   Built-in controls.

75. Which applies to vapour-compression refrigeration?

   1.   Refrigerant gas compressor may be multi-cylinder reciprocating piston or volute scroll.
   2.   Refrigerant always remains as a gas.
   3.   Refrigerant R12 is ozone friendly.
   4.   Refrigerant R22 is in domestic refrigerators.
   5.   Ammonia is not suitable as a refrigerant.

76. Which applies to vapour-compression refrigeration?

   1.   Compressed air.
   2.   Refrigerant vaporizes in the compressor.
   3.   Screw compressor.
   4.   Linear compressor.
   5.   Refrigerant condenses at low pressure and temperature.

77. Which applies to the vapour-compression refrigeration cycle?

   1.   Refrigerant liquid is pressurized by a centrifugal compressor.
   2.   Refrigerant liquid warms the inside of the building.
   3.   Reciprocating compressor increases refrigerant gas pressure.
   4.   Refrigerant thermostatic expansion valve stops and starts the flow of refrigerant to the compressor.
   5.   Refrigerant liquid evaporates fully at 25°C and 4 bar pressure in the evaporator.

78. What does COP of a refrigeration cooling system mean?

   1.   Convective operated pressure system.
   2.   Compressor operated performance.
   3.   Ratio of heat absorbed by refrigerant divided by power consumption of the compressor.
   4.   Number is always less than 1.0.
   5.   Ratio of the heat discharged in the condenser to the input power to the compressor.

79. Which applies to the lubrication of refrigeration compressors?

   1.   Reciprocating compressors produce oil carry-over into the refrigerant pipes.
   2.   Reciprocating compressors do not require lubrication.
   3.   Compressor lubricating oil never leaves the crankcase.
   4.   Piston rings do not let crankcase oil pass.
   5.   Refrigerant lubricates the compressor bearings.

80. What are refrigerants?

   1.   Combustible hydrocarbons.
   2.   Water.
   3.   Hydrocarbon oil.
   4.   Toxic.
   5.   Fluorinated hydrocarbons.

81. Which correctly describes the refrigeration cycle?

   1.   Thermostatic expansion valve regulates the rate of refrigerant flow into the evaporator to ensure superheated vapour enters the compressor.

2. Thermostatic expansion valve allows refrigerant liquid to expand.
3. Opening of the thermostatic expansion valve is controlled from a temperature sensor on the compressor discharge pipe.
4. Thermostatic expansion valve stops and starts the flow of refrigerant from a digital controller.
5. TEV is an evaporator isolating valve.

82. Which of these correctly describes a water-cooled refrigeration system?

1. Cold vapour leaves the compressor and enters a finned tube heat exchanger with axial flow cooling fans.
2. Low-pressure warm vapour condenses and rejects latent heat to the outside environment through a shell and tube heat exchanger.
3. Refrigerant vapour condenses at 40°C in an air-cooled heat exchanger.
4. Water-cooled condenser pump circulates water to a cooling tower.
5. Refrigerant evaporator is a direct expansion cooling coil in an air-handling unit.

83. Which does a cooling tower do?

1. Always remains completely clean as it is continuously washed with water circulation.
2. Never polluted with airborne contamination.
3. Operates without any energy input.
4. Collects atmospheric dust, debris and bird droppings.
5. Filters the condenser cooling water.

84. Which is a primary characteristic of a cooling tower?

1. Quiet operation.
2. Uses almost no water.
3. Potential source of water-based Legionella bacteria for outdoor air.
4. Compact unit usually installed within a chiller plant room.
5. Functions equally well in any outdoor climate.

85. Which is a primary characteristic for absorption refrigeration?

1. Absorbs heat from within the building whereas a vapour-compression system cools a water circulation system.
2. Has an absorption compressor.
3. Uses gas pressure to generate cooling.
4. Requires a source of primary heat energy.
5. Uses no electrical energy.

86. How is room ventilation rate measured?

1. Impossible to measure something that cannot be seen.
2. Can only be calculated from duct air flow rate measurement.
3. Found from releasing a non-toxic tracer gas into the room and measuring its rate of decay with a katharometer.
4. Measured quantity of tracer gas concentration in room remains constant when mechanical ventilation is switched off and measured with a thermo anemometer.
5. Tracer gas concentration measured with a carbon dioxide sensor and falls in a straight line graph when mechanical ventilation is switched off.

87. Which is correct about commissioning air-duct systems?

    1. Air-duct systems do not need to be inspected during commissioning.
    2. All air ducts must be internally cleaned prior to commissioning.
    3. All air ducts must be internally inspected with remote-controlled lamps and cameras before use.
    4. Rough internal projections, rivets and metal cuttings are removed by the commissioning technician.
    5. Air-duct systems are sealed in sections and pressure tested for an air-tightness standard compliance.

88. Which is correct about noise in air-conditioning ducts?

    1. Ducts are mounted on springs to isolate vibration from the building structure.
    2. Fan and motor are solidly bolted to a concrete plant base to isolate noise and vibration.
    3. Noise from sources within the building cannot enter air ducts and transfer elsewhere.
    4. Fans have a flexible air-tight fabric connection with air-conditioning ducts to stop transmission of vibration.
    5. Noise created in one room cannot travel through an air conditioning-duct and enter another room.

89. Which is not correct about chlorinated fluorocarbons?

    1. R22 commonly used in large refrigeration systems such as chilled-water plant.
    2. Contained within sealed refrigeration systems at below atmospheric pressure so never leaks into atmosphere.
    3. Used in halon fire extinguishing fluid.
    4. Non-toxic.
    5. Only exists at atmospheric pressure in gaseous form.

90. What happens to chlorinated fluorocarbons when released into the atmosphere?

    1. Dissolved by nearby water and rain.
    2. Harmlessly coexist in the atmosphere.
    3. Vaporized and dispersed by wind and rain.
    4. Degraded by ultraviolet solar radiation releasing chlorine into upper atmosphere that remains there for many years.
    5. Degraded by infrared solar radiation in the upper atmosphere releasing harmless oxides of chlorine, carbon and fluorine.

91. Which factor is included in SBS assessment?

    1. Poorly maintained mechanical equipment.
    2. Time of day.
    3. Shift work times.
    4. Overbearing management style over workforce.
    5. Illegal medication.

92. Which factor is not included in SBS assessment?

    1. Inadequately clean working environment.
    2. Staff not taking work breaks.
    3. Dust in workspace.
    4. Lighting glare.
    5. Tiredness.

93. Which symptoms manifest with sick building syndrome?

    1.   Upper respiratory infections.
    2.   High staff turnover not attributable to commercial factors.
    3.   Dry eyes.
    4.   Inability to retain staff in the building.
    5.   Inability to retain tenants in the building.

6 Hot- and cold-water supplies

Learning objectives

Study of this chapter will enable the reader to:

1. recognize the quality of water supplied to buildings;
2. explain pH value;
3. explain water hardness;
4. identify and apply appropriate water treatment methods;
5. decide appropriate applications for mains pressure and storage tank cold- and hot-water systems;
6. understand pressure-boosted systems for tall buildings;
7. apply economic instantaneous and storage techniques for hot-water provision;
8. understand primary and secondary pipe circulation systems;
9. calculate the heater power for hot-water devices;
10. understand demand units;
11. understand how cold- and hot-water pipe systems are designed;
12. be capable of carrying out basic cold- and hot-water pipe-sizing calculations;
13. understand pipe equivalent length;
14. understand pipe pressure loss calculation;
15. use CIBSE pipe-sizing data;
16. allocate sanitary appliances appropriate to building usage;
17. discuss the use of pipe materials and their methods of jointing;
18. have a basic understanding of how solar energy can be utilized in the provision of hot water in buildings.

Key terms and concepts

acidic and alkaline water 179; base exchange 180; centralized hot-water system 183; concentrating solar collector 198; corrosion 179; decentralized hot-water system 184; delayed-action ball valve 183; demand units 187; demineralization 180; dezincification 179; direct mains water 181; electrolytic action 195; equivalent length 188; flat-plate solar collector

Introduction

The convenience of piped water systems is likely to be taken for granted by those who have not been camping or caravanning. The provision of safe and hygienic water supplies is of paramount importance, and a considerable amount of engineering is involved in such provision.

The basics of water treatment are outlined, and then the ways in which water is distributed throughout buildings are discussed. The flows of water to sanitary appliances depend upon their frequency of use, which is not accurately predictable. The concept of pipe sizing with demand units leads to design calculations of water networks.

Calculation of the likely number of sanitary fittings to be installed is demonstrated. Water system materials and the application of solar heating are explained.

Water treatment

About 10% of the rainfall in the UK is used in piped services. Storage in reservoirs allows sedimentation of particulate matter, and then the water is filtered through sand and injected with chlorine for sterilization. A slow sand filter consists of a large horizontal bed of sand or a sand and granulated activated carbon sandwich. The carbon comes from coal and acts as a very efficient filter that traps microscopic traces of pesticides and herbicides. Water percolates down through the bed by gravity. Rapid sand filters have the raw water pumped through a pressurized cylinder that contains the filter medium. This filter material is either crushed silica, quartz or anthracite coal. Filtering removes metallic salts, bacteria and turbidity (muddiness). It also removes colouring effects, odours and particles, which affect the taste of the water. The naturally occurring pH value and the total dissolved salt concentration are virtually unaltered by the water supply authority.

Water quality varies with the local geology and can be classified as hard, soft, acidic or alkaline. Mineral salts of calcium and magnesium have soap-destroying properties and are considered in the evaluation of water hardness.

Temporary hardness is due to the presence of calcium carbonate, calcium bicarbonate and magnesium bicarbonate, which dissolve in water as it passes through chalky soil. These salts are deposited as scale on heat transfer surfaces during boiling, causing serious reduction in plant efficiency. They are known as carbonate hardness.

Permanent hardness is due to the presence of the non-carbonate salts calcium sulphate, calcium chloride, magnesium chloride and other sulphates and chlorides. Neutralization of these is achieved by means of chemical reactions.

Soft water contains up to 100 mg/l of hardness salts, as in Cornwall, and hard water contains as much as 600 mg/l, as in parts of Leicestershire. Acidic water is produced by contact with decomposing organic matter in peaty localities and normally occurs in soft-water regions. This water is very corrosive to steel, is plumbo-solvent and can cause dezincification of gunmetal pipe fittings.

The pH value denotes acidity or alkalinity due to the presence of free hydrogen ions in the water: acidic water, pH < 7; neutral water, pH = 7; alkaline water, pH > 7. Copper and plastic pipes and fittings can be used in acidic water regions. Hard-water areas are generally alkaline.

Water treatment for large boiler plants includes chemical injection to reduce corrosion from dissolved oxygen, and the pH value is raised to 11. Galvanized metal can be used where the pH value is 7.4 if the carbonate hardness is greater than 150 mg/l.

Users of large amounts of water may have treatment plant that removes or converts hardness salts to less harmful salts.

Base exchange

Raw water from the mains passes through a tank of zeolite chemicals where a base exchange takes place:

calcium carbonate + sodium zeolite → sodium carbonate + calcium zeolite

    (in water)   (in tank)   (in water)   (left in tank)

Similar base exchanges occur between the zeolites and other hardness salts in the raw water, turning them into non-scale-forming salts. On complete conversion of all the sodium zeolite, the filter bed is backwashed with brine (sodium chloride solution) which undergoes exchange with the calcium zeolite.

calcium zeolite + sodium chloride → sodium zeolite + calcium chloride

    (in tank)   (in flush water)   (in tank)   (flushed to drain)

The normal flow direction can then be resumed. The running cost of the system is limited to consumption of common salt, a small pump, periodic replacement of the zeolites and a small amount of maintenance work.

Steam boilers accumulate the salts passing through the treatment plant, and if they were allowed to become too numerous they would be carried over into the steam pipes and clog safety valves and pressure controllers. Either continuous or intermittent blow-down of boiler water to the drain is designed to control salt concentration. The high-pressure blow-down water is cooled before being discharged into the drains, and the heat is recycled.

Demineralization

Complete removal of mineral salts is very expensive, but it is essential for power station steam boilers, high-performance marine boilers and some manufacturing processes, where the presence of impurities is unacceptable. Raw water is passed through chemical filters in several stages to complete the cycle.

Reverse osmosis

Reverse osmosis is a filtration technique in which untreated water is pumped alongside a semi-permeable membrane in a pipe system. Clean water passes through the membrane. This method is used to produce drinking water in desert regions.

Solar distillation

Solar stills consist of glass-covered water troughs in which solar radiation evaporates the water, which then condenses on the cooler sloping glass roof and is collected in channels. This method can be used in hot locations.

Cold-water services

Mains water is used in two ways: direct from the main and as low-pressure supplies from cold-water storage tanks.

Mains supplies

At least one tap per dwelling and taps at suitable locations throughout large buildings are connected to the main for drinking water. The main also supplies ball valves on cold-water storage tanks and machines requiring a high-pressure inlet.

The economical use of water is important for safety, environmental and cost control reasons. The manual flush control of WCs and the tap operation of other appliances allows responsible usage. Urinals present a particular hygiene and water consumption contradiction. The user has no control over the flushing of water through the trough or bowl. The absence of flushing water leaves the urinal unpleasantly odorous and discoloured. Cleaning staff may counteract this by the excess dumping of deodorant blocks into the urinal. Perfumed toilet blocks are up to 100% para dichlorobenzene. Toilet-cleaning fluid contains phosphoric acid. These toxic chemicals are passed to the sewage treatment plant through the drain system. Uncontrolled flushing when the urinals remain unused, particularly overnight, results in wasteful water consumption and no benefit to the user. In the UK the supply of potable water plus the removal of waste water from consumers may cost around £1.50 per m^3 from a meter on the supply inlet pipe. An uncontrolled urinal cistern of 9 litre would flush, say, four times per hour, 24 h per day for 365 days in a year and consume 315 m^3 of water costing up to £500.

The installation of a water inlet flow control valve to a range of urinals will only allow flushing when appliances have been used, saving consumption. The valve may be operated from a passive infrared presence detector, discharge water temperature sensor or a variation in the water pressure within the same room. A short-term water flow to a WC or basin causes the stored water pressure within bellows to exceed the pressure in the pipeline. A diaphragm opens and allows water to flow to the urinal cistern until the accumulator pressure again equals the pipeline pressure; water flow can be adjusted to avoid wastage.

Low-pressure supplies

Static water pressures in tall buildings are reduced by storing water at various levels. Sealed storage tanks are used for drinking water. Open water tanks become contaminated with airborne bacteria and are only used for sanitary purposes. Cold-water services are taken to taps, WC ball valves, hot-water storage cylinders and equipment needing low-pressure supplies. A separate cold feed is taken to a shower or group of showers to avoid the possibility of scalding. Tanks are sized to store the total cold-water requirement for a 24-h period.

The minimum mains water pressure available in the street is 100 kPa (1 bar), which is 1 atmosphere gauge or 10 m height of water. The water supplier may be able to provide 300 kPa, or enough pressure to lift water to the top of a building 30 m high; however, allowance has to be made for friction losses in pipelines and discharge velocity, which effectively limits the vertical distance to between 2 and 6 storeys.

Separation of the contaminated water being used within the building for washing, flushing sanitary appliances, circulating within heating and air-conditioning cooling systems, evaporative cooling towers, ornamental fountains, agricultural irrigation or manufacturing processes from potable mains water is achieved by using the following:

1. a storage tank with ball valve (break tank);
2. a permanent air gap between the tap discharge and the contaminated water level (e.g. wash basin);
3. a single-seat non-return valve (check valve);
4. a double-seat check valve.

The Water Byelaws 1989 classify the risk of contamination from the building reaching upstream into the water main in three groups, each having its own protection (Table 6.1).

Cold-water storage tanks are expected to contain water of similar quality to that supplied from the main and so must be covered to exclude foreign matter, insects and light as well as being thermally insulated and not contaminating the stored water themselves. Tanks are generally not larger than 2 m long by 1 m wide by 1 m high, and pipe connections must ensure that water flushes through all of them to eliminate stagnation.

Servicing or isolating valves are located on the inlet to all ball valves on storage tanks and WC cisterns to facilitate maintenance without unnecessary water loss or inconvenience to the occupier. A servicing valve is required on all outlets from tanks of more than 15 l, that is, larger than a WC cistern.

The drinking and food-rinsing water tap at a kitchen sink must be connected to the water main before any water softener enters and a check valve is required between this tap and the softener.

Service entry into a building is via an underground pipe passing through a drain pipe sleeve through the foundations and rising in a location away from possible frost damage. An external stop tap near the boundary of the property is accessible from a brick or concrete pit. A ground cover of 760 mm is maintained over the pipe. A stop valve and drain tap are fitted to the main on entry to the building to enable the system to be emptied if the building is to be unoccupied during cold weather.

A water meter is the next pipe fitting. This has a rotary flow sensor, which is used to integrate the quantity of water that has passed. The cubic metres of water that are supplied, and charged for, are assumed to be discharged into the sewer. A separate charge is levied for the supply of potable water and for the acceptance of the contaminated discharge foul water. The consumer normally has no choice but to pay both the charges.

In tall buildings the pressure required to reach the upper floors can be greater than the available head, or pressure, in the mains. A pneumatic water-pressure-boosting system is used as shown

Table 6.1 Classification of contamination risks.

Class	Risk	Example of risk class	Type of protection
1	Serious danger to life	WC, bidet, urinal, agricultural or industrial process	Permanent air gap or break tank
2	May cause minor illness	Clothes, dishwashing and drinks vending machines, commercial water softeners	Permanent air gap, break tank or double check valve
3	May cause an unpleasant taste, odour or discolouration	Single outlet mixer taps and domestic water softeners	Any class 1 or 2 protection or a single check valve

6.1 Pneumatic water-pressure-boosting system for tall buildings.

in Fig. 6.1. Float switches in the storage tanks operate the pump to refill the system and minimize running times to reduce power consumption. A delayed-action ball valve on the cold-water storage tanks can be used. This delays the opening of the ball valve until the stored water has fallen to its low-level limit. System pressure is maintained by a small air compressor and pneumatic cylinder. The controller relieves excess pressure and switches on the compressor when the air pressure falls. During much of the day, water is lifted pneumatically at much lower cost than if it were pumped.

Cold-water storage to cover a 24-h interruption of supply (CIBSE, 1986) ranges from 45 l/person for offices to 90 l/person for dwellings and 135 l/person for hotels.

Hot-water services

Hot water can either be generated by the central boiler plant and stored, or produced close to the point of use by a more expensive fuel.

Central hot-water storage

The low-cost fuel used for the central heating plant is also used for the hot-water services boiler. This is located within the main boiler house and a large volume storage cylinder is employed. A small power input boiler is run almost continuously, winter and summer, under thermostatic control from the stored hot water. Primary circulation pipes are kept short and well-insulated.

This system can meet sudden large demands for hot water. Secondary circulation pipes distribute hot water to sanitary appliances. A pump is fitted in the secondary return; its function is to circulate hot water when the taps are shut and it does not appreciably assist draw-off rates from taps. Connections from the secondary flow to the tap are known as dead-legs and are limited to 5 m of 15 mm diameter pipe. This minimizes wastage of cold water in the non-circulating pipework when running a tap and waiting for hot water to arrive.

Decentralized system

The decentralized system is mainly for small hot-water service loads distributed over a large building or site where it would be uneconomic to use a central storage cylinder and extensive secondary pipework. Electricity or gas can be used in small storage or instantaneous water heaters located at the point of use. They are connected directly to the water main. Figures 6.2 and 6.3 show the operational features of gas-fired instantaneous and storage water heaters.

Mains-connected storage water heaters are protected from excess pressure and water temperature by a combined safety valve such as that shown in Fig. 6.4. On rise of mains water pressure, an internal spring relieves water to outdoors through the female-screwed pipe connection. On rise of water temperature within the stored volume, the wax thermal probe expands to open the relief valve. Manual testing of the spring valve during routine inspections is done by raising the lever to discharge water. When testing like this, collected debris from corrosion of the water

6.2 Gas-fired instantaneous water heater.

To taps

Flue

Cold-water inlet
from storage tank

Hot-water connection

Insulation material

Gas inlet

Thermostat

Spreader

D

Burner

6.3 Gas-fired storage water heater.

6.4 Mains water pressure safety valve.

storage tank or salt encrustation often causes the spring-controlled valve to remain cracked open and leak water, resulting in a valve replacement task.

Electric instantaneous heaters have power consumptions of up to 6 kW and produce water at 40°C and up to 3 l/min at 100% efficiency. Small electric storage heaters of 7 l are fitted over basins or sinks. Capacities of up to 540 l operate on the off-peak storage principle. Immersion heaters are controlled by time switches and thermostats and are connected in 3 kW stages.

The indirect hot-water system

The basic layout of the combined central heating and indirect hot-water service system is shown in Fig. 6.5. The cylinder is insulated with 75 mm fibre glass and should have a thermostat attached to its surface at the level of the primary return. Water is stored at 65°C, and when fully charged the thermostat closes the motorized valve on the primary return. This 'off' signal may also be linked into the pump and boiler control scheme to complete the shut-down when the central heating controls are satisfied.

Hot-water pipes are insulated with a minimum of 25 mm of insulation, as are tanks exposed to frost. The primary system feed and expansion tank has a minimum capacity of 50 l, and the cold-water storage tank has a capacity of at least 230 l.

Hot-water storage requirements at 65°C are as follows: office, 5 l/person; dwelling, 30 l/person; hotel and sports pavilion, 35 l/person (CIBSE, 1986).

6.5 Indirect hot-water storage system.

EXAMPLE 6.1

A dwelling has a 135 l hot-water storage indirect cylinder. The stored cold-water temperature is 10°C and the hot water is to be at 65°C. Calculate the necessary heat input rate to provide a 3-h recovery period from cold.

$$\text{Heat input rate} = \frac{\text{mass kg}}{\text{time h}} \times SHC \ \frac{\text{kJ}}{\text{kgK}} \times \text{temperature rise K}$$

$$= \frac{135 \ l}{3 \ h} \times \frac{1 \ \text{kg}}{1 \ l} \times 4.19 \ \frac{\text{kJ}}{\text{kgK}} \times (65 - 10) \ \text{K}$$

$$= \frac{135}{3} \times 4.19 \times (65 - 10) \ \frac{\text{kJ}}{h} \times \frac{1 \ h}{3600 \ s} \times \frac{\text{kWs}}{\text{kJ}}$$

$$= 2.881 \ \text{kW}$$

This can be rounded to 3 kW to allow for heat losses and is adequate for most dwellings with two to four occupants.

Pipe sizing

The demand for water at sanitary appliances is intermittent and mainly random but has distinct peaks at fairly regular times. The pipe sizes for maximum possible peak flows would be uneconomic. Few appliances are filled or flushed simultaneously. To enable designers to produce pipe systems that adequately match likely simultaneous water flows, demand units (*DU*) are used.

DU are dimensionless numbers relating to fluid flow rate, tap discharge time and the time interval between usage.

They are based on a domestic basin cold tap water flow rate of 0.15 l/s for a duration of 30 s and an interval between use of 300 s. This application is given a theoretical *DU* of 1.0 and other appliances are given relative values. Table 6.2 lists practical *DU*s.

The use of spray taps and small shower nozzles greatly reduces water consumption. Design water flow rates of 0.05 l/s for a spray tap, 0.1 l/s for a shower spray nozzle over a bath and 0.003 1/s per urinal stall can be used in place of *DU*. Figure 6.6 (CIBSE, 1986) is used to convert *DU* into water flow rates.

The design procedure for pipe sizing is as follows.

1. Draw the pipework layout on the building plans.
2. Mark the *DU* appropriate to each sanitary fitting on the drawing.

Table 6.2 Practical demand units (*DU*).

Fitting	Application		
	Congested	*Public*	*Private*
Basin	10	5	3
Bath	47	25	12
Sink	43	22	11
WC (13.5 l cistern)	35	15	8

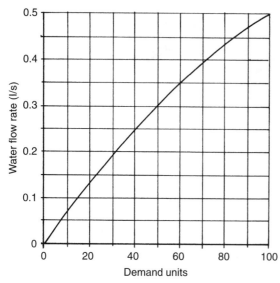

6.6 Simultaneous flow data for water draw-off points.

Table 6.3 Flow of water in copper tube of various diameters.

$\Delta p/EL$ (N/m³)	Water flow rate q (kg/s)				
	15 mm	22 mm	28 mm	35 mm	42 mm
1000	0.160	0.429	0.933	1.60	2.58
1500 v = 1.5	0.201	0.537	1.170	2.00	3.22 v = 3
2000	0.236	0.630	1.370	2.34	3.77
2500 v = 2	0.268	0.712	1.540	2.65	4.26
3000	0.296	0.787	1.710	2.92	4.70 v = 4

Note: v = water velocity (m/s).

Source: Reproduced from *CIBSE Guide* (CIBSE, 1986) by permission of the Chartered Institution of Building Services Engineers.

3. Sum all the *DU* along the pipework to the water source, which will be the storage tank or incoming water main.
4. Convert *DU* to water flow rates using Fig. 6.6.
5. Find the head of water *H* (in metres) causing the flow to each floor level.
6. Estimate the equivalent length (*EL*) of the pipe run to each floor in metres. This can be taken as the measured length plus 30% for the frictional resistance of bends, tees and the tap.
7. Find the index circuit. This is the circuit with the lowest ratio of *H* to *EL*.
8. Choose pipe sizes from Table 6.3 for the index circuit.
9. Determine the other pipe sizes from the *H/EL* figure appropriate to each branch of the index circuit.
10. Determine the water flow rate and head for a bronze-body hot-water service secondary pump, if one is required.

A notation system can be adopted to gather pipe-sizing data together on drawings, as shown in Fig. 6.7.

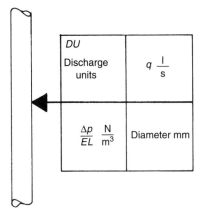

6.7 Notation for pipe-sizing data on drawings.

EXAMPLE 6.2

A domestic cold-water service system is shown in Fig. 6.8. The water main in the street is 50 m from the entry point shown and the supply authority provides a minimum static pressure of 20 m water gauge. The velocity energy and the frictional resistance at the ball valve amount to 2 m water gauge. Determine the pipe sizes.

6.8 Cold-water service pipe-sizing diagram for Example 6.2.

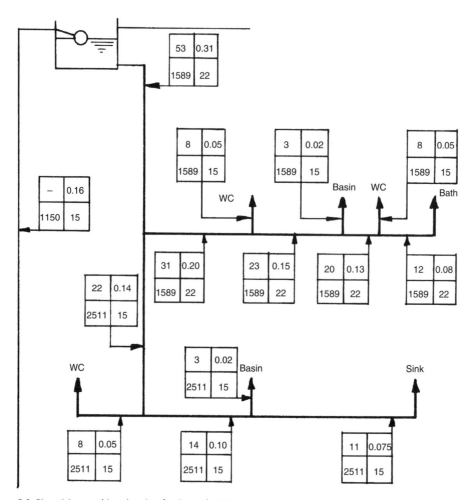

6.9 Pipe-sizing working drawing for Example 6.2.

The pipe-sizing data are shown in Fig. 6.9. Taps and ball valves will be 15 mm on domestic sanitary appliances and 22 mm on baths. *DU* values are taken from Table 6.2: WC, 8; basin, 3; sink, 11; bath, 12. These are entered into the appropriate boxes on the working drawing and then totalled back to the water main. Great accuracy is not needed in reading the water flow rates appropriate to each *DU*.

The heads of water causing flow are 4 m for the upper floor and 8 m for the lower floor. The measured length of pipe from the storage tank to the furthest fitting on the upper floor, the bath, is:

$$L_1 = (4 + 1 + 4 + 4 + 2 + 3 + 1) \text{ m} = 19 \text{ m}$$

and the equivalent length of the circuit to the bath is,

$$EL_1 = 1.3 \times 19 \text{ m}$$

$$= 24.7 \text{ m}$$

Similarly, the equivalent length for the lower floor circuit to the sink is:

$$EL_2 = 1.3 \times (4 + 1 + 3 + 1 + 4 + 10 + 1) \text{ m} = 31.2 \text{ m}$$

The head loss rate to the bath is:

$$\frac{H_1}{EL_1} = \frac{4 \text{ m}}{24.7 \text{ m}}$$

$$= 0.162 \text{ m head/m run}$$

The head loss rate to the sink is:

$$\frac{H_2}{EL_2} = \frac{8 \text{ m}}{31.2 \text{ m}}$$

$$= 0.2564 \text{ m head/m run}$$

The pipe circuit to the bath has the lowest H/EL figure; this is the index circuit. H_1/EL_1 is the available pressure loss rate, which drives water through the upper part of the system. Branches to other fittings on the same floor level can be sized from the same figure. All pipes below the upper floor are sized using the value of H_2/EL_2 that is appropriate to that circuit. Now,

$$H = \frac{\Delta p}{9807} \text{ m } H_2O$$

where Δp is the pressure exerted by a water column of height H m. Therefore

$$\frac{\Delta p_1}{EL_1} = \frac{H_1}{EL_1} \times 9807 \text{ N/m}^3$$

$$= 0.162 \times 9807 \text{ N/m}^3$$

$$= 1589 \text{ N/m}^3$$

and

$$\frac{\Delta p_2}{EL_2} = 0.2564 \times 9807 \text{ N/m}^3$$

$$= 2514.5 \text{ N/m}^3$$

These pressure loss rates are rounded and entered on Fig. 6.8. A different pressure loss rate could be calculated for each pipe but sizing them all on the index values is sufficiently accurate at this stage. Suitable pipe diameters are chosen from either Table 4.3 or Table 6.3, or the full data tables in CIBSE (1986). Notice that the bath connection size has been used rather than the 15 mm pipe, which would satisfy the design data for much of the upper floor branch. The lower water velocities produced will minimize noise generation in the pipework. The pressure loss rate causing flow in the water main is:

$$\frac{H_3}{EL_3} = \frac{\text{head available for overcoming pipeline friction}}{\text{equivalent pipe length}}$$

Now,

$$H_3 = \text{water main pressure} - \text{vertical lift} - \text{ball valve resistance}$$

$$- \text{water velocity pressure head at ball valve}$$

$$= 20 - (4 + 1 + 3 + 1) - 2 \text{ m}$$

$$= 9 \text{ m water guage}$$

and,

$$EL_3 = 1.3 \times (50 + 4 + 1 + 3 + 1)\,m = 76.7\,m$$

Therefore,

$$\frac{H_3}{EL_3} = \frac{9\,m\ water}{76.7\,m\ run} = 0.1173\,m\ water/m\ run$$

and,

$$\frac{\Delta p}{EL_3} = 0.1173 \times 9807\,N/m^3 = 1150.4\,N/m^3$$

While this pressure loss rate is available, a water main 15 mm in diameter would provide a flow of a little over 0.16 kg/s. Then, while the taps are shut, the storage tank would refill in:

$$230\,kg \times \frac{s}{0.16\,kg} \times \frac{1\,h}{3600\,s} = 0.4\,h$$

EXAMPLE 6.3

A hot-water service secondary system is shown in Fig. 6.10. Estimate the sizes of the pipes and specify the pump size.

6.10 Secondary hot-water service system for Example 6.3.

Taps on the lower floor are within the limit for dead-leg non-circulating pipes and a secondary return is shown for the group of appliances on the upper floor. Figure 6.11 is the working drawing. Water flow through the cold-feed pipe into the indirect cylinder is at the same rate as the expected outflow. The pressure loss rate to X from the cold-water storage tank is:

$$\frac{H_1}{EL_1} = \frac{3 \text{ m head}}{1.3 \times 24 \text{ m run}}$$

$$= 0.0962 \text{ m head/m run}$$

Then,

$$\frac{\Delta p_1}{EL_1} = 0.0962 \times 9807 \text{ N/m}^3$$

$$= 943.4 \text{ N/m}^3$$

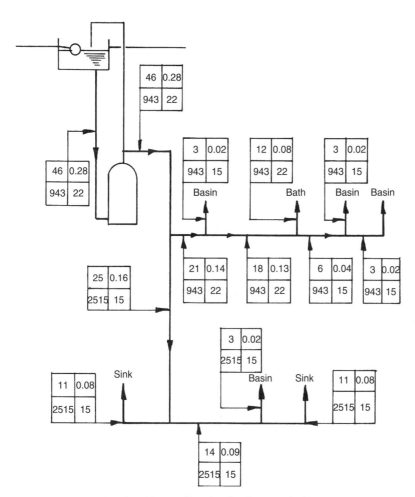

6.11 Hot-water service pipe-sizing working drawing for Example 6.3.

Table 6.4 Insulated pipe heat emission.

Pipe diameter (mm)	15	22	28	35	42
Heat emission (W/mK)	0.19	0.23	0.25	0.29	0.32

Source: Reproduced from *CIBSE Guide* (CIBSE, 1986) by permission of the Chartered Institution of Building Services Engineers.

The pressure loss rate to Y is:

$$\frac{\Delta p_2}{EL_2} = \frac{6}{1.3 \times 18} \times 9807 \text{ N/m}^3$$

$$= 2514.6 \text{ N/m}^3$$

The secondary return pipe is not intended to play an active part in water discharge from the taps but only to pass an amount of water relating to circulation pipe heat losses. Pipes insulated with 25 mm glass fibre have heat emissions as shown in Table 6.4.
The heat loss from the secondary circulation is,

$$\text{heat loss} = 13 \text{ m} \times 0.23 \frac{W}{mK} \times (65 - 20) \text{ K}$$

$$+ 13 \text{ m} \times 0.19 \frac{W}{mK} \times (65 - 20) \text{ K}$$

$$= 245.7 \text{ W} = 0.2457 \text{ kW}$$

The water flow rate necessary to offset this pipe heat loss while losing say 5°C between the outlet and inlet connections at the cylinder is:

$$q = \frac{0.2457}{4.19 \times 5} \text{ kg/s}$$

$$= 0.0117 \text{ kg/s}$$

By reference to Table 4.3, a pipe 15 mm or 22 mm in diameter carrying 0.0117 kg/s has a pressure loss rate of much less than 200 N/m^3. A gross overestimate of the pump head for this circuit is:

$$H = 26 \text{ m} \times 1.3 \times 200 \times \frac{1}{9807}$$

$$= 0.69 \text{ m water}$$

A pump that delivers 0.0117 kg/s at a head of 0.69 m would meet the requirement. Pump C in Fig. 4.16 would provide a far higher flow rate than this, and either a smaller pump would be used or the control valves would be partially closed to avoid the production of noise due to high water velocities.

Allocation of sanitary appliances

The recommended numbers of sanitary fittings are given in Table 6.5.

Table 6.5 Recommended allocation of sanitary fittings.

Building accommodation	No. of occupants	Male	Female	Urinals	Male	Female
Staff	1–100	1 + 1 per 25	1 + 1 per 14	1 + 1 per 25	1 + 1 per 25	1 + 1 per 14
	Over 100	+ 1 per 30	+ 1 per 20	+ 1 per 30	+ 1 per 30	+ 1 per 20
Transient public	1–200	1 per 100	2 per 100	1 per 50	1 per WC	1 per WC
	200–400	1 per 100	+ 1 per 100	1 per 50	1 per WC	1 per WC
	Over 400	+ 1 per 250	+ 1 per 100	1 per 50	1 per WC	1 per WC

Source: Reproduced from *IHVE Guide* (CIBSE, 1986 [IHVE, 1970]) by permission of the Chartered Institution of Building Services Engineers.

Table 6.6 Allocation of sanitary accommodation for Example 6.4.

Staff	Students
2 male WCs	4 male WCs
2 female WCs	4 female WCs
2 urinals	8 urinals
2 male basins	4 male basins
2 female basins	4 female basins

EXAMPLE 6.4

A new university building is to have 70 male and 20 female staff. The student population is considered to be 400 males and 300 females. Students are considered to be transient public. Recommend a suitable allocation of sanitary accommodation.

Using Table 6.5, we obtain the results given in Table 6.6. The accommodation is to be distributed around the site to ensure uncongested access and reasonable walking distances. A tall building would ideally have toilets on each floor, close to the stairways and lifts, so that all the pipework can run vertically in a service duct. At least one toilet for disabled people is to be included at suitable location, with appropriate access, in the above schedule.

Materials for water services

The materials used in hot- and cold-water systems are listed in Table 6.7. Corrosion protection is provided by ensuring that incompatible materials are not mixed in the same pipework system, by recirculating the water in central heating systems to reduce fresh oxygen intake, and by adding inhibiting chemicals to the water. Hot- and cold-water service systems are continually flushed with fresh water, making it necessary to use galvanized metal, copper or stainless steel.

Copper and galvanized steel should not be used in the same system because electrolytic action will remove the internal zinc coating and lead to failure. A galvanized metal cold-water storage tank can be successfully used with copper pipework as the low temperature in this region limits electrolytic action. Heat accelerates all corrosion activity.

Black mild steel is used in recirculatory heating systems, and an initial layer of mill scale, which is metal oxide scale formed during the high-temperature working of the steel during its

Table 6.7 Materials for hot- and cold-water systems.

Material	Application	Jointing
Lead	Elderly hot- and cold-water pipes up to 22 mm; water becomes contaminated during storage in the pipework and must not be consumed	Hot molten lead-wiped joints and swaged (flared) connections to copper pipe
Copper	All water services, gas and oil pipelines; 6–10 mm soft copper tube supplied in rolls for oil lines and microbore heating; semi-hard tube in various thicknesses in diameters of 15 mm upwards; hand- or machine-formed large radius bends are popular; aluminium finned pipe is used in convector heaters	Compression, manipulative, silver solder, bronze weld, flanged or push-fit ring seal using polybutylene fittings
Black mild steel	Indirect hot-water heating systems, pipework and radiators	British Standard pipe thread (BSPT), screwed and socketed, flanged or welded
Galvanized mild steel	Hot- and cold-water pipework on open systems and water mains; cold-water storage tanks and indirect cylinders	BSPT screwed, socketed, flanged or welded
Stainless steel	Hot- and cold-water pipework 15–28 mm; thickness and diameter correspond to those of semi-hard copper and can be bent in the same way; larger diameters are used for chemicals or for sterile fluids in hospital services	Compression, silver solder, flanged or push-fit ring seal using polybutylene fittings
Cast iron	Central heating radiators, boilers and centrifugal pump bodies	BSPT-screwed fittings and gasket joints
Brass	Pipe fittings and valves	BSPT screwed
Gunmetal	Pipe fittings and valves; pump impeller; also body of a secondary hot-water service centrifugal pump	BSPT screwed
Polybutylene	Cold- and hot-water pipes and fittings in 15 m and 22 mm diameters, withstanding 90°C at 3 bar (atm) pressure	Push-fit ring seal, compression fusion weld
Polyethylene	Underground mains cold water	Compression
Unplasticized poly vinyl chloride (UPVC)	Mainly cold-water services	Compression, solvent weld
Cast aluminium	Boiler body, central heating radiators with convection fins	BSPT screwed

manufacture, helps to slow further corrosion. Discoloration of the central heating water from rust to black during use shows steady corrosion. A black metallic sludge forms at low points after some years. Large hot-water and steam systems have the mill scale chemically removed during commissioning and corrosion-inhibiting chemicals are mixed with the water to maintain cleanliness and avoid further deterioration.

The formation of methane gas in heating systems during the first year of use is due to early rapid corrosion, and radiators need frequent venting to maintain water levels. Proprietary inhibitors should be added to all central heating systems. These control methods of corrosion are anti-bacterial. Without them, steel boilers and radiators can rust through in 10 years.

Solar heating

Solar heating can be employed to assist the generation of hot water in secondary storage systems with a consequent reduction in energy costs. The highly variable nature of solar radiation in the

UK produces a financial return on the capital invested in equipment of around 10% per annum when electrical water heating is used. Solar energy is used to provide:

1.   comfort heating through architectural design in a 'passive' system;
2.   comfort heating using collectors, with air as the heat transfer fluid, in an 'active' system;
3.   comfort heating and hot-water using collectors, with water as the heat transfer fluid, in an active system.

   Thermal storage of the collected energy is needed to balance the supply of solar heat with its times of use, usually when the supply has greatly diminished. The following methods of thermal storage are used:

1.   thick concrete or brick construction for passive systems, combined with large areas of south-oriented glazing;
2.   rock heat storage, heated by air recirculation;
3.   water tanks;
4.   phase change salts.

   Flat-plate solar collectors are the most popular as they can heat water to 30–40°C without the danger of boiling. They can be incorporated into the architectural design of a sloping roof and are reasonably cheap. They are used to preheat the water supplied to the hot-water service storage cylinder, as shown in Fig. 6.12. Only the simplest arrangement is indicated, where the cold-water

6.12 Solar collector to preheat a hot-water service system.

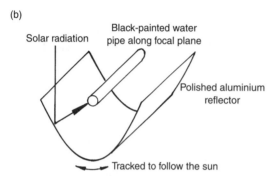

6.13 (a) Flat-plate solar collector; (b) parabolic trough concentrating solar collector.

storage tank supplies only the hot-water cylinder. A wide variety of system designs are in use, depending upon requirements. The solar pump switches off when the temperature sensors find no measurable water temperature rise. All the water then drains back into the storage tank to avoid frost damage. The system will be shut down during the coldest weather.

Concentrating solar collectors are used to generate water temperatures of 80°C and over for hot-water service heating or air-conditioning systems. They are usually driven by electric motors, through reduction gears, to track the sun's movement during each day so that the collecting pipe is kept in the focal plane of the parabolic mirror or polished aluminium reflector. Typical flat-plate and concentrating solar collectors are shown in Fig. 6.13.

Questions

1. Sketch and describe the earth's natural water cycle.
2. List and describe the sources of water and the methods used for its storage and treatment.
3. What pollutants are present in naturally occurring water and where do they come from?
4. Explain the terms 'temporary' and 'permanent' hardness, and list their characteristics.
5. State the use of service reservoirs and describe mains water distribution methods.
6. State the design parameters for cold-water mains and storage systems within buildings, giving particular information on protection from frost damage, suitability for drinking and protection of the mains against contamination from the building.
7. List the design parameters for hot-water service systems, giving typical data.
8. Sketch the layout of a water services system in a house, showing typical sizes of equipment and methods of control. Show how wastage of water is minimized.
9. Sketch and describe the methods used to generate hot water, noting their applications, economy and thermal performance.

10. Sketch and describe a suitable cold-water services installation for a 20-storey hotel where the mains water pressure is only sufficient to reach the fifth floor.

11. Draw the layout of an indirect hot-water system employing a central heating boiler and secondary circulation. Show all the pipework and control arrangements.

12. A small hotel is designed for 20 residents and 4 staff. Hot water to be stored at 65°C is taken from a cold-water supply at 10°C and heated during a 4-h period. Calculate the heat input rate required.

13. Explain the meaning and use of 'demand units'.

14. The cold-water service system in Fig. 6.9 has three identical sanitary fittings in the locations shown for each type. The pipe lengths shown are to be multiplied by 1.5 but vertical dimensions remain the same. Calculate the pipe sizes.

15. A cold-water storage tank in a house with five occupants is to have a capacity of 100 l per person and be fed from a water main able to pass 0.25 l/s. How long will it take to fill the tank?

16. A secondary hot-water service system has 55 m of 28 mm circulation pipework and 40 m of 15 mm circulation pipework. Water leaves the cylinder at 65°C and returns at 60°C. Air temperature around the pipes is 15°C. Pressure loss rate in the pipework is 260 N/m^3. The frictional resistance of the pipe fittings is equivalent to 25% of measured pipe length. Specify the head and flow rate required for the secondary circulation pump and choose a suitable pump. Use Table 4.3 and Fig. 4.9.

17. A bank building is to house 115 male and 190 female staff. Recommend a suitable allocation of sanitary accommodation.

18. Draw cross-sections through four different types of pipe joint used for water services, showing the method of producing a water seal in each case.

19. List the factors involved in the provision of a pipework system for the conveyance of drinking water within a curtilage. Comment on the suitability, or otherwise, of jointing materials, lead pipes and storage tanks in this context.

20. Describe the corrosion processes that take place within water systems and the measures taken to protect equipment.

21. Sketch and describe the ways in which solar energy can be used within buildings for the benefit of the thermal environment and to reduce primary energy use for hot-water production. Comment on the economic balance between capital cost and expected benefits in assessing the viability of such equipment.

22. List the types of sanitary appliance available and describe their operating principles, using appropriate illustrations. Comment on their maintenance requirements, water consumption, long-term reliability and materials used for manufacture.

23. How can water usage be minimized in domestic hot- and cold-water systems?

    1. Throttle pipe circuit balancing valves to minimize maximum water flow rates.
    2. Provide all water outlets from high-level storage tanks.
    3. Use spray taps on washbasins, low-flow shower heads, dual flush toilets and urinal occupancy sensors.
    4. Reduce washing and toilet facilities to below recommended numbers.
    5. Cannot be reduced as people have to use a certain amount for each facility.

24. The Demand Unit for domestic hot- and cold-water service outlets is based upon:

    1. Maximum sink water flow of 2.5 l/s at 1-h intervals.
    2. Total water flow required by a group of basin, sink, shower, bath and WC during average usage pattern.

3.   Demand Unit is just a fictitious number.
4.   Demand Unit is 1.0 for a domestic basin cold-water tap water flow rate of 0.15 l/s for a duration of 30 s and an interval between use of 5 min.
5.   100% usage in a typical house group of appliances.

25. What is reverse osmosis?

1.   Osmotic pressure difference cannot be reversed.
2.   A natural phenomenon.
3.   The opposite of water diffusion through a semi-permeable membrane to equalize the concentration of salt in solution.
4.   Dissolved salts migrating through a membrane, leaving desalinated water behind.
5.   Heat and pressure applied to salty water causes water molecules to pass through a fine membrane into a desalinated water stream on the other side.

26. Which of these is the correct basic safety provision for the public mains water supply system?

1.   Stop valve as mains water enters each building.
2.   Backflow preventer valve on each mains water pipe entering a building.
3.   Ball valve on a cold-water storage tank.
4.   Toilet flushing cistern.
5.   Air gap beneath every ball valve and tap outlet.

27. Should domestic potable cold water rise above the safe distribution temperature, what must be done?

1.   Redesign the distribution system and reinstall new pipes.
2.   Test water temperature, insulate pipes, drain and flush system, cool the mains water with a refrigeration system.
3.   Evacuate the building, drain and flush system.
4.   Nothing, it will soon cool down again.
5.   Test water temperature, find cause of heating, insulate or reroute pipes, drain and flush system and if necessary disinfect the system pipes and tanks.

28. Why is domestic hot water stored at high temperature?

1.   Maximizes the thermal efficiency of the water heater.
2.   Minimizes pipe sizes for flow rates.
3.   Minimizes storage quantity.
4.   It should not be as it can scald users.
5.   Kill Legionella bacteria in water.

29. What is part of the primary water treatment of public water supply systems in the UK?

1.   Aeration in outdoor reservoirs.
2.   Filtration by water percolating through the ground.
3.   Pumped filtration through granite chippings.
4.   Flocculation in gravel beds.
5.   Slow filtration by gravity flow through sand and activated carbon beds.

30. Which water description is correct about soft water?

1.   Occurs in granite rocky soil parts of the country.
2.   High pH value.

3.   Low suspended salt concentration.
4.   Total dissolved salt concentration of over 200 ppm.
5.   High soap-destroying capability.

31. Which water description is correct about hard water?

1.   Occurs in chalk soil parts of the country.
2.   Very low pH value.
3.   High suspended salt concentration and suspended vegetable matter.
4.   Total dissolved salt concentration below 50 ppm.
5.   Laundries consume minimal soap.

32. Which causes temporary hardness in water?

1.   Suspended solids.
2.   Acidic ground water sources.
3.   Precipitation of salts during storage in the building.
4.   Sulphates and chloride salts in the water.
5.   Dissolved carbonate salts.

33. Which is correct about permanent water hardness salts?

1.   Cannot be removed from water.
2.   Only removed by steam generation and condensation.
3.   Salts deposited on heat transfer surfaces during water boiling.
4.   Only removable by dosing ground water with acid.
5.   Due to presence of non-carbonate dissolved salts.

34. Which is not correct about water pH value?

1.   A measure of free hydrogen ions in water.
2.   pH of 7 means water is neither acidic nor alkaline.
3.   pH below 7 is always found in hard water.
4.   Steam boiler water is treated to a pH of 11.
5.   Acidic water from granite ground has a low pH.

35. Which is correct about water treatment for closed-circulation water systems?

1.   Not needed.
2.   Water purifies itself due to release of dissolved oxygen and salts during commissioning.
3.   Absence of fresh oxygen in closed system avoids need for corrosion treatment.
4.   There is no difference between the affects of pH value, acidity, alkalinity and corrosiveness of water.
5.   Treatment provided to combat electrolytic corrosion.

36. Which applies to the base exchange water treatment system?

1.   Calcium carbonate in public mains water chemically reacts with sodium zeolite in treatment tank.
2.   Calcium carbonate in public mains water is absorbed by zeolite salt in the treatment tank and remains there.
3.   Zeolite salts are consumed by incoming hardness salts and residue has to be removed for disposal.
4.   Zeolite salts filter out calcium carbonate and other hardness salts and must be disposed to waste water when fully clogged.

5. Zeolite salts destroy hardness salts in the public mains water supplied to steam boilers.

37. What is the meaning of low-pressure water supply main?

1. Building has a pressure reduction valve on the incoming public water main.
2. Street water main is at very low pressure.
3. There is no such thing as all public water supply systems function at well above atmospheric pressure.
4. All water services in buildings below four-storey height.
5. Mains water pressure supplies storage tanks that service the building's systems.

38. Which is the minimum allowable pressure of a water main in the street?

1. 100 bar.
2. 10 atmospheres.
3. 30 m water gauge.
4. 100 kPa.
5. 10 000 kPa.

39. When mains water pressure in the street is 250 000 N/m^2 and pipe friction and discharge losses are negligible, what building height can water lift to?

1. 250 m.
2. 9.807 m.
3. 26 m.
4. 12 storeys.
5. 25.49 m.

40. The measured pressure in the water main in the street is 3.5 b. To what height could the pressure of the water accommodate?

1. 35.6 m.
2. 0.35 m.
3. 3.5 m.
4. 356.8 m.
5. 35.0 m.

41. Which technical feature ensures a consumer building does not contaminate the public water supply pipework system?

1. Non-return valve on the supply pipe to each building.
2. Manual isolating valve at each building entry pipe.
3. Maintaining the building's piped water systems at a lower pressure with a pressure-reducing valve at entry from the public main.
4. Dosing all pipe systems in a building with biocide.
5. An air gap.

42. What is a water-pressure-boosting system?

1. Public mains water pumping station.
2. Gas-fired storage water heater to raise water pressure.
3. Sump pump to remove flood water.
4. Circulating pump for heating, chilled water or domestic hot water.
5. Pump to raise water to floors above where the public mains pressure can reach.

43. What does central domestic hot-water storage system mean?

    1. Hot-water storage tank in the middle of a building.
    2. Centrally heated storage calorifier.
    3. Domestic hot-water storage cylinder with pipework distribution to all hot-water taps.
    4. Gas-fired storage water heaters.
    5. Electrically heated domestic hot-water unit near each tap where the power is generated from the public system.

44. What is an instantaneous domestic hot-water unit?

    1. One where opening a hot-water outlet causes immediate flow.
    2. Gas-fired non-storage water heat exchanger.
    3. Something like a kettle.
    4. Electrically heated insulated storage water heater located beneath, alongside or above a hot tap outlet.
    5. Large domestic hot-water storage calorifier that cannot run out of hot water.

45. What is the health risk, if any, from stored domestic hot water?

    1. No health risk when public mains water is heated and stored as it always remains dosed with drinkable chemicals.
    2. Heating water does not raise its health risk.
    3. Only mains water that has come into contact with the atmosphere could become contaminated.
    4. Mains water bacteria always remain dormant.
    5. Legionella bacteria present in water grow rapidly between 20°C and 40°C.

46. How do we protect ourselves from domestic hot-water systems?

    1. Heat and store hot water at 42°C.
    2. Only use hot water at less than 45°C.
    3. Water stored at 65°C so all bacteria killed.
    4. Heat stored water to 65°C to kill bacteria, then wait for it to cool to 40°C prior to use.
    5. Heat stored water to 100°C to sterilize bacteria, then wait for it to cool to 40°C prior to use.

47. Which is the hotel domestic hot-water storage requirement in litres at 65°C, per person?

    1. 35
    2. 25
    3. 65
    4. 45
    5. 120

48. What is a domestic hot-water circulation system called?

    1. Personnel circulation.
    2. Hot-water pipe system.
    3. Tap circulation.
    4. Draw-off circulation system.
    5. Hot-water secondary flow and return.

49. A 1000 l domestic hot-water storage calorifier has a heating water rate of input of 50 kW. How long will it take to raise the water content from 10°C to 65°C if heat loss from the cylinder is negligible?

    1.   0.305 h.
    2.   3 days.
    3.   1.5 h.
    4.   1.28 h.
    5.   6.5 h.

50. Why are domestic hot- and cold-water distribution pipes not designed to always provide full flow when all taps are open?

    1.   But they are, have to be.
    2.   Users rarely open taps fully.
    3.   Pipe resistance always slows water flow, so full flow rate is never possible.
    4.   Demand for water at taps, showers and washing machines is randomly intermittent.
    5.   If too many taps open simultaneously, users simply have to wait a few seconds longer.

51. What does 'demand unit' stand for?

    1.   Amount of hot water needed to fill a standard basin in one minute.
    2.   Maximum demand for hot water from a group of taps.
    3.   Statistical assessment of the simultaneous flow in litre/s from a system of hot- or cold-water outlets.
    4.   Number relating to water flow rate, tap discharge time and interval between usage.
    5.   Total water flow rate demand in a system divided by the number of taps.

52. Which is the value of demand units?

    1.   Litre/s.
    2.   Total litres.
    3.   Time interval.
    4.   Number of water outlets.
    5.   Dimensionless.

53. Which is the appropriate temperature for shower, tap and bath water to avoid scalding?

    1.   36°C.
    2.   65°C.
    3.   50°C.
    4.   45°C.
    5.   41°C.

7 Soil and waste systems

Learning objectives

Study of this chapter will enable the reader to:

1. define the parts of waste and drain systems;
2. understand the type of fluid flow in a waste pipe;
3. explain the ventilation requirements of drainage pipes;
4. list and explain the ways in which the water seal can be lost from traps;
5. know the permitted suction pressure in a waste system;
6. understand air pressure distribution in a stack;
7. know how to connect waste pipes into a stack;
8. know the diameters, slopes and maximum permitted lengths of waste and drain pipes for above-ground systems;
9. know how to design waste and drain pipes for ranges of sanitary appliances;
10. design domestic, high-rise and commercial building waste installations;
11. understand and use discharge units for pipe sizing;
12. explain the uses of different materials and jointing methods;
13. understand how drain systems are tested;
14. explain how drain systems are maintained.

Key terms and concepts

access 208; air pressure 206; air static pressure 207; blockage 217; connection to stacks 207; discharge stack 207; discharge units 214; drain 208; flow surge 206; grease and residues 217; induced syphonage 209; inspection 216; lime scale 217; maintenance 217; materials and jointing 215; ranges of appliances 213; rodding 217; self-syphonage 206; sewer 207; slope 207; smoke 216; solid deposition 211; testing 216; trap 207; trap seal loss 209; vented systems 214; waste pipe 206; water 206; water seal 207.

Introduction

The terminology of drainage systems is outlined and then the characteristic flow within the pipework is explained. Understanding how fluid flows through waste and drain pipework is fundamental to correct design.

The potential for water seal loss in traps beneath sanitary fittings and the deposition of solids in long sloping drains is examined.

Various standard pipework layouts for above-ground systems are shown. The fluid flow through drain pipes is subject to diversity in timing and duration, as are the hot- and cold-water supplies to the same appliances. However, the characteristic flows into and out of the appliance are not the same, and the use of discharge units for drains is explained.

The materials and jointing methods used for pipework are demonstrated, as are the testing and maintenance procedures.

Definitions

The following terms are used.

Bedding: material around a buried pipeline assisting in resisting imposed loads from ground and traffic.

Benching: curved smooth surfaces at the base of manholes, which assist the smooth flow of fluids.

Combined system: a drainage system in which foul and surface-water are conveyed in the same pipe.

Crown: the highest point on the internal surface of a pipe.

Discharge stack: vertical pipe conveying foul fluid/solid.

Foul drain: a pipe conveying water-borne waste from a building.

Foul sewer: the pipework system provided by the local drainage authority.

Invert: the lowest point on the internal surface of a pipe.

Manhole: an access chamber to a drain or sewer.

Separate system: a drainage system in which foul and surface-water are discharged into separate sewers or places of disposal.

Stack: vertical pipe.

Subsoil drains: a system of underground porous or un-jointed pipes to collect groundwater and convey it to its discharge point.

Surface-water drain: a pipe conveying rain water away from roofs or paved areas within a single cartilage.

Surface-water sewer: the local authority pipework system.

Waste pipe: pipe from a sanitary appliance to a stack.

Fluid flow in waste pipes

The discharge of fluid from a sanitary appliance into a waste, soil or drain pipe is a random occurrence of short duration exhibiting a characteristic curve similar to that shown in Fig. 7.1.

Flows in waste pipes occur as surges, or plugs of fluid, which last for a short time. The pipe flows full at some time and a partially evacuated space appears towards the end of discharge, as shown in Fig. 7.2. Separation between the water attempting to remain in the P-trap and the plug falling into the soil stack causes an air pocket to form. The static pressure of this air will be subatmospheric. Air from the room and the ventilated soil stack bubbles through the water

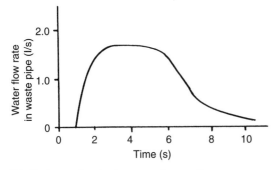

7.1 Discharge of water from a sanitary appliance.

7.2 Design of a basin waste pipe to avoid self-syphonage.

to equalize the pressures and a noisy appliance operation results. The inertia of the discharge may be sufficient to syphon most of the water away from the trap, leaving an inadequate or non-existent seal. The problem is avoided by using 32 mm basin waste pipes when the length is restricted to 1.7 m at a slope of 20 mm/m run, about 1°.

The sloping waste pipe can be up to 3 m long if its diameter is raised to 40 mm after the first 50 mm of run. This allows aeration from the stack along the top of the sloping section. Longer waste pipes with bends and steeper or even vertical parts have a 25 mm open vent pipe as shown in Fig. 7.3.

Vertical soil and vent stacks are open to the atmosphere 900 mm above the top of any window or roof-light within 3 m. Underground foul sewers are thus atmospherically ventilated. Water discharged into the stack from an appliance entrains air downwards and establishes air flow rates of up to a hundred times the water volume flow rate. Air flow rates of 10–150 l/s have been measured. The action of water sucking air into the pipe lowers the air static pressure, which is further reduced by friction losses.

Water enters the stack as a full-bore jet, shooting across to the opposite wall, falling and establishing a downward helical layer attached to the pipe surface. Restricted air passageways at such junctions further lower the air static pressure by their resistance to flow. Atmospheric pressure will be re-established at the base of the stack because of the flow of air into the low-pressure region. The falling fluid tends to fill the pipe near the base and positive air static pressures can be generated. Appliances connected to such a region may have their water seals intermittently forced out. Figure 7.4 shows the probable air static pressures during the simultaneous discharge of three appliances.

The pressure gradient shown in Fig. 7.4 can be drawn with the aid of data from experimental work (Wise, 1979). The maximum permitted pressure is −375 Pa as this is equivalent to the

7.3 Vented basin waste pipe.

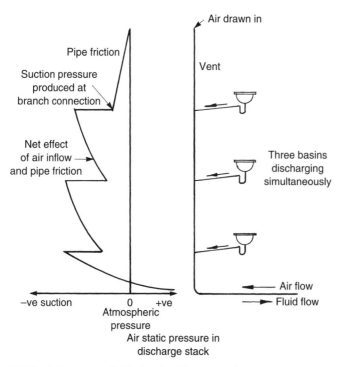

7.4 Air static pressure distribution in soil and vent pipes.

recommended trap depth of 75 mm water gauge for single-stack drain installations. When suction of this magnitude is applied to a 75 mm water seal, some of the water is sucked from the trap, leaving about 56 mm. This is sufficient to stop fumes entering the building.

Loss of water seal from a trap can occur through the following mechanisms.

1. Self-syphonage: this can be avoided by placing restrictions on lengths and gradients and venting long or steep gradients.
2. Induced syphonage: water flow past a waste pipe junction in a stack or along a sloping horizontal range of appliances can suck out the water seal. This is overcome by suitable design of the pipe diameters, junction layouts and venting arrangements.
3. Blow-out: a positive pressure surge near the base of a stack could push out water seals of traps connected in that region. Waste pipes are not connected to the lower 450 mm of vertical stacks, measured from the bottom of the horizontal drain.
4. Cross-flow: flow across the vertical stack from one appliance to another. Waste pipes are not connected to soil and vent pipes where cross-flow, particularly from WC branches, could be caused, as shown in Fig. 7.5.
5. Evaporation: this amounts to about 2.5 mm of seal loss per week while appliances are unused.
6. Wind effects: wind-induced pressure fluctuations in the stack may cause the water seal to waver out. The vent terminal should be sited away from areas subject to troublesome effects. Wind-tunnel tests using smoke as a tracer are performed for large developments.

7.5 Permitted stack connections avoiding cross-flow.

7. Bends and offsets: sharp bends in a stack can cause partial or complete filling of the pipe, leading to large pressure fluctuations. Foaming of detergents through highly turbulent fluid flow will aggravate pressure fluctuations. Connections to the vent stack before and after an offset equalize air pressures. A bend of minimum radius 200 mm is used at the base of a soil stack to ensure constant ventilation.

8. Surcharging: an underground drain that is allowed to run full causes large pressure fluctuations. Additional stack ventilation is required.

9. Intercepting traps: where a single-stack system is connected into a drain with an interceptor trap nearby, fluid flow is restricted. Additional stack ventilation is used.

10. Admission of rainwater into soil stacks: when a combined foul and surface-water sewer is available, it is possible to admit rainwater into the discharge stack. Continuous small rainwater flows can cause excessive pressure fluctuations in buildings of about 30 storeys. Flooding of the stack during a blockage would cause severe damage.

11. Pumped or pneumatically ejected sewage lifting: the discharge stack is gravity-drained into a sump, from where it is pumped into a street sewer at a higher level. A separate vent is used for the sump chamber and pumped sewer pipe to avoid causing pressure surges.

12. Capillary: lint or hair remaining in a trap may either block the capillary or empty it. Additional maintenance is carried out in high-risk locations.

13. Leakage: leakage can occur through mechanical failure of the joints or the use of a material not suited to the water conditions.

Figure 7.6 shows the principle of operation of an anti-syphon trap. When excessive suction pressure occurs in the waste pipe, some of the water in the trap is syphoned out. When the central ventilation passage becomes uncovered, it connects the inlet and outlet static air pressures. This returns the waste pipe to atmospheric pressure and the syphonage ceases. Sufficient water remains in the trap to maintain a hygienic seal.

7.6 Anti-syphon trap.

Drainage installations should remove effluent quickly and quietly, be free from blockage, and be durable and economic. They are normally expected to last as long as the building and be replaced only because of changed requirements or new technology. Blockages occur when the system is overloaded with solids, becomes frozen, suffers restricted flow at poorly constructed bends or joints, or has building material left inside pipe runs. Each section of discharge pipework must be accessible for inspection and internal cleaning.

Transport of solids from WCs is the controlling problem in the design, installation and maintenance of sloping drains. Swaffield and Wakelin (1976) showed that, to maintain the flow from a WC and avoid deposition of solids in the drain, the ratio of the length L of sloping drain (metres) to the gradient G must be:

$$\frac{L}{G} = 35^2$$

Pipe bends produce rapid deceleration of solids downstream, followed by velocity regain as the remaining flush water catches up with and accelerates the solids with minimal loss of inertia. When minimum gradients are used, solid deposition could occur at a bend. To avoid this, the equivalent length of a bend can be taken as 5 m of straight pipe in design calculations. Solid deposition can also occur at a top entry into a sewer. Branch connections should be at 45° to the horizontal.

EXAMPLE 7.1

A WC is to be connected to a 100 mm soil stack, which runs in a 250 mm deep service duct formed from a false ceiling. It is intended that the WC be 16 m from the stack. There is a 90° bend just before the drain enters the stack. Determine whether the proposed layout will be satisfactory. Figure 7.7 shows the intended arrangement.

7.7 Sloping drain in a false ceiling for Example 7.1.

The allowed fall will be approximately 150 mm, subject to free available passage. The equivalent length of the sloping drain is $(L + 5)$ m. Then the gradient is:

$$G = \frac{0.15}{L\,m}$$

The intended conditions are:

$$G = \frac{0.15}{16} = 0.009375$$

and,

$$L = 16 + 5 = 21 \text{ m}$$

Using $L/G = 35^2$ for an equivalent length of 21 m the gradient cannot be less than:

$$G = \frac{L}{35^2}$$
$$= \frac{21}{35^2}$$
$$= 0.01714$$

The intended gradient is less than the minimum allowable gradient and so the design must be modified.

Assuming that the WC can be brought nearer to the stack, how far away can it be? There are two conditions to be met:

$$G = \frac{0.15}{L} \tag{7.1}$$

$$\frac{L + 5}{G} = 35^2 \tag{7.2}$$

Substitute equation (7.1) into equation (7.2) to eliminate G:

$$(L + 5) \times \frac{L}{0.15} = 35^2$$

This can be rearranged to:

$$L^2 + 5L = 183.75$$
$$L^2 + 5L - 183.75 = 0$$

This is a standard quadratic equation of the form,

$$ax^2 + bx + c = 0$$

where $x = L$, $a = 1$, $b = 5$ and $c = -183.75$. The solution of the quadratic is:

$$x = \frac{-b \pm \sqrt{(b^2 - 4ac)}}{2a}$$

Therefore

$$L = \frac{-5 \pm \sqrt{[5^2 - 4(-183.75)]}}{2}$$

$$= -5 \pm \sqrt{\frac{760}{2}}$$

$$= 11.284 \text{ m}$$

Thus, if the WC is situated 11 m from the stack, the installed gradient of $0.15/11 = 0.014$ is steeper than that required by the design guide ($L/35^2$):

$$G = \frac{11 + 5}{35^2} = 0.013$$

and hence solid deposition should not occur.

Pipework design

The arrangement of pipework (Building Research Establishment Digest 205, 1977) for various sanitary appliances is shown in Figs 7.8–7.12.

Groups of appliances for dwellings are depicted in Figs 7.13 and 7.14. A pumped WC discharge unit, as shown in Fig. 7.15, enables the use of a 22 mm diameter copper pipe to run long distances, and upwards, to connect into the soil and vent stack at a convenient location.

7.8 Branch pipe to a WC.

7.9 Branch for a range of WCs.

7.10 Branch for a range of urinals.

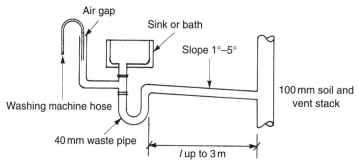

7.11 Branch from a sink or bath.

7.12 Branch discharge pipe for a range of up to 10 basins.

Discharge unit pipe sizing

The intermittency of discharge from appliances necessitates the use of discharge units that relate to the flow volume, flow time and interval between flows from sanitary fittings in a similar way to the demand units for water supplies to such fittings. Typical discharge units are as follows (domestic use): WC, 14; basin, 3; bath, 7; urinal, 0.3; washing machine, 4; sink, 6. A group consisting of WC, bath, sink and two basins has a value of 14 discharge units.

A 100 mm diameter stack can carry 750 discharge units, a 125 mm diameter stack can carry 2500 discharge units and a 150 mm diameter stack can carry 5500 discharge units.

7.13 Soil and vent stack in housing.

7.14 Drainage pipework for a bungalow.

7.15 Pumped WC discharge system.

Materials used for waste and discharge systems

The materials available for waste pipes and soil and vent stacks are listed in Table 7.1.

A clearance of 30 mm should be left between external pipes and the structure to allow free access and for painting. Secure bracketing to the structure is essential and allowance for thermal expansion should be made. Pipes passing through walls or floors should be sleeved with a layer of inert material to prevent the ingress of moisture into the building and provide the elasticity required for thermal movement. This is particularly important with plastics.

Table 7.1 Materials for waste and discharge pipework.

Material	Application	Jointing
Cast iron	50 mm and above vent and discharge stacks	Lead caulking with molten or fibrous lead; cold compound caulking
Galvanized steel	Waste pipes	BSPT screwed
Copper	Waste pipes and traps	Compression, capillary, silver solder, bronze weld or push-fit ring seal
Lead	Waste pipes and discharge stacks	Soldered or lead welded
ABS	Up to 50 mm waste and vent pipes	Solvent cement and push-fit ring seal
High-density polyethylene	Up to 50 mm waste and ventilating pipes and traps	Push-fit ring seal and compression fittings
Polypropylene	Up to 50 mm waste and ventilating pipes and traps	Push-fit ring seal and compression couplings
Modified PVC	Up to 50 mm waste and vent pipes	Solvent cement and push-fit ring seal
Unplasticized PVC	Over 50 mm soil and vent stacks; vent pipes under 50 mm	Solvent cement and push-fit ring seal
Pitch fibre	Over 50 mm discharge and vent stacks	Driven taper or polypropylene fitting with a push-fit ring seal

Note: ABS, acrylonitrile butadiene styrene; PVC, polyvinyl chloride.

Testing

Inspection and commissioning tests on drainage installations are carried out as follows.

Inspection

During installation, regular inspections are made to check compliance with specifications and codes. Particular attention is given to quality of jointing, security of brackets and removal of swarf, cement or rubble from inside pipe runs.

Air pressure

Prefabricated waste pipe systems will be factory tested before delivery to the site. The complete system will be tested on completion by filling the water seals and inserting air bags, expandable bungs, into the ends of stack pipes. A rubber hose is inserted into the vent stack through a WC water trap. The air pressure in the stack is hand-pumped up to 38 mm water gauge, measured on a U-tube manometer. This pressure must remain constant for 3 min without further pumping. Soap solution wiped onto joints will reveal leak locations.

Smoke

Existing stacks can be tested by the injection of smoke from an oil-burning generator or a smoke cartridge, provided that it will not cause damage to the drain materials. This is less severe than the air test, as smoke pressure remains low and damage from the test itself is less likely. Suitable warnings must be given to the occupants of the building.

Syphonage

The simultaneous discharge of several appliances should reveal a minimum remaining water seal of 25 mm in all traps. Discharge should take place quietly and smoothly.

Maintenance

Periodic inspection, testing, trap clearance, removal of rust and repainting should be a feature of an overall service maintenance schedule. Washers on access covers require occasional replacement. The use of chemical descaling agents, hand or machine-operated rodding and high-pressure blockage removal must be carefully related to the drainage materials and the skill of the operator.

Lime scale removal

Lime scale is found in hard-water areas. A dilute corrosion-inhibited acid-based descaling fluid is applied directly to scale visible on sanitary appliances and is then thoroughly flushed with clean water. The fluid is a mixture of 15% inhibited hydrochloric acid and 20% orthophosphoric acid.

Removal of grease and soap residues

A strong solution of 1 kg of soda crystals and 9 l of hot water is flushed through the system. The soda crystals are mixed with the hot water in a basin. When the soda is fully dissolved, the plug is released. This may be necessary frequently in commercially used appliances.

Blockage

A hand plunger may be sufficient but repeated blockage should be investigated. Hand rodding from the nearest access point can be performed using various tools as appropriate. A kinetic ram gun can be used for blockage in branch pipes. The impact of compressed air from the gun creates a shock wave in the water, which dislodges the solids. However, a blow-back from a stubborn blockage may injure the operator and damage the pipework and therefore the ram gun must be limited to the removal of soft materials. Coring and scraping mechanical tools can be used to remove hard lime scale in 100 mm pipes, provided that the materials will withstand the maintenance operation. A steel cutter is revolved by a flexible drive fed through the drain pipework.

EXAMPLE 7.2

Draw a suitable arrangement of a sanitary pipe system for a 22-storey block of flats with two groups of appliances on each floor. Each group consists of a WC, a bath, a sink and a basin, all sited close to the stack.

Discharge units carried by the stack $= (22 \times 2)$ flats $\times 14\,DU$

$$= 616\,DU$$

Therefore a 100 mm soil and vent stack can be used as 750 DU is the capacity limit. A typical floor layout is shown in Fig. 7.16.

7.16 Typical floor layout for two flats in Example 7.2.

7.17 Typical floor layout for Example 7.3.

EXAMPLE 7.3

A 14-storey office block is to have ranges of three WCs, three basins and three urinals on each floor. The WCs and urinals are situated on each side of the stack, but the common waste pipe from the basins is to be 5 m long and have four bends. The following discharge units can be assumed: urinal, 0.3; WC, 14; basin, 3. Draw a suitable discharge pipework arrangement and state the pipe sizes to be used.

Discharge units carried by the stack $= 14 \times (3 \times 0.3 + 3 \times 14 + 3 \times 3)$

$$= 727 \, DU$$

A 100 mm discharge and vent stack can be used in the arrangement shown in Fig. 7.17. An additional ventilating stack of diameter 40 mm is recommended in BS 5572: 1978, as shown.

Questions

1. List the ways in which an above-ground drainage installation satisfies its functional requirements.
2. Describe, with the aid of sketches, the ways in which the water seal can be lost from a trap and the precautions taken to avoid this happening.
3. Describe the types of fluid flow encountered in drainage pipework.
4. Outline the development of current drainage practice from the early Roman occupation of Britain.
5. State the meaning of the following terms: bedding; combined system; drain; sewer; manhole; separate system; stack; discharge pipe; vent.
6. Sketch the pipework layout for a typical group of sanitary appliances in a dwelling, where they are all connected into a stack. Show suitable pipe sizes, slopes and details of the connections at the stack.
7. A range of WCs is to be connected into a common branch pipe of outside diameter 125 mm fitted within a false ceiling 300 mm deep. It is intended that the furthest WC should be 18 m from the stack. The branch has a 90° bend between the last WC and the stack. Determine whether the proposed arrangement would be satisfactory. If it is not, calculate the maximum distance that could be allowed between the furthest WC and the stack.
8. State the meaning of the term 'discharge unit'. How many WCs can be connected into a discharge stack of diameter 100 mm?
9. Sketch sections through four types of joint used in drainage pipework to show their constructional features, method of providing a seal, and thermal expansion facility.
10. Sketch the installation of a vertical discharge stack within a plasterboard service duct in a house, clearly showing suitable dimensions and support.
11. Sketch and describe the methods of testing above-ground drainage installations.
12. Describe the maintenance work needed to support the efficient operation of drainage installations in residences, laundries, canteens and hotels.
13. Draw a suitable sanitary pipework installation for a 10-storey block of flats with two groups of appliances on each floor connected to one stack. Show pipe sizes and routes.
14. A 20-storey office block is to have ranges of 5 basins, 5 urinals and 5 WCs on alternate floors. Draw a suitable pipework installation, stating pipe sizes and slopes.
15. What is the principle design requirement for waste pipes from water appliances?

    1. Pipes not leaking.
    2. Must empty basin, sink or shower tray within 30 s.
    3. Avoidance of long horizontal pipes.
    4. Never reduce pipe diameter below waste outlet size.
    5. Maintain water seal at trap of each appliance.

16. What type of water flow occurs in the waste pipe from a basin?

    1. Turbulent.
    2. Laminar.
    3. Steady continuous stream along lower half of sloping pipes.
    4. Water swirls clockwise down vertical and along sloping pipes, adhering to the walls of the pipe due to the Coanda effect.
    5. Full-bore surge followed by dribbling.

17. Which of these is a problem that can occur in waste pipe systems?

    1. There are no problems in a correctly designed system.
    2. Pipes installed at too steep an angle create noise and suction problems.

      3.   Long horizontal waste pipes connecting to vertical stacks do not drain water away fast enough and cause flooding in baths and shower trays.

      4.   Self-syphonage from the trap due to inertia of water flow.

      5.   Self-syphonage due to leaks.

18. Which is a cause of induced syphonage in a waste pipe?

      1.   Positive back-pressure from downstream pipes.

      2.   Low atmospheric air pressure.

      3.   Waste pipes running full.

      4.   High atmospheric air pressure.

      5.   Insufficient ventilation of vertical stack.

19. How are waste pipe system syphonage risks reduced?

      1.   Adequate ventilation to open air.

      2.   Installing larger diameter waste pipes than normally recommended.

      3.   Connecting every waste trap individually to the vertical stack.

      4.   Maintaining air-tightness of the drainage system.

      5.   Regular internal cleaning of all waste and drain systems.

20. How are ground floor waste pipes connected into the sewer system?

      1.   Unvented direct connection.

      2.   Trapped pipe connects directly to a trapped outdoor gulley on sewer system.

      3.   Untrapped direct pipe connection to the foul sewer.

      4.   Trapped pipes at a steep gradient connect to the foul sewer.

      5.   Trapped pipes connect vertically to the foul sewer.

      6.   Trapped pipes at a shallow gradient connect to the foul sewer.

21. How are above-ground drain systems tested to ensure adequate water seals remain in traps during normal use?

      1.   Each sanitary fitting filled, drained and remaining trap seal depth measured.

      2.   All sanitary fittings in a group filled then drained simultaneously.

      3.   Each section of waste pipe system sealed and pumped to an air pressure of 150 mm water gauge for 10 min, then air and water released.

      4.   Water pumped up stack to test effectiveness of water seals in traps.

      5.   Air pressure in stack and waste pipes hand-pumped to 38 mm water gauge. This pressure maintained for 3 min without further pumping.

8 Surface-water drainage

Learning objectives

Study of this chapter will enable the reader to:

1. calculate rainfall run-off into surface-water drains;
2. calculate gutter water flow capacity;
3. choose appropriate gutter and rainwater down pipe combinations to create an economical design;
4. calculate and find gutter sizes;
5. assess methods for the disposal of surface-water;
6. calculate soakaway pit design.

Key terms and concepts

dimensions 224; fall 224; flow capacity 224; gutter 223; impermeability factor 222; interceptor chamber 226; rainfall intensity 221; rainwater down pipe 223; rainwater flow rate 222; road gulley 226; roof pitch 223; sewer 226; soakaway pit 226.

Introduction

Design calculations for roofs, gutters and ground drainage are presented along with practical exercises in suitable arrangements. It is important for the designer to maintain the closest contact with the architect during this process because of the required integration.

Flow load

In the UK, ground surface-water systems are designed on the basis of a rainfall intensity of 50 and 75 mm/h for roofs. The quantity of water entering a drain depends on the amount of evaporation into the atmosphere and natural drainage into the ground. The drain flow load is represented by the impermeability factor, and typical figures are shown in Table 8.1.

Table 8.1 Ground impermeability factors.

Nature of surface	Impermeability factor
Road or pavement	0.90
Roof	0.95
Path	0.75
Parks or gardens	0.25
Woodland	0.20

The drain water flow rate Q is given by:

$$Q = \text{area drained m}^2 \times \text{rainfall intensity } \frac{\text{mm}}{\text{h}} \times \text{impermeability factor}$$

EXAMPLE 8.1

Footpaths, roadways and gardens on a commercial estate cover an area of 75 000 m², of which 20% is garden and grassed areas. Estimate the surface-water drain flow load in litres per second. How many 15 l/s surface-water drain gulleys are needed?

From Table 8.1, impermeability factors are 0.9 for the roads and paths and 0.25 for gardens and grass. Therefore:

$$\text{Average impermeability} = 0.2 \times 0.25 + 0.8 \times 0.9$$
$$= 0.77$$

Hence,

$$Q = 75\,000 \text{ m}^2 \times 50 \frac{\text{mm}}{\text{h}} \times \frac{1 \text{ h}}{3600 \text{ s}} \times \frac{1 \text{ m}}{10^3 \text{ mm}} \times 0.77 \times \frac{10^3 \text{ l}}{1 \text{ m}^3}$$
$$= 802.1 \text{ l/s}$$

Number of drain gulleys needed

$$= 802.1 \frac{\text{l}}{\text{s}} \times \frac{\text{gulley s}}{15 \text{ l}}$$
$$= 54 \text{ gulleys}$$

Roof drainage

A rainfall intensity of 75 mm/h occurs for about 5 min once in 4 years. An intensity of 150 mm/h may occur for 3 min once in 50 years, and where overflow cannot be tolerated this value is used for design. The flow load Q for a roof is calculated from:

$$Q = A_r \text{ m}^2 \times 75 \frac{\text{mm}}{\text{h}} \times \frac{1 \text{ h}}{3600 \text{ s}} \times \frac{1 \text{ m}}{10^3 \text{ mm}} \times \frac{10^3 \text{ l}}{1 \text{ m}^3}$$
$$= A_r \text{ m}^2 \times 0.021 \text{ l/s}$$

where A_r is the surface area of a sloping roof of pitch up to 50° and no evaporation takes place, which is characteristic of a cold saturated atmosphere.

For a roof pitch of greater than 50°:

$$Q = 0.021 \times (1 + 0.462 \tan\theta) \times A_r \text{ l/s}$$

where θ is the roof pitch in degrees.

The flow capacity of a level half-round gutter is given by:

$$Q = 2.67 \times 10^{-5} \times A_g^{1.25} \text{ l/s}$$

where A_g is the cross-sectional area of the gutter in mm^2. For level gutters other than half-round the flow capacity can be found from:

$$Q = \frac{9.67}{10^5} \times \sqrt{\left(\frac{A_o^3}{W}\right)} \text{ l/s}$$

where A_o is the area of flow at the outlet, mm^2, and W is the width of the water surface, mm. The depth of flow in a level gutter discharging freely increases from the outlet up to a maximum at the still end. It can be assumed that the depth of flow at the outlet is about half that at the still end. Thus the depth at the outlet is half the gutter depth. A_o is found from the gutter cross-sectional area at half its depth. W will normally be the width across the top of the gutter.

A fall of 1 in 600 increases flow capacity by 40%. The frictional resistance of a sloping gutter may reduce water flow by 10%, and each bend can reduce this further by 25%. Water flow in down pipes is much faster than in the gutter and they will never flow full. Their diameter is usually taken as being 66% of the gutter width. Some typical gutter flow capacities are given in Table 8.2.

For calculation purposes, a roof is divided into areas served by a gutter with an end-outlet rainwater down pipe. If the whole roof is drained into a gutter with an outlet at one end, then the gutter carries water from the entire roof area. However, when a centre outlet is used, a smaller gutter size might be possible as it only carries half the flow load. The number and disposition of the down pipes is considered in relation to the gutter size, architectural appearance, cost and complexity of the underground drainage system.

Table 8.2 Typical flow capacities for a PVC half-round gutter at a 1 in 600 fall.

Nominal gutter width (mm)	Q (l/s)	
	End outlet	Centre outlet
75	0.46	0.76
100	1.07	2.10
125	1.58	2.95
150	3.32	6.64

Source: Reproduced from *1HVE Guide* (CIBSE, 1986 [IHVE, 1970]) by permission of the Chartered Institution of Building Services Engineers.

Rectangular gutter sizes can be found from:

$$A_o = \frac{WD}{2}$$

where D is the gutter depth (mm).

EXAMPLE 8.2

A sports centre roof of dimensions 15 m × 8 m is laid to fall to a PVC half-round gutter along each long side. Find an appropriate gutter size when the gutter is to slope at 1 in 600. Each gutter can have a centre or end outlet.

The flow load is:

$$Q = 0.021A_r \text{ l/s}$$
$$= 0.021 \times 15 \times 8 \text{ l/s}$$
$$= 2.52 \text{ l/s}$$

Each gutter will carry half of Q, that is, 1.26 l/s. For one gutter, the fall will increase the carrying capacity by 40% and friction will reduce it by 10%. The required gutter area can then be found from:

$$Q = 1.40 \times 0.9 \times 2.67 \times 10^{-5} \times A_g^{1.25} \text{ l/s}$$

Hence,

$$A_g^{1.25} = \frac{10^5 \times Q}{2.67 \times 1.4 \times 0.9}$$
$$= \frac{10^5 \times 1.26}{3.3642}$$
$$= 0.375 \times 10^5$$

Therefore,

$$A_g = (0.375 \times 10^5)^{1/1.25} \text{ mm}^2$$
$$= (0.375 \times 10^5)^{0.8} \text{ mm}^2$$
$$= 4563 \text{ mm}^2$$

For a half-round gutter:

$$A_g = \frac{\pi W^2}{8}$$

and hence,

$$W = \sqrt{\frac{8Ag}{\pi}}$$

$$= \sqrt{\left(\frac{8 \times 4563}{\pi}\right)} \text{ mm}$$

$$= 108 \text{ mm}$$

Thus a 125 mm half-round gutter with an end outlet would be used along each side. This can be checked with the data in Table 8.2. A smaller 100 mm gutter would be possible if a centre outlet was appropriate to the appearance of the building and the underground drain layout.

EXAMPLE 8.3

A roof sloping at 42° has a level box gutter 125 mm wide and 50 mm deep. The roof is 15 m long and 5 m up the slope. Calculate whether the gutter will adequately convey rainwater when the rainfall intensity is 75 mm/h. Recommend the outlet location.

The flow load is given by:

$$Q = 0.021 \times (1 + 0.462 \times \tan 42°) \times 15 \times 5 \text{ l/s}$$

$$= 3.24 \text{ l/s}$$

If an end outlet is used, the water depth at the outlet will be half the gutter depth, that is, 25 mm. Thus,

$$A_O = 125 \times 25 \text{ mm}^2$$

$$= 3125 \text{ mm}^2$$

The gutter flow capacity is expected to be:

$$Q = \frac{9.67}{10^5} \times \sqrt{\frac{3125^3}{125}} \text{ l/s}$$

$$= 1.69 \text{ l/s}$$

This is less than the imposed flow load and would produce overflow from the gutter. A centre outlet would have the effect of halving the flow load on the gutter. For a centre outlet,

$$\text{Flow load } Q = 0.50 \times 3.24 \text{ l/s}$$

$$= 1.62 \text{ l/s}$$

$$A_O = 3125 \text{ mm}^2 \text{ as previously calculated}$$

$$\text{Gutter capacity } Q = 1.69 \text{ l/s}$$

The solution is to use the centre outlet so the gutter capacity exceeds the calculated flow load from the roof.

Cast iron covers over drainage gulleys in roadways pass 20 l/s or more, depending upon surface-water speed, degree of flooding and blockage from debris.

Disposal of surface-water

Surface-water can be removed from a site by one or more of the following methods.

Sewer

Where the local authority agrees that there is adequate capacity, surface-water is drained into either a combined sewer or a separate surface-water sewer. Surface-water from garage fore-courts and car parks is run in open gullies to an interceptor chamber. Ventilation of explosive and poisonous petrol vapour is essential, as a concentration of 2.4% in air is fatal. It is illegal to discharge petrol, oil or explosive vapour into public sewers. The interceptor chamber is an under-ground storage tank of concrete and engineering bricks, which allows separation of the clean water from the oily scum remaining on its surface. It is intermittently pumped out and cleaned. The discharge drain to the sewer is turned downwards to near the bottom of the interceptor and three separate chambers are used in series.

Soakaway

Ground permeability is established using borehole tests to measure the rate of natural drainage within a curtilage. If running underground water is found, a simple rock-filled pit can be used. Slow absorption is overcome by constructing a perforated precast concrete, dry stone or brick pit, which stores the rainfall quantity. The stored volume is found from an assumed steady rainfall of 15 mm/h over a period of 2 h. This is exceeded around once in 10 years, so there may be occasional flooding for short periods. A soakaway pit is circular with its depth equal to its diameter.

Storage

An artificial pond or lake, or even an underground storage tank, will be necessary if the expected run-off from a curtilage is at a greater rate than could be accommodated by a sewer or watercourse.

Watercourse

The relevant local authorities may allow the disposal of surface-water into watercourses. Expected flow rates at both normal and flood water levels must be established.

EXAMPLE 8.4

Storage soakaway pits 2.25 m deep are to be employed for a tarmac-covered car park of dimensions 100 m × 30 m. Determine the number and size of the pits needed. Draw a suitable drainage layout for the car park.

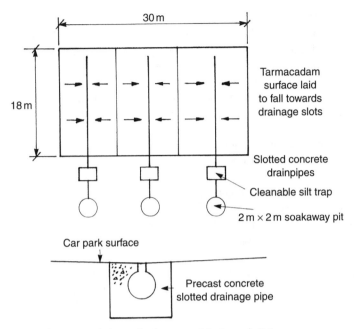

8.1 Surface-water drainage for the car park in Example 8.4.

$$\text{Volume to be stored} = 15 \ \frac{mm}{h} \times 2 \ h \times \frac{1 \ m}{10^3 \ mm} \times 100 \ m \times 30 \ m \times 0.9$$

$$= 81 \ m^3$$

Pits 2.25 m deep will have diameters of 2 m. Therefore:

$$\text{Pit volume} = \frac{\pi \times 2^2}{4} \times 2.25 \ m^3 = 7.07 \ m^3$$

and hence,

$$\text{number of pits} = \frac{81}{7.07} = 12$$

Figure 8.1 shows a suitable arrangement of precast concrete slot drainage channels and soakaway pits.

Questions

1. Explain how the rainwater flow load is calculated for roof and groundwater drainage system design.
2. A housing estate has footpaths and roads covering an area of 4000 m². Calculate the rainwater flow load and the number of drain gullies required.
3. Find the flow load for a flat roof of dimensions 20 m × 12 m.
4. Find the flow load for a roof of dimensions 10 m × 4 m with a pitch of 52°.
5. Calculate the flow capacity of a level half-round gutter 125 mm wide. Ignore friction.
6. Calculate the flow capacity of a level box gutter 200 mm wide and 150 mm deep when running full. Ignore friction.

7. A PVC half-round gutter 150 mm wide slopes at 1 in 600 and has an end outlet to a rainwater pipe. The water depth at the outlet is half the gutter depth. Assume that A_o is half the gutter cross-sectional area. Take W as the gutter width. Calculate the gutter flow capacity.

8. A house has two sloping roofs, each side of a ridge, of dimensions 15 m × 8 m. Calculate the flow load and determine the gutter and rainwater pipe design from Table 8.2.

9. The flat roof of a school is to be of dimensions 30 m × 20 m with a rectangular gutter on each long side and sloping at 1 in 600 to an outlet at each end. Calculate suitable dimensions for the gutter.

10. A pitched roof of dimensions 20 m × 5 m drains into a level box gutter 120 mm wide and 80 mm deep on one long side. The gutter has one end outlet. Calculate whether this is a satisfactory arrangement.

11. Describe the actions taken during the design and construction of a surface-water disposal system, stating what options are investigated.

12. Storage soakaway pits 2 m deep are to be used to dispose of rainwater from a roof of dimensions 10 m × 8 m. Determine a suitable size and number of pits.

13. List the techniques used for subsoil drainage systems.

14. Describe the features and maintenance requirements of surface-water drainage systems for car parking, garage forecourts and large paved areas in shopping centres.

9   Below-ground drainage

Learning objectives

Study of this chapter will enable the reader to:

1. understand the design principles for underground drainage pipework;
2. use discharge units;
3. use a pipe-sizing chart;
4. calculate flow capacity;
5. calculate loads on buried pipelines;
6. identify materials and jointing methods;
7. understand sewage-lifting requirements;
8. know testing procedures;
9. carry out a design assignment;
10. explain the principles of below-ground drain layout;
11. know the location and types of access fitting.

Key terms and concepts

access 230; access chamber 231; bedding 230; design velocity 232; discharge units 232; drain diameter 230; excavation 230; fluid flow rate 233; gradient 230; gravity flow 229; gulley 232; manhole 230; materials and jointing 234; pneumatic ejector 235; pumped sewage 235; rodding 231; rodding eye 231; sewage lifting 235; testing 236.

Introduction

Below-ground drainage systems are designed to operate without the input of energy, wherever possible, to be reliable and to require little, if any, maintenance. Their layout has to be such that drains are not subject to undue stress from foundations or traffic and are fully accessible for occasional clearance.

Design calculations can be made on the basis of flow rates, utilizing discharge units, gradients, pipe material and pipe diameter. Stress loads, pipe materials and jointing, sewage lifting and testing are discussed, and a design layout assignment is given.

Design principles

Sanitary discharge services operate by gravity flow and require no energy input. Parts of buildings or sites that are below the sewer invert require a pump to raise the fluid. These operate intermittently to minimize electrical power consumption. Drains are laid to fall at an even gradient, which produces a self-cleaning water velocity so that potential deposits are accelerated and floated downstream. Large drops in drain level are accommodated in a back-drop manhole, rather than a lengthy steep slope, in order to minimize excavation.

Pipes are laid in a series of straight lines between access points used for inspection, testing and cleaning. Branch connections are made obliquely in the direction of flow. There is a preference against running drains beneath buildings owing to possible settlement and the potential cost of later excavation to make repairs, and because the drain invert is lower than the floor damp-proof membrane and rising drains need a waterproof seal against groundwater.

Selected trench bedding and backfill material is used to provide continuous pipe support, to spread imposed ground loads due to the weight of soil and passing traffic, to protect drains from sharp objects and other services, as well as to divert stresses imposed by building foundations. Temporary boards are used to protect exposed drain trenches during construction work. Figure 9.1 shows a typical foul drain system from a single stack.

Access provision

Blockages may happen, as drain systems are likely to be in place for a hundred years or more and demands upon them continue to increase. Cleanability is an essential feature of good design. Good health depends upon satisfactory drainage. Domestic drains are likely to be located less than 1.5 m below ground level, at a maximum gradient of 1:40 and 100 mm in diameter, possibly increasing to 150 mm in diameter at the downstream end of an estate or large building.

Vitrified clay or PVC pipe and fittings are often used, with flexible joints to accommodate ground movement and thermal expansion due to variations in fluid temperature. Brick, concrete,

9.1 Foul drainage installation.

PVC or glass-reinforced plastic (GRP) access chambers are used. All changes in direction are either through 135° or large radius-swept bends.

Access points are provided for removing compacted material and for using rigid rods to clear blockages in the direction of flow, even though flexible water-jetting techniques are currently available and it is possible to clear obstructions from either direction. Air-tight covers are desirable to avoid access points allowing a health hazard or flooding.

Figure 9.2 shows the types of access.

Rodding eye: a 100 mm diameter drain pipe extended from any depth to ground level to allow rodding in the downstream direction.

Shallow access chamber: a removable threaded cap on a branch fitting to allow access in either direction located such that the distance from ground level to drain invert is less than 600 mm to facilitate reaching into the drain.

Sealed inspection chamber: a 600 mm deep, 500 mm diameter chamber for access to screwed caps on drain junctions.

Open-channel inspection chamber: a 600 mm deep, 500 mm diameter access chamber with benched smooth surfaces for drain junctions.

9.2 Types of access to below-ground drainage.

Manhole: the main access point for an operative wearing breathing apparatus to climb down steps to any depth; a 1 m deep manhole is 450 mm², and a 1.5 m deep manhole has dimensions of 1200 mm × 750 mm or 1050 mm diameter, and a cover 600 mm².

Gulley: ground-level connection point for various waste pipes and the below-ground 100 mm diameter drain providing a water trap against sewer gas and allowing debris removal and rodding access; it may have a sealed lid or open grating.

The first access point close to the building is either a gulley, a removable WC or a shallow access chamber just áfter the base of the internal drainage stack. It is not necessary to fit access points at every change in drain direction, but pipe junctions are made with access chambers. The maximum spacings between access points are 12 m from the start of the drain to the first access, 22 m from a rodding eye to a shallow access chamber, 45 m from a rodding eye to an access chamber or manhole and 90 m between manholes. Figure 9.3 demonstrates a typical housing estate drain layout. Careful integration with the surface-water drainage system is necessary as falls to the sewers are preconditioned by the sewer inverts, and the two drains may run within the same trenches and cross each other where branch connections are made.

Pipe diameters for surface-water pipework are based on the flow loads discharging from each down pipe. Those for foul drains are found from the discharge units in each stack. Flows in underground drains are found by totalling calculated flow rates along the route of the collecting drain run. Discharge units are converted into flow rate using Fig. 9.4. Pipe sizes and fluid velocities can be read on Fig. 9.5 from the calculated flow rate and desired gradient as appropriate to the maximum allowable fall available on the site.

9.3 Typical site layout showing access.

9.4 Simultaneous flow data for foul- and surface-water drains (reproduced from data in *CIBSE Guide* (CIBSE, 1986) by permission of the Chartered Institution of Building Services Engineers).

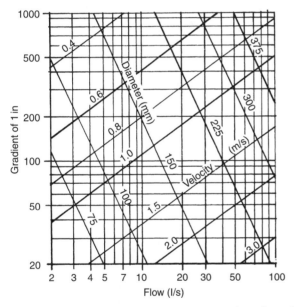

9.5 Sizes for two-thirds full spun precast concrete drains (reproduced from *IHVE Guide* [IHVE, 1970], now superseded, by permission of the Chartered Institution of Building Services Engineers [CIBSE, 1986]).

EXAMPLE 9.1

The flow from a 100 mm stack is equivalent to 750 discharge units and is to run underground for 30 m before entering the foul sewer at a depth 375 mm lower than at the building end. Find a suitable diameter for a spun precast concrete drain so that it will not be more than two-thirds full at maximum flow rate.

From Fig. 9.4, note that the horizontal scale is 0–9000, 750 discharge units are equivalent to a flow rate of 7 l/s. The gradient of the drain is given by:

$$\text{Gradient} = \frac{0.375}{30} = 1 \text{ in } 80$$

From Fig. 9.5, a flow rate of 7 l/s at a gradient of 1 in 80 requires a 150 mm diameter drain. The fluid velocity will be 1.05 m/s. Check this result with a simple calculation of carrying capacity. Drains run at a maximum of two-thirds full to provide continuous ventilation throughout their length and avoid creating large suction pressures on water traps in sanitary fittings in buildings. This will only be an approximation to published chart data.

Pipe flow capacity $Q = \dfrac{2}{3} \times Av$

$$= \frac{2}{3} \times \pi \times \frac{0.15^2}{4} \text{ m}^2 \times 1.05 \, \frac{\text{m}}{\text{s}} \times \frac{10^3 \text{ l}}{1 \text{ m}^3}$$

$$= 12.3 \text{ l/s, so chart result is confirmed}$$

It is worth noting that, at this gradient, a 100 mm drain will carry only 5.3 l/s. A 100 mm drain would carry 7 l/s if its gradient was increased to 1 in 48, but this may be impractical. A 150 mm drain is the next available size up from the design point on the chart but it is grossly oversized for the present duty. At 1 in 80 it can carry a flow rate of 15 l/s, which corresponds to 2200 discharge units. This represents a possible future increase in discharge unit capacity of

$$\frac{2200 - 750}{750} \times 100\% = 193\%$$

Where drain use may be increased by additional site or area development, this is a useful advantage.

Materials for drainage pipework

Traditionally, glazed vitrified clay (GVC) pipes have been used because they represent an efficient use of UK national resources. The finished internal surface of GVC pipes offers less frictional resistance to flow than that of concrete pipes and is resistant to chemical attack and abrasion. Rigid joints consist of a socket and spigot cemented together. The brittle nature of such pipe runs has led to the introduction of flexible joints, which can withstand ground movement due to thermal and moisture variations and settlement of buildings. Plastic and rubber sealing ring joints allow up to 5° of bending and longitudinal expansion and contraction. Pipe sizes range from 75 to 750 mm in diameter.

Spun concrete drain pipes of diameter up to 1.83 m with oval cross-sections, which maintain flow velocity at periods of low discharge, are used. Plastic sleeves with rubber sealing rings give joints flexibility and a telescopic action.

Asbestos cement pressure pipes in lengths of up to 4 m have been used because of their lower weight. Flexible sleeve joints with rubber ring seals are used. Diameters from 100 to 600 mm are produced.

Pitch fibre pipes are formed by impregnating wood fibre with pitch. They are lightweight and can be used for some drainage applications. Lengths of 2.5 m are easily handled and can be hand sawn. Push taper joints are made using a hand-operated chamfering tool. Pipelines have flexibility and require well-selected backfill and careful protection during site work. Hot fluid or chemical discharges may lead to the early collapse of the pipe from ground pressure. Plastics are used for bends and other pipe fittings. Diameters are in the range 75–200 mm.

Cast iron drain runs are used for overground sections and where the ground movement might otherwise cause fracture. Pipework beneath buildings can either be cast iron encased in concrete or short lengths with flexible joints. Rigid socket and spigot joints are caulked with tarred yarn and then filled with hot lead or lead wool.

Plastics have increasingly replaced naturally occurring materials owing to their low weight and high degree of prefabrication. Complete systems from the sanitary appliance to the sewer, using one supplier and material, are common. Such materials are derived from crude oil and their higher cost needs to be compensated by reduced site time. Smooth bore drain systems can be assembled with minimum skill and they are highly resistant to corrosion. Thermal expansion is greater, and telescopic joints are used. Short-term discharges from some appliances, for example, some types of washing machine, can be at temperatures of 80°C or higher. Polypropylene and acrylonitrile butadiene styrene, ABS, pipes are suitable for the high-temperature applications.

Sewage-lifting pump

Where sanitary appliances discharge into drain pipework that is below the foul sewer invert, a collecting sump, pump and fluid level controller are used in the manner shown in Fig. 9.6. Either a large clearance centrifugal pump, driven by a 440 V three-phase electric motor, or a pneumatic ejector is used.

The storage chamber is sized to accommodate several hours of normal discharge so that the pump only runs for short periods and total electrical power consumption need not be high. Duplicate pump sets ensure a continuity of service during breakdowns and maintenance.

The pneumatic ejector collects the discharge in a steel tank containing a float. At the upper fluid level, the float operates a change-over valve, which admits air from a compressor and

9.6 Sewage-lifting pump.

storage vessel. The incoming compressed air drives the sewage into the outlet drain at the higher level. Non-return valves are fitted to the inlet and outlet pipes to stop the possibility of reverse flow.

Both types of sewage-lifting equipment have open vent pipes to ensure that back pressures are not imposed upon the soil stack.

Testing

In addition to the inspection and smoke tests previously described, a water test can be applied to underground drainage pipelines. A test is carried out before, and sometimes after, backfilling. Drain runs are tested between manholes. The lower end is sealed with an expandable plug. A temporary upstand plastic or aluminium pipe is connected at the higher manhole. Water is admitted to produce a static head of 1.4–2.4 m and maintained for an hour or more. Some drainage materials will absorb water, and the initial water level is replenished from a measuring cylinder or jug. The maximum allowable water loss is 1.0 l/h for 10 m of 100 mm diameter pipe or pro rata for other diameters and lengths.

An air test can be conducted in a similar manner: a static pressure of 75 mm water gauge on a U-tube manometer should be maintained for a period of 5 min without further pumping.

Questions

1. List the principal requirements for an underground drainage installation.
2. Explain, with the aid of sketches, the differences between the following types of drain and sewer system: separate, combined and partially separate.
3. A 100 mm discharge stack connects to a drain laid to a 1 in 80 gradient and is expected to carry 500 discharge units. Find a suitable diameter for the drain.
4. A building has four 100 mm discharge stacks, which connect into a common underground drain. The stacks have discharge unit values of 400, 500, 600 and 700. Find the diameters of each part of the collecting drain if it has a gradient of 1 in 100.
5. A 100 mm PVC drain runs for 30 m to connect between a discharge stack and the sewer. If its gradient is 1 in 80 and it commences its route with minimum ground cover under a garden, what will its invert be at the sewer connection if the ground is level?
6. There are 236 houses on a new development. Each has a group of sanitary appliances with a discharge unit value of 14. The common drain is laid at a gradient of 1 in 100. Find the diameter of the common drain and the maximum possible number of houses that it could serve.
7. Sketch and describe the types of bedding used for drains, giving an application for each.
8. Under what circumstances may drains and sewers become damaged during their construction and service periods?
9. Describe, with the aid of sketches, the materials and jointing techniques used for below-ground drain systems.
10. Sketch and describe the operation of a sewage pumping installation. Draw the details of the construction of the below-ground chambers.
11. State the performance criteria for tests on below-ground drain systems.
12. Design a below-ground drainage system for the Pascal Sports Club shown in Fig. 9.7. The foul sewer is at an invert of 2 and 25 m to the right of the east wall of the club. Only one connection is allowed to be made to the 300 mm diameter sewer, and this is to its upper half.

9.7 Pascal Sports Club.

It will be necessary to design the above-ground waste pipework from all sanitary appliances in order to optimize the gulley positions, the 100 mm diameter pipe routes and the location of the one ventilation stack at the high point of the whole system. Minimize the use of underfloor pipework, all of which must be 100 mm in diameter and fully accessible.

Modifications can be made to the building to construct above-ground service ducts to accommodate hot- and cold-water pipes as well as wastes and drains.

A 100 mm diameter rainwater down pipe is located 500 mm from each external corner of the building on the north and south sides. These connect to 100 mm diameter below-ground drains, which run to the surface-water sewer alongside the foul sewer. Ensure that both drain systems are fully integrated and separated by a bedding of at least 100 mm thick shingle or broken stones of maximum size 5–10 mm. Access to the surface-water pipework is of the same standard as that to the foul pipework.

The last access prior to the sewer for both drains should be a manhole. There is no manhole at the junction of the drain and sewer. The shower rooms will have trapped floor gulleys that connect to the foul drain.

No model solution is provided as the design should be discussed with tutor and colleagues, and reference should be made to manufacturers' guides.

13. How are below-ground drains tested?

1. Filling with water and pumping the pressure up to 30 m water gauge for an hour.
2. Sealing ends of completed system, hand pumping air pressure up to 150 mm water gauge for an hour without further pumping.
3. Prior to backfilling trench, subject completed drain system to a static water height of 2.4 m for an hour.
4. Internal camera survey.
5. Watching for leaks prior to backfilling trench.

14. The flow capacity of a two-thirds full below-ground drain at a shallow gradient is approximated from the formula, $Q = (2/3)(\pi d^2 / 4)$ V m³/s. Which is the flow capacity of a 100 mm diameter drain when the water velocity is 0.9 m/s?

   1.   7.07 l/s.
   2.   9.4 l/s.
   3.   2.35 l/s.
   4.   1.5 l/s.
   5.   4.712 l/s.

15. A 20 m run of below-ground drain falls 175 mm. Which is the correct gradient?

   1.   1.75
   2.   8.75
   3.   0.114
   4.   1:175
   5.   1:114

10 Condensation in buildings

Learning objectives

Study of this chapter will enable the reader to:

1. identify the moisture content of humid air by its vapour pressure;
2. understand dew-point temperature;
3. identify the sources of moisture within a building;
4. understand the flow and storage characteristics of moisture flows found in habitable building;
5. explain the causes of condensation;
6. discuss the damage which can be caused by condensation;
7. calculate vapour diffusion resistance;
8. calculate vapour flow;
9. calculate air vapour pressure;
10. calculate air dew-point temperature;
11. understand atmospheric pressure terms;
12. use the e^x calculator function;
13. use the $\log_e$ calculator function;
14. be able to convert from e^x to $\log_e$ forms;
15. calculate and draw thermal temperature gradients through structures;
16. calculate and draw dew-point temperature gradients through structures;
17. identify condensation zones within structures;
18. discuss surface and interstitial condensation;
19. understand where to install thermal insulation and vapour barriers in relation to condensation risk and thermal and structural integrity requirements.

Key terms and concepts

atmospheric pressure 240; change of phase 240; condensation 240; dew-point 240; dew-point temperature gradient 251; diffusion 242; dry- and wet-bulb air temperatures 240; exponential and logarithmic functions 244; moisture flow 242; moisture production 240;

Introduction

Condensation risk is analysed during design of a building, when retrofit measures such as additional thermal insulation, double glazing or ventilation control are being considered or where damage from condensation has been discovered. Anti-condensation measures are linked to temperature control systems and ventilation provision in that they determine the size of plant and resulting operating costs.

The fundamentals of air and water vapour mixtures are introduced and then the moisture diffusion properties of building materials are analysed. A convenient form of equation to enable the air dew-point temperature to be found using a student's scientific calculator is derived and this saves the need to refer to charts or tables.

Thermal and dew-point temperature gradients are calculated, allowing moisture flow rate to be found. The rate of moisture deposition within the structure can then be assessed for its damage potential.

Sources of moisture

Air is a mixture of dry gases and water vapour. The water vapour exists in the form of finely divided particles of superheated steam at the air dry-bulb temperature. Total atmospheric pressure consists of the sum of the partial pressures of the two main constituents.

Typically, one standard atmosphere exerts a pressure of 1013.25 mb at sea level. When the air conditions are 25°C d.b. and 20°C w.b., the standard atmosphere is made up of 993.08 mb dry gas and 20.17 mb water vapour pressure. If this air is allowed to come into contact with a surface at a temperature of 17.6°C, the air becomes saturated with moisture and can no longer support all the water in its vapour state. This temperature is known as the air dew-point t_{dp}, and is shown on a sketch of the CIBSE psychrometric chart in Fig. 5.4. Further data on the properties of humid air can be obtained from the *CIBSE Guide* (CIBSE, 1986). Figure 10.1 is a reduced psychrometric chart that may be used to find data. It is reproduced by permission of the Chartered Institution of Building Services Engineers.

The sources of water vapour in an occupied building are as follows:

1. people, upwards of 0.7 kg per 24 h;
2. cooking;
3. washing, bathrooms, drying clothes;
4. humidifiers and open water surfaces;
5. animals (dogs exhale more moisture than people produce overall);
6. combustion of paraffin (the complete combustion of 1 kg of C_9H_{20} produces 1.41 kg of water vapour).

Porous structural surfaces, furniture and fabrics within the building absorb moisture and then release it into the internal atmosphere when the temperature and humidity allow this. Some moisture travels through the structure and evaporates externally unless it is prevented from doing so by an impervious layer or vapour barrier. The majority of internally produced humidity is removed by ventilation.

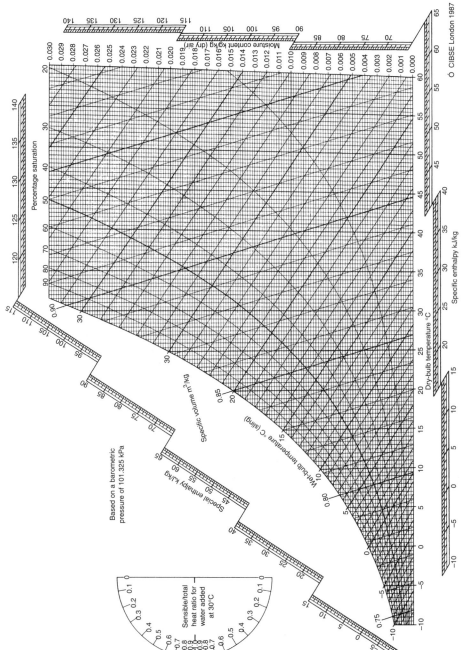

10.1 Psychrometric chart (reproduced by courtesy of the Chartered Institution of Building Services Engineers).

The warm internal atmosphere is able to hold more moisture than the cooler external air; thus the partial pressure of the vapour, that is, the vapour pressure p_s, is higher inside than outside. This vapour pressure difference causes mass transfer of moisture out of the building through the porous structure and by the ventilation air flow. When dense materials such as precast concrete or impervious barriers form a major part of the construction, moisture removal from the normal habitation is slow and condensation occurs on cold surfaces. Brickwork or masonry that is already saturated has the same effect.

Condensation and mould growth

Condensation and mould may readily form on window sills and in the corners of rooms, where the surface temperatures are lower compared with large areas. Gloss paint stops the absorption of moisture into an otherwise porous plaster or wooden component and water droplets form on the surface. The production of surface condensation in heated rooms is generally avoided in structures with a U value lower than 1.4 W/m^{2}K.

Dampness in the timber of roofs, not caused by the ingress of rain, may be due to condensation on low-temperature surfaces. A well-insulated flat plaster ceiling may produce these low-temperature conditions in the roof construction. Natural ventilation through gaps between the tiles and roofing felt is normally sufficient to stop rot and damp patches on the plasterboard. Well-sealed roofs, with boarding under the felt, should have their ventilation increased by means of openings in the soffit of the eaves after extra thermal insulation. Humid air enters the roof space through gaps in the fiat plaster ceiling around access hatches and pipes. Sealing these substantially reduces condensation risk. Roof insulation should stop at the wall head and not be pushed into the eaves, or ventilation will be restricted (Saunders, 1981).

Condensation forms within a structure or inside a solid material wherever the temperature falls below the dew-point of the moist air at that location. This is known as interstitial condensation. Vapour diffusion through building materials is calculated in a similar manner to the calculation of heat flow.

Vapour diffusion

The flow of water vapour through a porous building material or composite slab is analogous to the flow of heat through the structure. Convection currents transfer heat and moisture at the fluid–solid boundary. Conduction heat transfer is similar to vapour diffusion through a porous material, and its resistance to moisture flow varies with density, as does thermal resistance but in the opposite sense.

The mass flow rate of moisture through a composite structure consisting of a number of plane slabs and surfaces in series is

$$G = \frac{p_{s1} - p_{s2}}{R_v}$$

where G is the mass flow rate of vapour (kg/m^{2}s), R_v is the vapour resistance of the structure (Ns/kg) and P_{s1} and P_{s2} are the vapour pressures on surfaces 1 and 2 on each side of the slab (N/m^2 or Pa). The total vapour resistance R_v is given by

$$R_v = r_v \times l$$

where R_v is the total vapour resistance of a slab of homogeneous material (GN s/kg), r_v is the vapour resistivity of a material (GN s/kgm), and l is the thickness of material (m). Surface films and air cavities have only slight resistance to the flow of vapour and they are not normally included. Some typical values of vapour resistivity are given in Table 10.1. The complete resistance to the flow of vapour through some typical vapour barrier films is given in Table 10.2.

Further data are available in the reference source and in BRE Digest 369, February 1992, Building Research Establishment. Note that the values quoted in BRE Digest 369 are in MN s/g m (mega newton seconds per gram metre) for vapour resistivity and MN s/g for the vapour resistance of films. These units of measurement are the same as GN s/kgm (giga newton seconds per kilogram metre) and GN s/kg as both the numerator and denominator have been reduced by 1000 times.

Values of vapour pressure are available in CIBSE (1986) but can be calculated with sufficient accuracy from the following curve fit to the saturation conditions data:

$$p = 600.245 \exp (0.0684 \, t_{dp}) \text{ Pa}$$

where t_{dp} is the air dew-point temperature from the psychrometric chart (°C) and $\exp(x) = e^x$ where $e = 2.71828$ is the exponential operator. Tabulated vapour pressures are in millibars (mb), and since 1 bar $= 100\,000$ N/m$^2 = 100\,000$ Pa, 1 mb $= 100$ N/m$^2 = 100$ Pa.

Table 10.1 Vapour resistivity.

Material	Vapour resistivity (GN s/kgm)
Brickwork	40
Dense concrete	200
Aerated concrete	30
Glass fibre wool	10
Foamed urea formaldehyde	30
Foamed polyurethane, open cell	30
Foamed polyurethane, closed cell	1000
Foamed polystyrene	500
Hardboard	520
Insulating fibreboard	20
Mineral fibre wool	6
Plaster	50
Plywood	520
Wood wool/cement slab	15
Wood	50

Source: Reproduced from *CIBSE Guide* (CIBSE, 1986) by permission of the Chartered Institute of Building Services Engineers.

Table 10.2 Vapour resistances of films.

Material	Vapour resistance (GN s/kg)
Aluminium foil	Over 4000
Double layer Kraft paper	0.35
Gloss paint	8
Interior paint	3
Polythene film, 0.1 mm	200
Roofing felt	4 (and up to 100)

Source: Reproduced from *CIBSE Guide* (CIBSE, 1986) by permission of the Chartered Institute of Building Services Engineers.

EXAMPLE 10.1

State the vapour pressure for air at 25°C d.b., 20°C w.b. in Pascals.

The vapour pressure for air under these conditions is 20.17 mb. Thus:

$$p_s = 20.17 \text{ mb} \times \frac{100 \text{ Pa}}{1 \text{ mb}} = 2017 \text{ Pa}$$

EXAMPLE 10.2

Calculate the vapour pressure for saturated air at 25°C d.b., 20°C w.b. from the dew-point.

The dew-point is $t_{dp} = 17.6°C$. Therefore,

$$p_s = 600.245 \times \exp(0.0684 \times 17.6) \text{ Pa}$$

Scientific calculators have an e^x function, and in this calculation,

$$x = 0.0684 \times 17.6 = 1.20384$$

Consequently,

$$p_s = 600.245 \times e^{1.20384}$$

Having left 1.20384 in the displayed x register, execute the e^x function, producing 3.33289, and multiply by 600.245 to obtain,

$$p_s = 2000.6 \text{ Pa}$$

This is less than 1% different from the tabulated vapour pressure and is of sufficient accuracy bearing in mind the other figures involved in the problem.

The dew-point temperature can be found from the saturation vapour pressure by rearranging the equation:

$$p_s = 600.245 \times \exp(0.0684 \, t_{dp}) \text{ Pa}$$

to give,

$$\frac{p_s}{600.245} = \exp(0.0684 \, t_{dp})$$

This is a logarithmic equation of the form:

$$y = e^x$$

where y is the number whose logarithm to base e is x. Logarithms to base e are called natural logarithms and are expressed as follows:

$$\log_e y = x \quad \text{or} \quad \ln y = x$$

EXAMPLE 10.3

Compute the natural logarithm of 2 and then raise e to the power of this logarithm.

Enter 2 into the calculator x display and press the ln key; the answer is 0.6931. Therefore:

$$\ln 2 = 0.6931 \quad \text{or} \quad \log_e 2 = 0.6931$$

Now, as $x = 0.6931$ is displayed in the calculator, execute e^x. This results in

$$e^{0.6931} = 2$$

Thus it is seen that e^x is the antilogarithm of $\ln x$, and the two expressions $y = e^x$ and $\ln y = x$ are interchangeable to suit the problem. Thus for:

$$\frac{p_s}{600.245} = \exp(0.0684\, t_{dp})$$

we can write:

$$\ln\left(\frac{p_s}{600.245}\right) = 0.0684\, t_{dp}$$

and,

$$t_{dp} = \frac{1}{0.0684} \times \ln\left(\frac{p_s}{600.245}\right) {}^\circ C$$

EXAMPLE 10.4

Calculate the dew-point for saturated air with a vapour pressure of 2000.6 Pa.

$$t_{dp} = \frac{1}{0.0684} \times \ln\left(\frac{2000.6}{600.245}\right) {}^\circ C$$

$$= \frac{1}{0.0684} \times \ln 3.33$$

$$= 17.6^\circ C$$

An alternative general equation is stated in BRE Digest 369 for the calculation of vapour pressure:

$$p_s = 0.6105 \times \exp\left(\frac{17.269 \times t_{dp}}{237.3 + t_{dp}}\right) \text{ kPa}$$

This is less convenient to use when a dew-point temperature is to be calculated from a known vapour pressure. The curve-fit equation that has been demonstrated here is of sufficient accuracy for most manual estimations of condensation risk.

Typical values of vapour pressure are given in Table 10.3. These will accommodate some applications without reference to tables or equations.

EXAMPLE 10.5

Calculate the total vapour resistance of a cavity wall constructed from 13 mm plaster, 100 mm aerated concrete, 40 mm mineral wool, 10 mm air space and 105 mm brickwork.

For each layer:

$$R_v = r_v \times l$$

For the whole structure:

$$\sum (R_v) = \sum (r_v \times l)$$

From Table 10.1 the vapour resistivities are plaster 50, aerated concrete 30, mineral wool 6, brickwork 40. The surface films and air space have no resistance to the flow of vapour. Material thicknesses are used in metres.

$$\sum (R_v) = 50 \times 0.013 + 30 \times 0.1 + 6 \times 0.04 + 40 \times 0.105$$

$$= 8.09 \text{ GN s/kg}$$

Table 10.3 Vapour pressures.

Air condition t_a (°C d.b.)	% saturation	Dew-point t_{dp} (°C)	Vapour pressure (Pa)
−5	100	−5	402
−3	80	−5.6	381
0	80	−2.7	489
1	80	−1.8	526
2	80	−0.9	565
5	80	1.9	699
10	80	6.7	984
14	60	6.5	965
15	60	7.4	1030
20	50	9.4	1182
22	50	11.3	1339

Source: Reproduced from *CIBSE Guide* (CIBSE, 1986) by permission of the Chartered Institute of Building Services Engineers.

EXAMPLE 10.6

The cavity wall in Example 10.5 is to be used for a dwelling exposed to an external environment of $-1°C$ d.b., 80% saturation, where the heating system is designed to maintain the internal air at 22°C and 50% saturation. If the wall has a surface area of 110 m^2, find the moisture mass flow rate taking place through the wall.

From the CIBSE psychrometric chart, the internal and external air dew-point temperatures are found to be:

internal $t_{dp} = 11.5°C$

external $t_{dp} = -3.5°C$

The internal air vapour pressure is:

$$p_{s1} = 600.245 \times \exp(0.0684 \times 11.5) \text{ Pa}$$
$$= 1318.09 \text{ Pa}$$

and the external air pressure is,

$$p_{s2} = 600.245 \times \exp[0.0684 \times (-3.5)] \text{ Pa}$$
$$= 472.45 \text{ Pa}$$

Using $R_v = 8.09$ GN s/kg from Example 10.5, we obtain the moisture mass flow rate through the wall as:

$$G = \frac{p_{s1} - p_{s2}}{R_v} \text{ kg/m}^2\text{s}$$
$$= (1318.09 - 472.45) \frac{\text{N}}{\text{m}^2} \times \frac{\text{kg}}{8.09 \text{ GN s}} \times \frac{1 \text{ GN}}{10^9 \text{ N}}$$
$$= 1.045 \times 10^{-7} \text{ kg/m}^2\text{s}$$

Temperature gradient

Heat flows through a structure from an area of high temperature to one of lower temperature. Homogeneous materials have a linear temperature gradient through their thickness, as shown in Fig. 10.2. Temperature drops 1–2 and 3–4 are caused by the internal and external surface film resistances. To determine the surface and intermediate temperatures, the overall rate of heat flow through the whole structure is equated with the individual heat flows in each slab:

$$Q_f \text{ W} = U \frac{\text{W}}{\text{m}^2\text{K}} \times A \text{ m}^2 \times (t_1 - t_5) \text{ K}$$

This same rate of heat flow Q_f also passes through the internal surface film, the concrete and the external surface film. Therefore:

$$Q_f = \frac{1}{R_{si}} \frac{\text{W}}{\text{m}^2\text{K}} \times A \text{ m}^2 \times (t_1 - t_2) \text{ K}$$

10.2 Temperature gradient through a solid construction.

and,

$$Q_f = \frac{\lambda}{l} \frac{W}{m^2 K} \times A\, m^2 \times (t_2 - t_4)\, K$$

and,

$$Q_f = \frac{1}{R_{so}} \frac{W}{m^2 K} \times A\, m^2 \times (t_4 - t_5)\, K$$

The heat flow rate Q_f can easily be evaluated from U, t_{ai} and t_{ao}. The only unknowns in the other equations are the second temperatures t_2 and t_4. An intermediate temperature t_3 can be calculated at half the concrete thickness by using $l/2$:

$$Q_f = \frac{2\lambda}{l} \frac{W}{m^2 K} \times A\, m^2 \times (t_2 - t_3)\, K$$

If the wall area is taken as 1 m^2, then:

$$Q_f = \frac{1}{R_{si}}(t_1 - t_2)$$

and,

$$t_2 = t_1 - Q_f R_{si}$$

Similarly:

$$Q_f = \frac{\lambda}{l}(t_2 - t_4)$$

and,

$$t_4 - t_2 - \frac{l}{\lambda} Q_f$$

Also:

$$Q_f = \frac{1}{R_{so}}(t_4 - t_5)$$

and,

$$t_5 = t_4 - Q_f R_{so}$$

Calculating the outdoor air temperature t_5 is a check on the accuracy of the calculations and method. It should agree with the original value used in finding Q_f to within $\pm 1\%$.

To find t_3 use:

$$Q_f = \frac{2\lambda}{l}(t_2 - t_3)$$

Hence:

$$t_3 = t_2 - \left(\frac{l}{2\lambda}Q_f\right)$$

EXAMPLE 10.7

Calculate the temperature gradient through a medium-weight concrete block wall 100 mm thick. The internal and external air temperatures are 20°C d.b. and −1°C d.b.

From Table 3.1 thermal conductivity $\lambda = 0.51$ W/mK, from Table 3.2 $R_{si} = 0.12$ m²K/W and from Table 3.3 $R_{so} = 0.06$ m²K/W. Then the thermal transmittance is given by:

$$U = \frac{1}{R_{si} + (l/\lambda) + R_{so}}$$

$$= \frac{1}{0.12 + (0.1/0.51) + 0.06} \text{ W/m}^2\text{K}$$

$$= 2.66 \text{ W/m}^2\text{K}$$

For a wall area of 1 m²:

$$Q_2 = 2.66 \times [22 - (-1)] \text{ W}$$

$$= 61.16 \text{ W}$$

Using the numbered locations in Fig. 10.2:

$$t_1 = 22°C, t_5 = -1°C$$

$$t_2 = 22 - 61.16 \times 0.12 = 14.66°C$$

$$t_4 = 14.66 - \frac{0.1}{0.51} \times 61.16 = 2.67°C$$

and,

$$t_5 = 2.67 - 61.16 \times 0.06 = -1°C$$

which agrees with the input data. Also,

$$t_3 = 14.66 - \left(\frac{0.1}{2 \times 0.51} \times 61.16 \right) = 8.66°C$$

which should be the temperature midway between t_2 and t_4 that is $(14.66 \pm 2.67)/2 = 8.67°C$, which it is.

EXAMPLE 10.8

Calculate the temperature gradient through a cavity wall consisting of 13 mm lightweight plaster, 100 mm lightweight concrete block, 40 mm mineral fibre slab, 10 mm air space and 105 mm brickwork. Internal and external air temperatures are 22°C and 0°C.

From Table 3.1, the thermal conductivities are as follows: plaster, $\lambda_1 = 0.16$ W/mK; concrete, $\lambda_2 = 0.19$ W/mK; mineral fibre, $\lambda_3 = 0.035$ W/mK; brickwork, $\lambda_4 = 0.84$ W/mK. From Tables 3.2, 3.3 and 3.4, $R_{si} = 0.12$ m²K/W, $R_{so} = 0.06$ m²K/W and $R_a = 0.18$ m²K/W. Then

$$U = \left(R_{si} + \frac{l_1}{\lambda_1} + \frac{l_2}{\lambda_2} + \frac{l_3}{\lambda_3} + R_a + \frac{l_4}{\lambda_4} + R_{so} \right)^{-1}$$

$$= \left(0.12 + \frac{0.013}{0.16} + \frac{0.10}{0.19} + \frac{0.04}{0.035} + 0.18 + \frac{0.105}{0.84} + 0.06 \right)^{-1} \text{ W/m}^2\text{K}$$

$$= 0.45 \text{ W/m}^2\text{K}$$

For a wall area of 1 m²:

$$Q_f = 0.45 \times (0.22 - 0) = 9.9 \text{ W}$$

Using the notation from Fig. 10.3, temperatures are calculated as follows:

$$t_2 = 22 - (0.12 \times 9.9) = 20.81°C$$

$$t_3 = 20.81 - \left(\frac{0.013}{0.16} \times 9.9 \right) = 20.01°C$$

$$t_4 = 20.01 - \left(\frac{0.01}{0.19} \times 9.9 \right) = 14.8°C$$

$$t_5 = 14.8 - \left(\frac{0.04}{0.035} \times 9.9 \right) = 3.49°C$$

$$t_6 = 3.49 - (0.18 \times 9.9) = 1.71°C$$

$$t_7 = 1.71 - \left(\frac{0.105}{0.84} \times 9.9 \right) = 0.47°C$$

$$t_8 = 0.47 - (0.06 \times 9.9) = 0.12°C$$

10.3 Temperature notation for Example 10.8.

A slight error of 0.55% has occurred as a result of rounding all the results to two decimal places. Notice the larger temperature drops across the two main insulating materials.

Dew-point temperature gradient

Moist air passing through the structure from the high internal air vapour pressure to the lower external air vapour pressure will form a gradient of vapour pressures. The vapour pressure at any location is calculated from the mass flow rate of vapour G and vapour resistances in the same manner as for the thermal gradient.

The moist air dew-point temperature is calculated for each of these vapour pressures and another temperature gradient is drawn on the structural cross-section. If the thermally produced structure temperature equals or falls below the local air dew-point, then condensation will commence at that location. This information is used to decide whether a wall or roof will remain dry and whether a vapour barrier should be installed. The vapour barrier is fitted on the warm side of any zone of interstitial condensation.

Once the internal and external air vapour pressures, total vapour resistance and mass flow rate of vapour are known, the equation:

$$G = \frac{p_{si} - p_{s2}}{R_v}$$

can be written as,

$$p_{s2} = p_{s1} - R_v G$$

Note that this is of the same form as:

$$t_2 = t_1 - RQ_f$$

Surface air films and cavities offer negligible resistance to the flow of moisture.

EXAMPLE 10.9

Calculate the dew-point gradient for the cavity wall in Example 10.8. Determine whether surface or interstitial condensation will take place. Internal and external percentage saturations are 50 and 100 respectively.

Referring to Fig. 10.2,

$$p_{s1} = p_{s2}$$
$$p_{s5} = p_{s6}$$
$$p_{s7} = p_{s8}$$

Thus,

$$p_{s3} = p_{s1} - (\text{vapour resistance of plater} \times G)$$

where p_{s3} and p_{s1} are the vapour pressures on each side of the plaster and,

$$\text{vapour resistance of plaster} = r_v \times l$$

Similar equations are written for the other materials, with appropriate resistivities r_v and thicknesses l. The resistivities are plaster 50, concrete 30, mineral wool 6 and brickwork 40. From the CIBSE psychrometric chart, for internal air at 22°C d.b., 50% saturation, $t_{dp} = 11.5$°C, and for external air at 0°C d.b., 100% saturation, $t_{dp} = 0$°C. Then,

$$p_{s1} = 600.245 \times \exp(0.0684 \times 11.5) \text{ Pa}$$
$$= 1318.09 \text{ Pa}$$
$$p_{s8} = 600.245 \times \exp(0.0684 \times 0) \text{ Pa}$$
$$= 600.245 \text{ Pa, because } (e^0 = 1)$$

The vapour resistance R_v for this wall was calculated in Example 10.5:

$$R_v = 8.09 \text{ GN s/kg}$$

The mass flow of vapour per m² of wall area is:

$$G = (1318.09 - 600.245) \frac{\text{N}}{\text{m}^2} \times \frac{\text{kg}}{8.09 \text{ GN s}} \times \frac{1 \text{ GN}}{10^9 \text{ N}}$$
$$= 88.7 \times 10^{-9} \text{ kg/m}^2\text{s}$$

The vapour pressure and corresponding dew-point temperature can now be calculated for each numbered point through the wall:

$$p_{s3} = 1318.09 \frac{N}{m^2} 50 \times 0.013 \frac{GN\,s}{kg} \times \frac{10^9\,N}{1\,GN} \times \frac{88.7}{10^9} \frac{kg}{m^2 s}$$

$$= (1318.09 - 57.7)\,N/m^2$$

$$= 1260.4\,Pa$$

$$t_{dp3} = \frac{1}{0.0684} \times \ln\left(\frac{p_{s3}}{600.245}\right)^{\circ}C$$

$$= \frac{1}{0.0684} \times \ln\left(\frac{1260.4}{600.245}\right)^{\circ}C$$

$$= 10.85^{\circ}C$$

$$p_{s4} = 1260.4 - 30 \times 0.1 \times 88.7\,Pa$$

$$= 994.3\,Pa$$

$$t_{dp4} = \frac{1}{0.0684} \times \ln\left(\frac{994.3}{600.245}\right)^{\circ}C$$

$$= 7.38^{\circ}C$$

$$p_{s5} = 994.3 - 6 \times 0.04 \times 88.7\,Pa$$

$$= 973\,Pa$$

$$t_{dp5} = \frac{1}{0.0684} \times \ln\left(\frac{973}{600.245}\right)^{\circ}C$$

$$= 7.06^{\circ}C$$

$$p_{s7} = 973 - 40 \times 0.105 \times 88.7^{\circ}C$$

$$= 600.5^{\circ}C$$

$$t_{dp7} = \frac{1}{0.0684} \times \ln\left(\frac{600.5}{600.245}\right)^{\circ}C$$

$$= 0.006^{\circ}C \text{ shows small calculation inacuracy}$$

$$= 0^{\circ}C \text{ the significant value}$$

The dew-point temperature gradient ends with the input data and is superimposed upon a scale drawing of the thermally induced gradient shown in Fig. 10.4.

Owing to the high thermal resistance and very low vapour resistance of the mineral fibre slabs fixed in the cavity, under the design conditions as stated the material temperature drops below the moist air dew-point temperature midway through its thickness. Interstitial condensation will occur within the mineral fibre, air space and external brickwork. Ventilation of the remaining wall cavity allows evaporative removal of the droplets. Variation of internal and external air conditions will limit the duration of such temperature gradients. Periods of condensation will be very intermittent. Solar heat gains to the external brickwork will raise the temperature of the structure and help to reduce condensation periods. The walls most at risk are those always shaded from direct sunlight and having reduced wind exposure due to the proximity of nearby buildings.

10.4 Temperature gradients in the wall in Example 10.9.

When condensation takes place, a change of phase occurs as the water vapour turns into liquid. The calculations of water vapour transfer end at this discontinuity. The reason for undertaking the calculations was to establish if and where condensation is formed. The whole thickness of the layer where liquid forms is likely to be dampened owing to capillary attraction within porous solid material. The quantity of condensation that may be formed during a 60-day winter period can be assessed from the expected average air conditions. This aids the prediction of the physical damage that may be caused.

Installation note

Added thermal insulation can cause conditions where condensation will take place owing to the lowered structural temperatures. The installation of a vapour barrier raises the overall vapour resistance and may be able to keep the dew-point gradient below the thermally produced temperatures.

1. Materials with a low thermal resistance but a high vapour resistance, such as aluminium foil, sheet plastic, roofing felt and gloss paint, are placed on the warm side of the structure.
2. Materials of high thermal and vapour resistance can be placed anywhere.
3. Materials of moderate vapour resistance but high thermal resistance should be placed on the cold side of the structure. Such materials will require the addition of a weather-resistant coating when applied to the outside of a building.

Thermal or acoustic insulation applied to industrial roofs can increase the possibility of the occurrence of condensation. Alternative schemes are shown in Figs 10.5 and 10.6 (Saunders, 1981).

10.5 Cold-deck roof.

10.6 Warm-deck roof.

In the cold-deck design, the roofing sheets remain at little more than the external air temperature. Ventilation through gaps at the eaves, junctions where the sheets overlap and holes around steel supports provides passageways for the ingress of moist air. Condensation on the underside of the cold deck will cause water to run down the steelwork and wet the ceiling. It is unlikely that a tight seal can be made between the vapour barrier and the steel supports to avoid this happening. Before insulation of an existing roof, its underside is maintained at above the local dew-point except during severe weather or when the heating plant is off.

The warm-deck design improves the weather resistance of the roof and raises the original underside surface temperature even further. Interstitial condensation is unlikely because of the vapour barriers.

EXAMPLE 10.10

Calculate the external air temperature that will cause condensation to form on the underside of a factory roof constructed from 10 mm corrugated asbestos cement sheet. Internal air conditions near the roof are 23°C d.b., 30% saturation. The thermal conductivity of asbestos cement sheet is 0.4 W/mK. The roof has a severe exposure.

From the CIBSE psychrometric chart, the internal air dew-point is 5°C. From Tables 3.2 and 3.3, $R_{si} = 0.1$ m^{2}K/W and $R_{so} = 0.02$ m^{2}K/W. Let the unknown external air temperature be t_o. Then for a roof area of 1 m^2:

$$Q_f = U (23 - t_o) \text{ W/m}^2$$

For the external surface film:

$$Q_f = \frac{1}{R_{si}}(23-5) \text{ W/m}^2$$

$$= \frac{1}{0.1} \times 18 \text{ W/m}^2$$

$$= 180 \text{ W/m}^2$$

For the roof:

$$U = \frac{1}{R_{si} + (l/\lambda) + R_{so}}$$

$$= \frac{1}{0.1 + (0.01/0.4) + 0.02}$$

$$= 6.9 \text{ W/m}^2$$

Now, rate of heat flow through roof sheets = rate of heat flow through internal surface film.

$$6.9 \frac{\text{W}}{\text{m}^2 \text{ K}} \times (23 - t_o) \text{ K} = 180 \frac{\text{W}}{\text{m}^2}$$

$$26.0 = 23 - t_o$$

$$t_o = -3.09°\text{C}$$

Questions

1. Describe how the following forms of condensation occur: temporary, permanent and interstitial.
2. List the sources of moisture in buildings.
3. What is the purpose of installing a vapour barrier and what effect does it have on the dew-point temperatures within a structure?
4. Discuss the use of thermal insulation in reducing the likelihood of condensation in walls and roofs.
5. State examples of thermal insulation increasing the risk that condensation occurs.
6. List the actions that could be taken to reduce the water vapour input to a dwelling.
7. Discuss the use of heating and ventilation in combating condensation problems.
8. Why might prefabricated concrete buildings suffer more from condensation than other constructions?
9. What sources of moisture would you look for when consulted about mould growth on a building?
10. Describe the constituent parts of the atmospheric pressure.
11. What drives water vapour from one area to another?
12. Describe the way in which moisture is alternatively stored and released by porous building materials.
13. What is the flow of vapour through a composite structure analogous to?
14. State the conditions under which water vapour will condense on or within a construction.
15. Calculate the temperature gradients through the following structures. Internal and external air temperatures are to be taken as 21°C d.b. and −1°C d.b. Assume that $t_a = t_e$. The answers should be expressed as the surface or interface temperatures in descending order

from the warm side. Outside surfaces are taken as having a high emissivity and normal exposure. All air spaces are ventilated. The thermal conductivity of glass is 1.05 W/mK.

(a) 6 mm single-glazed window.
(b) 6 mm double-glazed window.
(c) Cavity wall of 15 mm dense plaster, 100 mm lightweight concrete block, air space and 105 mm brick.
(d) An industrial roof of 10 mm asbestos cement corrugated sheet which has been given an external coating of 50 mm phenolic foam. The thermal conductivity of asbestos sheet is 0.4 W/mK.

16. A shop window consists of 6 mm plate glass in an aluminium frame. The display area air temperature is expected to be 15°C d.b. and to have a dew-point of 7°C. Find the external air temperature that will start to produce condensation on the inside of the window. The window has normal exposure.

17. If a double-glazed window is to be fitted in the shop in Question 16, what could the external air temperature drop to before condensation starts?

18. A hospital ward is to be maintained at 24°C d.b. and 80% saturation. The air dew-point is 20.5°C. Thermal insulation is to be added to the inside of the existing wall to avoid surface condensation when the external air temperature falls to −5°C d.b. The U value of the original wall is 1.9 W/m^{2}K. Calculate the thickness of insulation material required if its thermal conductivity is 0.06 W/mK.

19. Calculate the thermal transmittance, temperature and dew-point gradients through a flat roof consisting of 25 mm stone chippings, 10 mm roofing felt, 150 mm aerated concrete, 75 mm wood wool/cement slabs, a ventilated 50 mm air space and 12 mm plasterboard. The roof has a sheltered exposure. Internal air conditions are 22°C d.b., 50% saturation. External conditions are 2°C d.b., 80% saturation. The stone chippings have a high emissivity when weathered and offer no resistance to the flow of water vapour. Plot a graph of the two temperature gradients and find if condensation is likely to occur.

20. Calculate the thermal transmittance, temperature and dew-point gradients through a wall consisting of 15 mm dense plaster, 100 mm medium-weight concrete blockwork, 40 mm glass fibre quilt, a ventilated 10 mm air space and 105 mm brickwork. The wall has a severe exposure. The average internal air conditions are 14°C d.b., 60% saturation when the external conditions are 1°C d.b., 80% saturation. Plot a graph of the two temperature gradients and find if condensation is likely to occur.

21. Where are indoor condensation problems most likely found?

    1. Hot dry ambient air locations.
    2. Hot humid ambient air locations.
    3. Temperate maritime climates, such as the UK, where outdoor air humidity remains high.
    4. Below zero ambient air locations.
    5. In any building anywhere.

22. Why does condensation occur within a building?

    1. Users leave taps running, baths full, water evaporates and deposits onto room surfaces, making air moist.
    2. Kettles, cooking, fish tanks and open bowls of water evaporate more water vapour into room than ventilation can remove.
    3. Porous building materials provide pathways for cold moisture to ingress a warm building and make indoor surfaces damp.

4. Evaporated water within the building meets surfaces at below dew-point temperature.
5. Cyclic variation of indoor surface temperatures always produces below dew-point locations.

23. What always combats condensation problems within occupied buildings?

1. Thermal insulation.
2. Impervious building materials.
3. Removing open water surfaces.
4. Air conditioning.
5. Heating and ventilation.

24. From where does water vapour originate within a building?

1. Atmospheric rain.
2. Wind-driven atmospheric humidity.
3. Occupants and their activities.
4. Lack of sufficient natural and mechanical ventilation.
5. Refrigeration systems and food storage.

25. What is the relationship of building materials to moisture mass transfer?

1. There is none, building materials do not leak water.
2. Good design and construction removes all moisture issues.
3. Modern building materials have zero permeability.
4. Porous structural materials absorb and pass moisture.
5. Vapour barriers isolate brick, concrete and thermal insulation materials from moisture generated within a building.

26. What drives moisture flow through a structure?

1. Vapour pressure difference in Pascals.
2. Air temperature difference in °C d.b.
3. Percentage saturation difference.
4. Air moisture content difference in kg H_2O/kg dry air.
5. Wet-bulb air temperature difference in °C w.b.

27. Which is correct about condensation and mould growth?

1. Always occurs in buildings over 20 years old.
2. Cannot happen with current design standards.
3. Readily forms in the UK in room corners, on window sills, in cupboards on external walls and within structures having an overall thermal transmittance of over 1.4 W/m^{2}K.
4. Must be removed, the surface gloss painted and outdoor air ventilation minimized.
5. Impervious external surface materials need to be matched with an outdoor vapour barrier to stop moisture flowing into the wall, floor or roof structure.

28. Vapour diffusion into a structure that then condenses is called what type?

1. Adiabatic.
2. Complex.
3. Leakage.
4. Interstitial.
5. Intermediate.

29. What is happening in a building during vapour diffusion?

   1. Odours and gases produced indoors slowly percolate through the building structure to outdoors.
   2. Steam from water boiling, cooking and hot-water washing become absorbed into furniture, furnishings, carpets and surface plaster unless removed by ventilation.
   3. Internally sourced water vapour migrates through porous building structures.
   4. Low indoor vapour pressure drives moisture towards higher outdoor moist air vapour pressure.
   5. Liquid water passes through the building structure.

30. In the context of condensation in building, what does r_v symbolize?

   1. Vapour resistivity of a material.
   2. Vapour resistivity of a structure.
   3. Resistance to vapour flow of a structure.
   4. Volumetric resistivity of a structure.
   5. Volumetric resistance of a material to moisture transfer.

31. Which is true?

   1. 10^3 kN s/gm $=$ GN s/kgm.
   2. 10^3 N s/g $=$ MN s/g.
   3. MN s/gm $=$ GN s/kgm.
   4. 10^6 N s/kg $=$ MN s/g.
   5. kN s/kg $=$ MN s/g.

32. What is the vapour resistivity of a structural material measured in?

   1. kN/m^{2}s.
   2. MN s.
   3. kN s/kg.
   4. GN s/kg m.
   5. kN kg/m^{2}s.

11 Lighting

Learning objectives

Study of this chapter will enable the reader to:

1. explain the use of day and artificial lighting;
2. use lux and lumen per square metre;
3. state normal lighting levels;
4. explain general and task illumination;
5. understand permanent supplementary artificial lighting of interiors (*PSALI*);
6. understand artificial lighting terminology;
7. assess the importance of the maintenance of lighting installations;
8. calculate the room index;
9. find utilization factors;
10. discuss the problem of glare;
11. calculate the number of lamps needed to achieve a design illumination level;
12. calculate lamp spacing for overall design;
13. calculate the electrical loading produced by the lighting system;
14. consider how luminaires should be oriented in relation to room layout and visual tasks to be performed;
15. understand the use of air-handling luminaires;
16. understand lamp colour-rendering and colour temperature;
17. know the range of available lamp types;
18. apply appropriate lamp types to designs;
19. understand the working principles of lamp types and their starting procedures.

Key terms and concepts

artificial and natural illumination 261; *BZ* classification 264; colour-rendering 269; colour-rendering index 269; colour temperature 269; daylight factor 262; efficacy 264; electrical power consumption 268; glare 267; heat generation 262; illuminance 261; lamp types 269; *LDL* 262; lighting cost 272; light loss factor 265; lumen 265; luminaire 269; luminance

Introduction

Artificial illumination for both functional and decorative purposes is a major consumer of primary energy, and developed civilizations have become used to very high illumination standards with consequently high electricity consumption. The use of daylight is encouraged in order to reduce fuel consumption for lighting but this occurs at the expense of heating and cooling energy consumption at the building outer envelope, which is in contact with the external environment. A compromise solution is inevitable, and the building services engineer is at the centre of the calculations needed to minimize total energy consumption for all usages.

The factors involved in determining illumination requirements are discussed in relation to lighting levels for various tasks and the possible use of daylight. Lighting terms are introduced as are glare considerations. The lumen design method is demonstrated for office accommodation.

Lamp colour-rendering is discussed, and the use of luminaires with air-conditioning systems. Lamp types, their uses and control arrangements are explained.

Natural and artificial illumination

Natural illumination by penetration of direct solar and diffuse sky visible radiation requires correctly designed passive architecture. Large glazed areas may provide sufficient day lighting at some distance into the building but can also cause glare, overheating and high heating and cooling energy costs.

The other extreme of vertical narrow slot windows limits energy flows while causing very unequal lighting levels near the room's perimeter. Reflected illumination from other buildings, particularly from those having reflective glazing or metallic architectural features, may cause annoyance. A careful consideration of all the, largely conflicting, variable elements is necessary if a comfortable internal environment is to be produced.

Artificial lighting is provided to supplement daylight on a temporary or permanent basis. Local control of lights by manual and/or automatic switches aids economy in electricity consumption. The colours rendered by objects on the working plane should match the colours under daylight. The working plane may be a desk, drawing board or display area.

Illumination intensity, illuminance, measured in lux on the working plane, is determined by the size of detail to be discerned, the contrast of the detail with its background, the accuracy and speed with which the task must be performed, the age of the worker, the type of space within which the task is to be performed and the length of time continuously spent on the task. The working plane is the surface being illuminated. Other areas are lit by overspill from it and by reflections from other room surfaces. Table 11.1 gives some typical values for illuminance commonly encountered and used for design.

Higher levels of illuminance may be provided for particularly fine detail tasks at the area of use by local, or task, illumination: for example, up to 3000 lx for inspection of small electronic components and 50 000 lx on a hospital operating table. Bright sunlight provides up to 100 000 lx. Local spot lighting for display purposes and exterior illumination are used to accentuate particular features of the working plane.

Permanent supplementary artificial lighting of interiors (*PSALI*) has become common in modern office accommodation, shops and public buildings. Figure 11.1 shows the constituents of the overall design illumination.

Table 11.1 Typical values of illuminance.

Application	Illuminance (lx)
Emergency lighting	0.2
Suburban street lighting	5
Dwelling	50–150
Corridors	100
Rough tasks with large-detail storerooms	200
General offices, retail shops	400
Drawing office	600
Prolonged task with small detail	900

11.1 Permanent supplementary artificial lighting.

The heat generated by permanent lighting can be extracted from the light fitting (luminaire) by passing the ventilation extract air through it, thus raising the air temperature to 30–35°C, and then supplying this heated air to perimeter rooms in winter. Further air heating with finned tube banks and automatic temperature control would be part of a normal ventilation or air-conditioning system. This can be termed a heat reclaim system incorporating regenerative heat transfer between the outgoing warm exhaust air and incoming cold fresh air.

The penetration of daylight into a building can be enhanced with north-facing roof-lights, skylights having motorized louvres which are adjusted to suit the sun's position and weather conditions, or mirrored reflectors which direct light rays horizontally into the building. An example is shown in Fig. 11.2.

Localized task illumination can be provided in addition to a background lighting scheme but may not necessarily produce a reduction in total power consumption. Tests (Ellis, 1981; McKenna et al., 1981) of combinations of overall lighting plus task illumination revealed a preference for two 40 W white fluorescent lamps 1200 mm long plus indirect background lighting.

Desk-mounted lamps produce a range of up to 9:1 in illuminance values across the working surface and form strong contrasts between surrounding and working surfaces which may result in discomfort glare. Direct dazzle from unshaded lamps, reflected glare, shadows around objects and hands, and heat radiation can cause discomfort.

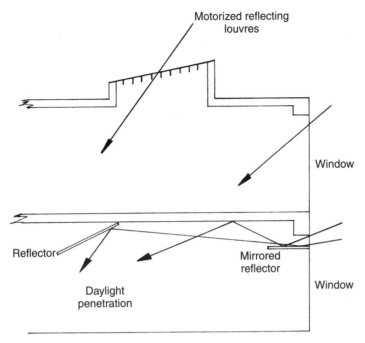

11.2 Use of daylight reflectors.

Lighting is, obviously, a visual subject and it may be treated as such for design purposes. It is not necessarily helpful to study lighting only as applied mathematics. Designs are an artistic and engineering combination of architecture, interior design, decoration, illumination functionality, economical use of electrical energy, maintainability, safety, environmental health, controllability, prestige and the overall and specific requirements of the user. Architects and engineers coordinate their skills to create an acceptable visual and technical solution.

To consider the design of lighting from a visual standpoint, acquire two or more cardboard boxes. These are to be used to represent rooms (W. Burt, Manchester University). The dimensions are unimportant as the principles of lighting apply to volumes of any size. They can be used to evolve answers to Questions 15–21 at the end of this chapter. This may be seen as a crude method of higher education; however, the use of scale models is a well-established artistic and scientific discipline. Computer-generated illumination plots are available for a similar design purpose but at much higher cost. The interior of the shoebox, or similar, can be covered or painted with dark or light colours. The source of light for the interior of the box is the indoor artificial illumination or daylight through a window of the user's location. Cut windows or roof-lights into the box that are in proportion to the room design that is to be modelled. Each window should be hinged so that different combinations of window opening, and night-time, can be reproduced. Battery lamps could be utilized for the illumination arrangements. Rectangular slots in the flat ceiling of the box can represent linear fluorescent luminaires. Small circular holes in the ceiling would model spot lighting. Advanced users can place objects within the scaled room to establish the effects upon desks, furniture, partitions and artifacts. Cut a small viewing window in the end of the box. Use this aperture to observe and sketch the areas of light and shade within the room. Try various combinations of surface colour, day, artificial and permanent supplementary artificial lighting. Figure 11.3 is an example of an observed lighting pattern in a model room that has both side windows and a representation of recessed tubular fluorescent luminaires in the ceiling.

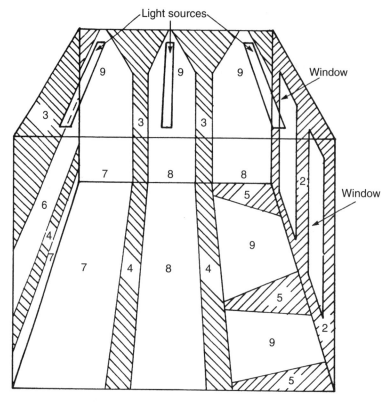

Light sources

Window

Window

Window

Shadow and brightness on a
scale of dark 1 to bright 10

11.3 Observed illumination pattern.

Definition of terms

Some of the terms that are used in lighting system design are as follows.

BZ classification: British Zonal classification of 1–10 for the downward light emitted from a luminaire. The *BZ* class number relates to the flux that is directly incident upon the working plane in relation to the total flux emitted. *BZ*1 classification is for a downward directional luminaire. A *BZ*10 describes a luminaire that directs all its illumination upwards so that the room is illuminated by reflection from the ceiling.

Contrast: the difference in the light and dark appearance of two parts of the visual field seen simultaneously.

Daylight factor: the ratio of the natural illumination on a horizontal plane within the building to that present simultaneously from an unobstructed sky, discounting direct sunlight. A standard figure of 5000 lx is adopted for the external illuminance in the UK.

Efficacy: the luminous efficacy is the lamp light output in lumens per watt of electrical power consumption.

Glare: the discomfort or impairment of vision due to excessive brightness.

Glare index: a calculated numerical scale for discomfort glare.

Illuminance: the luminous flux density at a surface in lumens per square metre, l/m^2, lux. The surface is normally the working plane.

Illumination: the process of lighting an object or surface.

Light meter: a current-generating photocell which is calibrated in lumens per square meter, lux.

Lumen, lm: SI unit of luminous flux. It is the quantity of light emitted from a source or received by a surface. A 100 W tungsten filament lamp emits around 1200 lm.

Lux, lx: SI unit of illuminance; $1 lx = 1 lm/m^2$.

Light loss factor, LLF: the overall loss of light from the dirtiness of the lamp (0.8), luminaire (0.95) and the room surfaces (0.95). Clean conditions LLF may be 0.7 but 0.5 when equipment and room become soiled. Preferred to maintenance factor.

Luminous intensity, I (candela): the power of a source or illuminated surface to emit light in a given direction.

Luminaire: the complete apparatus that contains the lamp, the light emitter and the electrical controls.

Maintenance factor, MF: an allowance for reduced light emission due to the build-up of dust on a lamp or within a luminaire. Normally 0.8 but 0.9 if the lamps are cleaned regularly or if a ventilated luminaire is used. Light loss factor is preferred.

Utilization factor, UF: the ratio of the luminous flux received at the working plane to the installed flux.

Maintenance

A planned maintenance schedule will include regular cleaning of light fittings and the lamp to ensure the most efficient use of electricity. Ventilated luminaires in air-conditioned buildings remain clean for quite long periods as the air flow through the building is mechanically controlled and filtered. The lamp also operates at a lower temperature, which prolongs its service and maximizes light output.

Because of gradual deterioration of the light output from all types of discharge lamps after their design service period, lamp efficacy could fall to half its original figure. Phased replacement of lamps after 2 or 3 years maintains design performance and avoids breakdowns. Figure 11.4 shows a typical illuminance profile for a tubular fluorescent light fitting with 6-monthly cleaning and a 2-year lamp-replacement cycle.

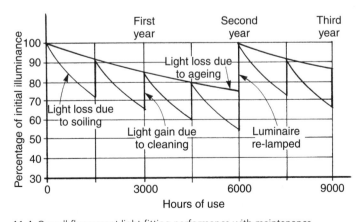

11.4 Overall fluorescent light fitting performance with maintenance.

Utilization factor

The utilization factor is provided by the manufacturer and takes into account the pattern of light-distribution from the whole fitting, its light-distributing efficiency, the shape and size of the room for which it is being designed and the reflectivity of the ceiling and walls. Values vary from 0.03, where purely indirect distribution is employed, the room has poorly reflecting surfaces and all the light is upwards onto the ceiling or walls, to 0.75 for the most energy-efficient designs. Spot lighting can have a utilization factor of nearly unity.

The configuration of the room is found from the room index:

$$\text{room index} = \frac{lW}{H(l + W)}$$

where l is the room length (m), W is the room width (m) and H is the height of the light fitting above the working plane (m).

The ability of a surface to reflect incident light is given by its luminance factor from BS 4800 : 1972. Sample values are given in Table 11.2.

Utilization factors for a light fitting comprising a white metal support batten and two 58 W white fluorescent lamps 1500 mm long (New Streamlite by Philips Electrical Limited) are given in Table 11.3. The data refer to bare fluorescent tubes suspended under the ceiling as used in commercial buildings. Enclosing the fitting with a plastic diffuser to improve its appearance

Table 11.2 Luminance factors for painted surfaces.

Surface	Typical colour	Luminance factor range (%)
Ceiling	White, cream	70–80
Ceiling	Sky blue	50–60
Ceiling	Light brown	20–30
Walls	Light stone	50–60
Walls	Dark grey	20–30
Walls	Black	10
Floor	—	10

Table 11.3 Utilization factors for a bare fluorescent tube fitting with two 58 W 1500 mm lamps (%).

Luminance factors		Room index								
Ceiling	Walls	0.75	1	1.25	1.5	2	2.5	3	4	5
70	50	48	53	59	64	71	75	79	83	86
70	30	40	46	51	57	64	69	73	78	82
70	10	35	40	46	51	59	64	68	74	78
50	50	43	48	52	57	63	67	70	74	76
50	30	37	41	46	51	57	62	65	70	73
50	10	33	37	42	46	53	58	61	67	70
30	50	39	42	46	50	55	59	61	65	67
30	30	34	37	42	46	51	55	58	62	65
30	10	30	33	38	42	48	52	55	59	62

usually lowers the utilization factor. This fitting has a *BZ6* rating, operates at 240 V, consumes 140 W and has a power factor of 0.85 and a running current of 0.68 A.

Glare and reflections

Disability glare is when a bright light source prevents the subject from seeing the necessary detail of the task. Veiling reflections can be formed on windows and visual display unit (VDU) screens from nearby lamps. A limiting glare index is recommended for each application, for example, general office 16, and this can be calculated (CIBSE, 1986).

To maximize contrast on the working plane, luminaires should be placed in rows parallel to the direction of view. The rows should be widely spaced to form work areas between them.

The zone of the ceiling that would cause glare or veiling reflections can be viewed with a mirror on the working plane from the normal angle of work. A luminaire or direct sunlight should not appear in the mirror.

Lumen design method

The number of light fittings is found from the total lumens needed at the working plane and the illumination provided by each fitting using the formula:

$$\text{number of fittings} = \frac{\text{lux} \times \text{working plane area m}^2}{LDL \times UF \times MF}$$

where *LDL* is the lighting design lumens produced by each lamp, *UF* is the utilization factor and *MF* is the maintenance factor. A typical high output luminaire with two fluorescent lamps, 1500 mm long, emits 5100 lm measured at 2000 h of use. This is known as its lighting design lumens (*LDL*).

EXAMPLE 11.1

A drawing office 16 m × 11 m and 3 m high has a white ceiling and light-coloured walls. The working plane is 0.85 m above the floor. 5100 lm double-lamp luminaires are to be used and their normal spacing-to-height ratio *SHR* is 1.75. Calculate the number of luminaires needed and draw their layout arrangement. Find the electrical power consumption of the lighting system.

From Table 11.2 the luminance factors are 70 for the ceiling and 50 for the walls. A high standard of maintenance will be assumed, giving a maintenance factor of 0.9. The lighting design lumens is taken as 5100 lm for the whole light fitting.

From Table 11.1 the illuminance required is 600 lm/m^2. The height *H* of fittings above the working plane is

$$H = (3 - 0.85) \text{ m}$$

$$= 2.15 \text{ m}$$

$$\text{room index} = \frac{lW}{H(l + W)} = \frac{16 \times 11}{2.15 \times (16 + 11)} = 3.03$$

From Table 11.3, for a room index of 3,

utilization factor $= 79\% = 0.79$

$$\text{number of fittings} = 600 \ \frac{\text{lm}}{\text{m}^2} \times \frac{16 \text{ m} \times 11 \text{ m}}{0.79 \times 0.9} \times \frac{\text{luminaire}}{500 \text{ lm}} = 29.12$$

The ratio of the spacing S between rows to the height H above the working plane is:

$$SHR = \frac{S}{H} = 1.75$$

Therefore,

$$S = 1.75 \ H$$
$$= 1.75 \times 2.15 \text{ m}$$
$$= 3.76 \text{ m}$$

If it is assumed that windows are along one long side of the office and that rows of luminaires will be parallel to the windows, this will produce areas between rows where drawing boards, desks and VDU terminals can be sited to gain maximum benefit from side day lighting without glare and reflection. The perimeter rows of luminaires are spaced at about half of S, 1.74 m, from the side walls.

Three rows of 10 luminaires are required, as shown in Fig. 11.5, giving 30 luminaires and a slightly increased illuminance.

The electrical power consumption of each luminaire is 140 W. For the room the power consumption will be 30×140 W, that is, 4200 W, which is:

$$\frac{4200 \text{ W}}{16 \text{ m} \times 11 \text{ m}} = 23.86 \text{ W/m}^2 \text{ floor area}$$

11.5 Arrangement of the luminaires in a drawing office in Example 11.1.

11.6 Ventilated luminaire.

Air-handling luminaires

Luminaires that are recessed into suspended ceilings are ideally placed to be extract air grilles for the ventilation system. The heat generated is removed at its source and the lamp can be maintained at its optimum operating temperature to maximize light output and colour-rendering properties. Dust build-up should also be less in an air-conditioned building where all the circulating air is filtered in the plant room. Figure 11.6 shows the air flow through a luminaire that has ventilation openings and mirrored reflectors.

Up to 80% of the electrical energy used by the light fitting can be absorbed by the ventilation air as it passes through. Air flow rates are around 20 l/s through a 1500 mm twin tube fluorescent luminaire and a temperature increase of about 8°C is achieved. Extract air at about 30°C can be produced and the heat it contains can be reclaimed for use in other parts of the building.

Colour temperature

Colour temperature is a term used in the description of the colour-rendering property of a lamp. Colours of surfaces under artificial illumination are compared with the colours produced by a black body heated to a certain temperature and radiating in the visible part of the spectrum between the ultraviolet and infrared bands. Any colour that matches that shown by the heated black body is said to have a colour temperature equal to the temperature of the black body. A candle has a colour temperature of 2000 K and blue sky has a colour temperature of 10 000 K. Correlated colour temperature is that temperature of the heated black body at which its colour most closely resembles that of the artificial source.

The colour-rendering index Ra8 is used to compare the colour-rendering characteristics of various types of lamp. Eight test colours are illuminated by a reference source, which is a black body radiator of 5000 K correlated colour temperature or 'reconstituted' daylight if more than 5000 K is needed. These eight colours are then illuminated by the test lamp. The average of the colour differences produced between the source and test lamps provides a measure of the colour-rendering properties of the test lamp. An Ra8 of 50 corresponds to a warm white fluorescent lamp. An Ra8 of 100 would be produced by a lamp that radiated identically with the reference source.

Lamp types

A summary of lamp types, their performances and applications is given in Table 11.4.

General lighting service (GLS) tungsten filament lamps are inexpensive, give good colour matching with daylight and last up to 2000 h in service. They can be controlled by variable-resistance dimmers and are used in a supplementary role to higher-efficacy illumination equipment. Tungsten halogen spot or linear lamps have a wide variety of display and floodlighting applications.

Table 11.4 Lamp data.

Lamp	Lamp designation BS colour	Efficacy (lm/W)	Colour temperature	Colour-rendering index Ra8	Application
Incandescent	GLS, tungsten	18	2900	100	Interiors
Incandescent	GLS, tungsten–halogens	22	3000	100	Interior displays, outdoors
Fluorescent	White	80	3500	50	Industrial
Fluorescent	Natural	85	4000	85	Commercial
Fluorescent	Warm white	85	3000	85	Social
Discharge lamps					
Low-pressure sodium	SOX	183	2000	—	Roads, car parks, floodlighting
High-pressure sodium	SON	112	2250	29	Floodlighting, exteriors, large hall interiors
Mercury fluorescent	MBF	50	4300	47	Roads, floodlights, factory interiors
Metal halide	MBI	85	4400	70	Industrial interiors, floodlighting
Mercury-blended tungsten fluorescent	MBTF	25	3700	—	Road, floodlights, factory interiors

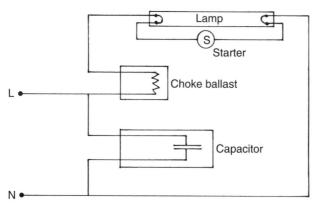

11.7 Switch-start circuit diagram for fluorescent and discharge lamps.

Miniature fluorescent lamps (SL) can be used as energy-saving replacements for GLS lamps. Folded-tube and single-ended types are available. A typical folded lamp, SL18 18 W, produces 900 lm at 100 h, has a correlated colour temperature of 2700 K, an Ra8 of 80 and a service period of 5000 h. Its lumen output is equivalent to a 75 W GLS filament lamp.

Low-pressure mercury-vapour-filled fluorescent tubular lamps (MCF) are the most common. The tube diameter is 38 mm. Electrical excitation of the mercury vapour produces radiation, which causes the tube's internal coating to fluoresce. The colour produced depends on the chemical composition of the internal coating. High-efficacy 26 mm diameter lamps (TLD) are filled with argon or krypton vapour and have a phosphor internal coating. A circuit diagram for a switch-start fluorescent lamp is shown in Fig. 11.7.

The glow-switch starter S has bimetallic electrodes, which pass the lamp electrode preheating current when starting cold. Upon becoming warmed, the bimetallic electrodes move into contact and establish a circuit through the lamp electrodes. Making this contact breaks the starter circuit, whose electrodes cool and spring apart after about a second, subjecting the lamp to mains voltage. Twin lamp and starterless controls are used in new installations to minimize running costs.

Luminaires are designed to run in air of 5–25°C d.b., and their service period will be reduced at higher temperatures. Circuit-breakers or high rupturing capacity (HRC) fuses, rather than rewirable fuses, should be used for circuit protection. All fluorescent luminaires make an operating noise, which may be noticeable against very low background noise levels. Some radio interference is inevitable but this diminishes with distance from the radio set.

Low-pressure sodium discharge lamps emit light from electrically excited sodium vapour in a glass discharge tube. High-pressure sodium discharge lamps comprise a discharge tube of sintered aluminium oxide containing a mixture of mercury and sodium vapour at high pressure.

Mercury fluorescent lamps consist of electrically excited mercury vapour in a quartz discharge tube, which emits ultraviolet radiation, an infrared component from the mercury arc, and visible light. A phosphor internal coating fluoresces to produce the desired colour. Metal halide lamps contain metal halides in a quartz discharge tube, which gives a crisp white light. MBTF lamps are the mercury fluorescent type with additional tungsten filaments, which need no control gear and give light immediately. Other discharge lamps incorporate current-limiting ballasts and power-factor-correcting capacitors in a similar arrangement to that in Fig. 11.7.

EXAMPLE 11.2

A windowless office is to be illuminated for 15 h per day, for 6 days per week for 50 weeks per year. The floor is 20 m long and 12 m wide. An overall illumination of 450 lx is to be maintained over the whole floor. The total light loss factor for the installation is 70%. The designers have the choice of using 100 W tungsten filament lamps, which have an efficacy of 12 lm/W and need replacing every 3000 h, or 65 W tubular fluorescent warm white lamps, which have an initial output of 5400 lm and are expected to provide 12 000 h of service. The room layout requires an even number of lamps. Electricity costs 8 p/kWh. The tungsten lamps cost £1 each while the fluorescent tubes cost £10 each. Compare the total costs of each lighting system and make a recommendation as to which is preferable, stating your reasons.

For the lighting system,

$$\text{lighting hours} = 15 \times 6 \times 50 = 4500 \text{ h/year}$$

$$\text{floor area} = 20 \text{ m} \times 12 \text{ m} = 240 \text{ m}^2$$

$$\text{installed lumens} = \frac{450 \times 240 \times 100}{70}$$

$$= 154\,286 \text{ lm}$$

For the tungsten lamps:

$$\text{input power} = \frac{154\,286}{12 \times 1000}$$

$$= 12.857 \text{ kW}$$

$$\text{number of lamps} = \frac{12.857 \times 1000}{100}$$

$$= 129 \text{ lamps}$$

Next even number is 130 lamps.

$$\text{installed power} = \frac{130 \times 100}{1000}$$

$$= 13 \text{ kW}$$

$$\text{electricity cost} = \frac{13 \times 4500 \times 8}{100}$$

$$= \pounds 4680/\text{year}$$

The average annual cost for replacing the lamps can be found by multiplying the number of installed lamps by the anticipated hours of use, dividing by the lamp manufacturer's rated average life hours and then multiplying by the replacement cost for each lamp. In this case,

$$\text{lamp cost} = \frac{130 \times \pounds 1 \times 4500}{3000}$$

$$= \pounds 195/\text{year}$$

These tungsten lamps expire within a year, so there will be annual expenditure. A new lighting system that has lamps providing reliable service for more than a year will not produce replacement expenditure in the first year or two. The owner needs to budget for the eventual replacement costs by assessing the average annual cost. A planned maintenance programme will have lamps replaced, and luminaires cleaned, prior to expiry. This work may be performed out of normal working hours for the building to avoid disturbance to its normal functions.

$$\text{Total annual cost} = \pounds(4680 + 195) = \pounds 4875/\text{year}$$

For the fluorescent lamps

$$\text{number of lamps} = \frac{154\,286}{5400}$$

$$= 29 \text{ lamps}$$

Next even number is 30 lamps

$$\text{input power} = \frac{30 \times 65}{1000}$$

$$= 1.95 \text{ kW}$$

$$\text{electricity cost} = \frac{1.95 \times 4500 \times 8}{100}$$

$$= \pounds 702$$

$$\text{lamp cost} = \frac{30 \times £10 \times 4500}{12\,000}$$

$$= £112.5\text{/year}$$

total annual cost $= £(702 + 112.5) = £814.5$/year

Both methods of lighting require at least the annual cleaning of lamps and luminaires.
The reasons for using fluorescent lamps are as follows.

1. They produce an annual cash saving of $£(4875 - 814.5) = £4060.5$/year.
2. They only need changing $(12\,000/4500)$ after 2 years 8 months of use.
3. Tungsten lamps need changing $(12 \times 3000/4500)$ every 8 months.
4. There is less heat emission from fluorescent lighting, particularly radiant heat, so the air-conditioning cooling load is lower.
5. They give better colour rendering.
6. They give better diffused light distribution.

Control of lighting services

The energy that is consumed by artificial lighting systems is both an expensive use of resources and a high monetary cost. A minimum level of illumination may be desirable for the security of personnel or monitoring of the building and its contents for the detection of intruders. The changes from day lighting to full artificial lighting and then to security illumination can be achieved with manual and automatically timed operation of switches. This usually leaves unoccupied areas illuminated. A light-sensitive photocell can be used to detect illumination level and an automatic controller may be programmed to reduce the use of the electrical lighting system. The presence of the occupants, or an intruder, can be detected by passive infrared, acoustic, ultrasonic or microwave-radar-based systems.

The detector and control system needs to be sufficiently fast-acting and sensitive so that the occupant is not stranded within a darkened room and suffers injury or fear. It is equally important that only those lights that are actually needed are switched on, and not for the entire room or space to be illuminated when only one person enters to use a small area. Local control of the light switching may be preferable to a system that is operated from a remote energy management system computer. The local system should be faster in operation and will be less subject to distribution system or computer breakdown.

The design of a control scheme for an occupied space may include a minimum number of luminaires, which are switched on from a time switch or by the occupant to provide safe access. Groups of luminaires that are near windows may be controlled from local photocell detectors to ensure that the perimeter lighting remains off as long as possible. The internal parts of the space may be operated from automatic presence detectors. Data on the length of operation of each lighting unit can be transmitted to the computer-based building management system so that real-time usage and electrical power consumption can be recorded. Timed-off controllers avoid lights left on excessively after working hours and when nobody is present.

Questions

1. Explain, with the aid of sketches, how interiors can be illuminated by daylight. State how natural illumination is quantified.
2. State the relationship between the visual task and the illuminance required, giving examples.

3. Sketch and describe how supplementary artificial lighting is used to achieve the desired illuminance.
4. Discuss the use of localized task-illumination systems in relation to the illumination level provided, reflection, energy conservation, shadows and user satisfaction.
5. Define the terms used in lighting system design.
6. Draw a graph of illumination provided versus service period for an artificial illumination installation to show the effect of correct maintenance procedure.
7. Calculate the room index for an office 20 m × 12 m in plan, 3 m high, where the working plane is 0.85 m above floor level.
8. State the luminance factors for a room having a cream ceiling and dark grey walls.
9. Find the utilization factor for a bare fluorescent tube light fitting having two 58 W, 1500 mm lamps in a room 5 m × 3.5 m in plan and 2.5 m high. The working plane is 0.85 m above floor level. Walls and ceiling are light stone and white respectively.
10. Sketch satisfactory arrangements for natural and artificial illumination in modern general and drawing offices, a library and a lounge. Comment particularly on how glare and reflections are controlled.
11. A supermarket of dimensions 20 m × 15 m and 4 m high has a white ceiling and mainly dark walls. The working plane is 1 m above floor level. Bare fluorescent tube light fittings with two 58 W, 1500 mm lamps are to be used, of 5100 lighting design lumens, to provide 400 lx. Their normal spacing-to-height ratio is 1.75 and total power consumption is 140 W. Calculate the number of luminaires needed, the electrical loading per square metre of floor area and the circuit current. Draw the layout of the luminaires.
12. Discuss how the use of air-handling luminaires improves the performance of the lighting installation and makes better use of energy.
13. Explain how the rendering of colours by illumination systems is measured.
14. Compare the energy efficiency and colour-rendering of different lamp types, stating suitable applications for each.
15. Use the cardboard box small-scale models of rooms to investigate the visual design of lighting systems.

    (a) Cut different shapes and locations of windows and roof-lights such that they all have the same open area.
    (b) Colour the internal surfaces differently by means of dark, light, removable and reflective sheets of materials.
    (c) Cut slots and holes into the ceiling to model different designs of strip fluorescent and filament lighting layouts; replaceable ceilings with different designs are helpful.
    (d) View the interior of the room under various day lighting and artificial lighting arrangements.
    (e) Make three-dimensional sketches of what you see of the lighting layouts produced, showing the shading. Write the lighting level found on each area on a scale of 1 (dark) to 10 (bright).

16. Using the models created for Question 15, answer the following.

    (a) What is the effect of *quantity* of daylight on the *quality* of the day lighting system created?
    (b) What effect do the colours of the room interior surfaces have on the quality of the lighting produced?
    (c) What colours should the room surfaces be?

    Justify your views in relation to the use of the room, its maintenance costs and design of the decoration.

17. Using the models created for Question 15, answer the following.

   (a) What patterns of illumination are produced on the end walls by differently spaced rows of strip lighting?
   (b) What are the best spacing arrangements between rows of strip or circular lamps? These depend upon what is being illuminated, so state the objectives of the lighting design first.

18. Create an approximate scale model of the interior of a room that is known to you. Experiment with three combinations of day lighting and artificial lighting to find the best overall lighting scheme for the tasks to be performed in the room.

19. Write a technical report to explain how the reflectance of room surfaces, the location, dimensions and shape of glazing, the spacing of rows of luminaires and their height above the working plane are related to the efficient use of electrical energy in the overall lighting design.

20. Put small boxes and partitions into a scale model of a room to represent furniture, desks, horizontal and vertical working planes. Carry out an experimental investigation of the problems that arise for the lighting designer.

21. State how the lighting design can be made to feature particular parts of the interior of the building and the parts that should be featured for safety and appearance reasons.

22. Analyse the costs of these competitive lighting systems and recommend which is preferable, stating your reasons.

   A heavy engineering factory is to be illuminated for 15 h per day, for 5 days per week for 50 weeks per year. The floor size is 120 m long and 80 m wide. An overall illumination of 250 lx is to be maintained over the whole floor. The overall light loss factor for the installation is 63%. The designers have the choice of using 150 W tungsten–halogen lamps, which produce 2100 lm and need replacing every 2000 h, 80 W tubular fluorescent lamps, which produce 6700 lm and are expected to provide 12 000 h of service, and 250 W high-pressure sodium lamps, which produce 27 500 lm and are expected to last for 24 000 h. The lighting layout needs an even number of lamps. Electricity costs 7.2 p/kWh. The tungsten lamps cost 90p each, the fluorescent tubes cost £10.50 each and the sodium lamps cost £61 each. Replacing any lamp takes two people 2 min and their combined labour rate is £17 per hour. The hire cost of scaffolding is £120 per 8-h day.

23. A lecture theatre is to be illuminated for 8 h per day, for 5 days per week for 30 weeks per year. The floor is 32 m long and 16 m wide. An overall illumination of 350 lx is to be maintained over the whole floor. An even number of lamps is to be used.

   The utilization factor for the installation is 0.73 and the maintenance factor is 0.7.

   The designers have the choice of using 100 W tungsten filament lamps, which have luminous efficacy of 10 lm/W and need replacing every 2000 h, 100 W quartz halogen low-voltage lamps in reflectors, which have an efficacy of 95 lm/W and provide 23 000 h use, and 65 W tubular fluorescent lamps, which have an efficacy of 57 lm/W and are expected to provide 7500 h of service. Electricity costs 8 p/kWh. The tungsten lamps cost 85 p each, the halogen cost 29 p each and the fluorescent tubes cost £11.25 each. Compare the total costs of each lighting system and make a recommendation as to which is preferable, stating your reasons.

24. When 900 lm fall onto a 2.0 m² surface from a fluorescent lamp, and a light meter finds that 360 lm are reflected, the luminance of the surface is:

   1.   0.71
   2.   0.4
   3.   0.2

   4.   Insufficient information.
   5.   1.25

25. Which is illumination intensity?

   1.   Luminosity.
   2.   Light level.
   3.   Lumens.
   4.   Lux.
   5.   Lumen per watt.

26. Illumination intensity provided is related to which?

   1.   Contrast between detail and background on the working plane.
   2.   Cleanliness of luminaires.
   3.   Task lighting needed.
   4.   Lighting colour temperature.
   5.   Provision of shadow-free lighting.

27. Which illuminance levels are correct? More than one correct answer.

   1.   Small electronic component inspection 3000 lx.
   2.   Corridor 250 lm/m^2.
   3.   Bright sunlight 100 000 lx.
   4.   Rough tasks 500 lm/m^2.
   5.   Hospital operating theatre table 10 000 lx.

28. Which is lighting design?

   1.   A primarily scientific application.
   2.   Engineering design.
   3.   A unique combination of technology and visual effects.
   4.   Entirely a mathematical application.
   5.   Best achieved by an iterative design technique.

29. Efficacy of a lamp means:

   1.   Efficiency at converting colour temperature into illuminance on the working plane.
   2.   Percentage of lamp globe or tube surface area that emits light.
   3.   Proportion of lamp light output directed towards the working plane.
   4.   Lumens light output per watt of electrical input power to lamp.
   5.   Luminous efficacy is total illuminance produced divided by lamp electrical input power including all control equipment consumption.

30. Light system glare means:

   1.   Whether discomfort is created in a particular viewing direction.
   2.   Impairment of vision due to excessive brightness.
   3.   Sedentary position includes direct view of sunshine.
   4.   Florescent indoor lighting systems do not cause glare.
   5.   All lighting causes glare when viewed directly.

31. What is an air-handling luminaire?

   1.   A sealed light fitting inside an air-handling unit to allow servicing work.
   2.   A light fitting open to the room air allowing cooling.

3. Luminaire passing air returning to the ductwork system from the conditioned room.
4. Lamp designed for low-temperature operation.
5. Sealed luminaire to keep out moisture.

32. What is meant by lamp colour temperature?

1. Description of colour-rendering property.
2. Colour shift in surface caused by the type of lamp.
3. Colour of the lamp when viewed directly.
4. Infrared fraction of the sun produced by the lamp.
5. Calculated average temperature from each of eight wavelength bands in the light produced by a lamp.

33. Which are correct descriptions of lighting system sensor operation? More than one correct answer.

1. Sensor detects light level in room and switches on rows of luminaires to maintain set lux level.
2. All sensor types are used by the building management computer system, BMS, to switch rows of luminaires on and off.
3. Groups of luminaires switched on from a sensor detecting occupancy within controlled space.
4. Microwave sensor detects use of electrical equipment within room and switches lights on.
5. Microwave sensor detects any small movement within controlled space and switches luminaires on and then off when no movement is detected for a set time interval.

34. Which is correct about lighting?

1. Light and heat energy are mutually convertible.
2. Light radiation is the same as heat radiation.
3. In the absence of humans or animals, there would be no light.
4. Light energy only exists within a visible spectrum.
5. Low-power laser beams are for heat energy.

35. Which is correct about units for lighting?

1. W/m^2.
2. Light intensity is 5.67×10^{-8} W/m^2K^4.
3. Lumens.
4. Flux in Webber/m^2.
5. Light intensity increases with distance from source.

36. Which is correct about lighting?

1. Light intensity varies inversely with the square of the distance from the source.
2. Light intensity varies inversely with the distance from the source.
3. Light intensity reduces with the distance from the source.
4. Light intensity varies inversely with the square root of the distance from the source.
5. Light intensity varies inversely with the distance cubed from the source.

37. Which is the recommended design illumination for office work?

1. 40 lx.
2. 400 lx.

3. 800 lm/m^2.
4. 2600 lx.
5. 200 lm/m^2.

38. Which is a valid technical reason for burning primary energy resources to provide artificial lighting?

1. Overcomes affect of dull sky.
2. Makes outside of building appear more impressive.
3. Human task illumination.
4. Good colour-rendering of exteriors.
5. Increases lighting available during daytime.

39. What has been the most significant benefit to humans since ancient ages?

1. Air conditioning in hot climates.
2. Refrigeration.
3. Central heating and cooling providing comfort.
4. Computers, mobile telephones and television.
5. Indoor electric lighting.

40. What is supplementary daylight?

1. Manually operated blinds or curtains.
2. Mechanically operated shading devices.
3. Perimeter fluorescent lighting.
4. Artificial lighting temporarily or permanently used to stabilize variations in day lighting.
5. Artificial lighting having the same colour-rendering temperature as daylight.

41. Which of these correctly describe *BZ* classification?

1. *BZ*0 means equal spherical illuminance.
2. *BZ*10 means light projection all downwards.
3. *BZ*5 projects only sideways and is used for wall-mounted luminaires.
4. *BZ*10 has a luminaire directing all light output upwards providing only reflection off the ceiling.
5. *BZ* defines the solid angle projection from a lamp in steradians.

42. What is illuminance?

1. Proportion of available light flux received on working plane.
2. Lux received on working plane.
3. Colour brightness index.
4. Luminous flux intensity reflected from a surface.
5. Brightness of perceived detail relative to background.

43. What is a light meter?

1. Easily portable data logger.
2. 1 lm/m^2 projected for 1 m.
3. Current-generating photocell calibrated in lux.
4. Current-generating photocell that activates an alarm in a BMS security control room.
5. Photocell that receives a laser beam from an emitter as part of a security, smoke or heat detection system.

44. What is a lumen?

    1. Unit of lighting not presently in use.
    2. 1000 candela/m^2.
    3. SI unit of light output or received.
    4. A directional measurement of light.
    5. Lighting power of a source.

45. What are lux?

    1. Total light output from a source.
    2. A measurement of glare.
    3. 31.4 candela.
    4. 1 lm/m^2
    5. A measurement of reflected light.

46. What does luminaire mean?

    1. A part of the lamp control system.
    2. The complete light fitting.
    3. A unit of light output flux.
    4. A unit of reflected light.
    5. Those parts of the lighting fitting other than the lamp.

47. Which is not true about fluorescent lamps?

    1. Low cost and reliable.
    2. Poor colour-rendering compared to daylight.
    3. Visible flicker may be observed from 50 Hz single-phase.
    4. Energy efficiency around 85 lm/wt.
    5. Colour temperature 4000 K and Ra8 of 85.

48. Where are sodium discharge lamps used?

    1. Low-pressure SOX sports facilities.
    2. Not used for street lighting.
    3. High-pressure SON produces industrial indoor white lighting.
    4. Where good colour-rendering is preferred.
    5. Regular replacement installations.

49. How does a fluorescent lamp work?

    1. Fluorescing powder-filled tube is electrically heated.
    2. Fluorescent lining of tube becomes heated.
    3. Vapour within tube fluoresces.
    4. Vapour within tube is electrically charged.
    5. Tube interior coating fluoresces when irradiated.

50. Which is a feature of an energy-efficient lighting control system?

    1. BMS reports maintained of all lighting system usage.
    2. All rows of luminaires remain switched on during working day as frequent starting uses more energy.
    3. Lamp deterioration increases with frequent switching, so should remain on continuously.
    4. Occupancy sensors are programmed to keep lights on for half an hour after occupants leave.
    5. Timed switch-off controller minimizes lighting use after occupation ceases.

12 Gas

Learning objectives

Study of this chapter will enable the reader to:

1. calculate the flow of gas required by an appliance;
2. identify gas pressures;
3. know how to measure gas pressure;
4. calculate gas pressures;
5. calculate gas pressure drops in pipelines;
6. calculate the equivalent length of a gas distribution pipe system;
7. use gas-pipe-sizing tables;
8. choose suitable gas pipe sizes;
9. describe gas service entry;
10. understand the working of a gas meter;
11. identify the space requirement for gas meters;
12. explain the flue requirements for gas appliances;
13. describe gas flue systems;
14. understand fan-assisted flue systems;
15. apply appropriate gas flue systems to designs;
16. understand how gas combustion is controlled and regulated;
17. know how gas systems are operated safely.

Key terms and concepts

density 281; efficiency 281; equivalent length 284; flue systems 286; gas burner controls 289; gas flow rate 281; gas meter 285; gas service entry 285; gross calorific value 281; ignition and safety controls 289; manufactured gas 281; natural gas 281; pipe size 285; pressure 282; pressure drop 283; pressure governor 289; specific gravity 281; U-tube manometer 282; ventilation 285.

Introduction

Gas services are provided to most buildings, and safety is of paramount importance. Economy is also important, and the versatility and controllability of gas are appreciated. It is used for heating, hot-water production, refrigeration for small and large cooling loads, electrical power generation, cooking and decorative heating.

Gas is converted from its primary fuel state into useful energy at the point of use. Its distribution energy loss is accounted for in the standing charge to the final consumer and, although it is charged for, does not appear to be related to the load as for water pipes or electricity cables. Gas is conveyed in a one-way pipe system and is not returned to the supplier as are electricity and water.

The use of natural gas and, ultimately, an artificially produced substitute from coal and oil is a highly efficient use of primary resources, and all efforts are directed at continuing this trend.

This chapter introduces the calculation of gas flow rate into a load and the gas pressure measurement that is used to monitor the flow rate and condition. The sizing of pipework depends on the gas pressure of the incoming service and that required by the final gas-burning appliance in order to maintain the correct combustion rate. Incoming gas service provisions, metering and particular flue systems are explained. Gas is more suitable than any other fuel for low-level flue gas discharge provided that sufficient dilution with fresh air is available. Typical methods of gas burner control and pressure reduction are explained.

Gas pipe sizing

The gas flow rate required for an appliance can be found from the manufacturer's literature or calculated from its heat output and efficiency:

$$\text{appliance efficiency } \eta = \frac{\text{heat output into water or air}}{\text{heat input from combustion of gas}} \times 100\%$$

Most gas appliances have an efficiency of 75%.

The gross calorific value (GCV) of natural gas (methane) is 39 MJ/m^3. If the appliance heat output is SH kW, then the gas flow rate Q required is

$$Q = \frac{SH \text{ kW}}{\eta} \times \frac{1 \text{ kJ}}{1 \text{ kWs}} \times \frac{1 \text{ MJ}}{10^3 \text{ kJ}} \times \frac{1}{GCV} \frac{m^3}{MJ} \times \frac{10^3 \text{ l}}{1 \text{ m}^3}$$

$$= \frac{SH}{\eta \times GCV} \text{ l/s}$$

The maximum allowable gas pressure drop due to pipe friction between the gas meter outlet and the appliance will normally be 75 Pa. Gas pressure in a pipe is measured with a U-tube water manometer as shown in Fig. 12.1. The water in the U-tube manometer is displaced by the gas pressure until the gas and water pressures in the two limbs are balanced. The atmosphere acts equally on both limbs under normal circumstances and all pressures are measured above atmospheric. These are called gauge pressures.

Methane has a specific gravity (SG) of 0.55 and so its density ρ is:

$$\rho_{gas} = SG \times \rho_{air}$$

Open to atmosphere

Rubber pipe

Test cock

H mm

Gas pipe

Datum-

Gas pressure P Pa

U-tube manometer

12.1 Measurement of gas pressure.

The standard density of air at 20°C d.b. and 1013.25 mb barometric pressure is 1.2 kg/m³. Thus

$$\rho_{gas} = 0.55 \times 1.2 \text{ kg/m}^3$$
$$= 0.66 \text{ kg/m}^3$$

The water column H in the manometer shows the net effect of the gas pressure P and the different density of the column H of gas in the right-hand limb. At the datum, the pressure exerted by the left-hand column equals the pressure exerted by the right-hand column.

$$P_{water} = P - \text{pressure of column } H \text{ of gas}$$

The pressure at the base of a column of fluid is $\rho g H$ where $g = 9.807$ m/s². The density of water ρ_w is 10^3 kg/m³. Therefore

$$\rho_w g H = P - \rho_{gas} g H$$
$$P = \rho_w g H - \rho_{gas} g H$$
$$= g H (\rho_w - \rho_{gas})$$

Thus

$$P = 9.807 \, \frac{m}{s^2} \times H \text{ mm} \times \frac{1 \text{ m}}{10^3 \text{ mm}} \times (1000 - 0.66) \, \frac{kg}{m^3}$$
$$= \frac{9.807 \times 999.34}{1000} - H \, \frac{kg}{ms^2} \times \frac{1 \text{ Ns}^2}{1 \text{ kgm}} \times \frac{1 \text{ Pa m}^2}{1 \text{ N}}$$
$$= 9.801 H \text{ Pa}$$

A sufficiently good approximation is $P = gH$ Pa, where H is in millimetres of water gauge.

EXAMPLE 12.1

A gas-fired boiler has a heat output of 280 kW and an efficiency of 82%. Calculate the flow rate of gas required.

$$Q = \frac{SH}{\eta \times GCV} = \frac{280}{0.82 \times 39} = 8.76 \text{ l/s}$$

EXAMPLE 12.2

Calculate the gas pressure if a water manometer shows a level difference of 148 mm.

$P = 9.807 \times 148 \text{ Pa}$

$= 1450 \text{ Pa} \times \dfrac{1 \text{ kPa}}{10^3 \text{ Pa}}$

$= 1.45 \text{ kPa}$

EXAMPLE 12.3

A gas-fired warm-air unit requires a gas pressure of 475 Pa at the burner test cock. What reading will be found on a water manometer?

$P = gH \text{ Pa}$

and so

$$H = \frac{P}{g} = \frac{475}{9.807} = 48.5 \text{ mm}$$

Gas pressure may be expressed in millibars (mb) as

1 bar $= 100\,000$ Pa

1 bar $= 100$ kPa

$1 \text{ mb} = \dfrac{100 \text{ kPa}}{1000} = 0.1 \text{ kPa}$

1 mb $= 100$ Pa

The gas pressure loss rate $\Delta p/EL$ along a pipeline is needed to find the pipe diameter. EL is the equivalent length of the measured straight pipe, bends and fittings. An initial estimate can be made for the flow resistance of pipe fittings by adding 25% to the straight lengths L m of pipe.

EXAMPLE 12.4

Calculate the pressure loss rate in a gas pipeline from the meter to a water heater. The pipe has a measured length of 34 m.

$$EL = (1.25 \times 34)\, m = 42.5\, m$$

The allowable pressure loss Δp is 75 Pa. Thus

$$\frac{\Delta p}{EL} = \frac{75\, Pa}{42.5\, m} = 1.76\, Pa/m$$

EXAMPLE 12.5

A pressure loss rate of 4.2 Pa/m is to be used for sizing a gas pipeline in a house. Calculate the maximum length of pipe that can be used if the resistance of the pipe fittings amounts to 20% of the installed pipe run.

$$\frac{\Delta p}{EL} = \frac{75\, Pa}{EL\, m} = 4.2\, Pa/m$$

Thus

$$EL = \frac{75}{4.2} = 17.9\, m$$

and

$$EL = 1.2\, l$$

Hence

$$l = \frac{17.9}{1.2} = 14.9\, m$$

This is the maximum length of run.

The gas pressure of the incoming service will be up to 5 kPa and this is reduced by a governor at the meter inlet to give 2 kPa in the installation within the building. For large gas-burning equipment the gas pressure may have to be increased with a booster. A boosting system comprises an electrically driven reciprocating compressor, a high-pressure storage tank and automatic pressure and safety controls. Pipe diameters can be found from Table 12.1.

EXAMPLE 12.6

Find the pipe size required for a gas service carrying 1.75 l/s and having an allowable pressure loss rate of 5.1 Pa/m.

Table 12.1 Flow of methane (natural gas) in copper pipes.

Δp/EL (Pa/m)	Gas flow rate Q (l/s) for pipe diameters of			
	15 mm	22 mm	28 mm	32 mm
1	0.08	0.31	0.69	1.22
2	0.16	0.47	1.05	1.84
3	0.21	0.59	1.33	2.34
5	0.29	0.81	1.8	3.15
7	0.35	0.98	2.2	3.83
10	0.44	1.21	2.7	4.71

Source: Reproduced from *IHVE Guide* (CIBSE, 1986 [IHVE, 1970]) by permission of the Chartered Institution of Building Services Engineers.

From Table 12.1, using the nearest pressure drop below 5.1 Pa/m, in this case 5 Pa/m, a 28 mm copper pipe can carry 1.8 l/s and would be suitable.

Gas service entry into a building

The gas service pipe from the road main should slope up to the entry point to the building, at right angles to the road main and entering the building at the nearest convenient place. Ground cover of 375 mm is maintained and new pipework is made of plastic. When old steel services are renewed, the plastic pipe can be run inside the steel.

A meter compartment can be built into the external wall in housing installations and the service clipped to the wall under a cover. This facilitates meter reading without entry to the property. Computer monitoring of energy meters using a telephone link to the supply authority will eventually replace manual reading.

Where the meter compartment is inside the building, the service should pass through the foundations in a pipe sleeve, plugged to stop the ingress of moisture and insects but allowing for some movement. A 300 mm^2 pit is provided in a concrete floor to allow the service to rise vertically to the meter. The pit can subsequently be filled with concrete.

The meter compartment must not be under the only means of escape in the event of a fire in a building where there are two or more storeys above the ground floor unless it is located in an enclosure having a minimum fire resistance of half an hour.

Gas service pipes, meters and appliances should always be in naturally ventilated spaces, as dilution with outside air is the best safety precaution against the accumulation of an explosive mixture with air. Early detection of leaks is essential, but ventilation assists the dilution of leaks. Gas detectors can be provided as an additional precaution.

Domestic credit meters pass up to 10 l/min, 0.17 l/s, and are 212 mm wide, 270 mm high and 155 mm deep. Their overall space requirement is approximately double the width and height measurements for pipework, valve and filter. Industrial meters have flanged steel pipework up to 100 mm in diameter and a bypass to allow uninterrupted gas flow in the event of meter breakdown. A 500 l/s meter is 2 m wide, 2.25 m high and 1.6 m deep. Due allowance must be made for doorways and access for replacing the meter during the building's use. A separate meter room is recommended, which should be secure, accessible, illuminated and weatherproof with no hot pipes or surfaces.

Manufactured town gas came from the conversion of coal or oil. It had a high hydrogen content and flame speed but its cross-calorific value was half that of methane. In future, substitute natural gas (SNG) may be manufactured from hydrocarbons as indigenous reserves become exhausted. SNG will come from the chemical conversion of coal, tar sand or crude oil and will have characteristics similar to those of methane.

Gas pipes or meters should usually be spaced 50 mm from electrical cables, conduits, telecommunications cables or other conductors. Electric and gas meters may be accommodated in a single compartment if a fire-resistant partition separates them.

Positive displacement mechanically operated meters are used as the billing meter. These meters have three compartments with a horizontal valve plate near the top of the casing and a vertical division plate in the lower section. Bellows formed by a metal disc surrounded by a leather diaphragm are located on each side of the division plate to measure the gas flow. Rotary meters may be used for downstream gas flow metering for energy management purposes and these may be logged by a BMS.

Flue systems for gas appliances

Gas appliances can be flued by a wide variety of methods, as the products of combustion are mainly water vapour, carbon dioxide, nitrogen and oxygen, at a temperature of about 95°C after the draught diverter. The function of the draught diverter is to discharge flue products into the boiler room during a down-draught through the chimney. Such reverse flows occur infrequently for a few seconds during adverse wind conditions. Diversion ensures that the correct combustion process is not interrupted. It stops the pilot flame from being blown out, with consequential shutdown of the appliance until manual ignition is arranged. The draught diverter also dilutes the products of combustion by entrainment of room air into the flue. A carbon dioxide concentration of 4% by volume is found in the secondary flue after the diverter. The primary flue pipes are those before the diverter. Flue systems are described below.

Brick chimney

New masonry chimneys must be lined with vitrified clay or stainless steel pipe. Existing chimneys may incorporate a stainless steel flexible flue liner, which can be pulled through an existing chimney with a rope and rounded plug. The liner has the same diameter as the appliance flue outlet, often 125 mm for domestic appliances, and is built into the top of the chimney with a plate to form a sealed air space between the liner and the brickwork. This acts as thermal insulation to maintain flue gas temperature. If the flue gases were allowed to cool to below about 25°C condensation of the water vapour would occur and deterioration of the metal and brickwork would reduce serviceability. Asbestos cement or glazed earthenware pipes can be built into new chimneys for protection of the brickwork. A cowl is fitted to the flue to reduce the ingress of rain and the possibility of down-draughts.

Free-standing pipe

Figure 4.22 shows a typical free-standing pipe flue, which can be used for a gas appliance. The pipe will be either asbestos cement or double-walled stainless steel with thermal insulation between the inner and outer pipes. A flue pipe taken through a roof is fitted with a lead slate to weatherproof the junction. The terminal should be 600 mm from the roof surface and clear of windows or roof-lights. An internal flue from a small domestic appliance can be connected to a ridge terminal. An externally run asbestos cement flue pipe has a branch tee junction at its emergence through the wall. A 25 mm copper drain pipe takes condensation to a drain gulley.

Balanced flue

Figure 12.2 shows the balanced-flue system used for boilers, warm-air units, convectors and water heaters. It is used for appliance ratings up to around 30 kW. External wind pressure is applied equally to the combustion air inlet and the flue gas outlet parts of the combined terminal. The only pressure difference causing air flow through the appliance is that caused by combustion. The flue terminal should not be underneath a window or within 0.5 m of a doorway or openable window. It should not be located in a confined corner where external air flow might be restricted. Fan-assisted balanced flues have been used and these allow more flexibility in siting the appliance further away from the terminal. Balanced-flue appliances are also called room-sealed appliances.

Se-ducts and U-ducts

Room-sealed appliances in multi-storey flats are connected to a vertical precast concrete shaft extending from the fresh air inlet grille at ground level to a terminal on the roof. Combustion air is taken from the duct by each heater and its flue products are passed into the shaft. The duct is sized so that sufficient ventilation is provided for the whole installation. With a U-duct a separate combustion air inlet duct takes air from the roof downwards to the lowest appliance, and then the upward duct acts in the same manner as the Se-duct.

Shunt duct

Precast concrete wall blocks, 100 mm wide, with a rectangular flue passage are built into partition walls or the inner leaf of a cavity wall. A continuous flueway is formed for each heater, often a gas fire, to ceiling level. An asbestos pipe then connects to a ridge terminal. Several flues built into a wall side by side are called a shunt duct system.

Fan-diluted flue

Fan-diluted flues are mainly used in commercial buildings where a conventional flue pipe and terminal could not be used or would be unsightly: for example, in a shopping precinct. Fresh air enters a galvanized sheet metal duct, which passes through the boiler plant room and discharges back into the atmosphere. A centrifugal or axial flow fan in the duct is started before the boilers

12.2 Balanced-flue gas appliance.

are ignited and an air flow switch cuts the burners off in the event of fan failure. Secondary flue pipes from the boilers are connected into the duct on the suction side of the fan. Dilution of the combustion products takes place and the discharge air from the system may contain as little as 0.5% carbon dioxide and be down to 30°C. Any condensing moisture is carried by the high-velocity air stream and is dispersed as steam into the atmosphere.

The air inlet and discharge louvres should be positioned on the same external wall to balance wind pressures on each. The discharge can be made into a shopping arcade or covered walkway to make use of the available heat. Careful fan selection is essential to avoid creating a noise nuisance. Figure 12.3 shows a fan-diluted flue installation for one boiler.

Boosted flue

A domestic boiler may have a booster centrifugal fan fitted into its flue pipe to allow a long horizontal run or even a downwards run. The pipe diameter can be smaller than the boiler flue outlet diameter and the fan pressure rise is used to overcome the frictional resistance. A typical installation is shown in Fig. 12.4.

12.3 Fan-diluted flue (reproduced by courtesy of Airdelle Products, High Wycombe).

12.4 Boosted flue (reproduced by courtesy of Airdelle Products, High Wycombe).

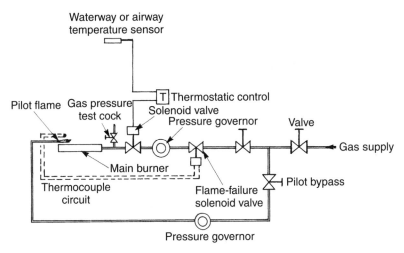

12.5 Gas burner controls.

Ignition and safety controls

Natural gas is burnt in an aerated burner in which half the air needed for combustion is entrained into the gas pipeline by a nozzle and venturi throat. This premixed gas and air goes to the burner, which is often a perforated plate through which the mixture passes. Further mixing occurs above the plate and the flame is ignited by a permanently lit pilot jet. A sheet of clear blue flame is established over the top of the burner plate or matrix. Large gas boilers, over 45 kW, use forced-draught burners in which gas and air are blown under pressure into a swirl chamber where the flame is established, with a fair amount of noise.

Gas burner control of appliances under 45 kW is achieved as follows. The pilot flame, which is ignited manually or with a piezoelectric spark, heats a thermocouple circuit whose electrical voltage and current holds open the flame-failure solenoid valve. In the event that the pilot flame is extinguished, the flame-failure solenoid becomes de-energized because the thermocouple is no longer heated, and the main gas supply is stopped. As long as the pilot is alight, the control thermostat in the boiler waterway or warm-air unit outlet duct is able to ignite the burner by opening its own solenoid valve. Figure 12.5 shows the diagrammatic arrangement of the control system for a gas burner of less than 45 kW. A combination gas valve is used to incorporate some of these functions into one unit.

The governor maintains a constant gas pressure to the burner by means of a synthetic nitrile rubber diaphragm which rises when the inlet pressure is increased. The diaphragm is connected to a valve, which closes when the diaphragm rises. This action increases the resistance to gas flow and maintains the outlet gas pressure at the set value. An adjustable spring is used to set the downstream pressure appropriate to the gas flow rate required by the burner.

Questions

1. A gas-fired water heater has a heat output of 30 kW at an efficiency of 75% and a gas pressure of 1225 Pa. Calculate the gas flow rate required at the burner and the reading on a U-tube manometer in millimetres water gauge at the outlet from the pressure governor.
2. Express gas pressures of 55 mm H_2O, 350 N/m^2, 75 Pa, 1.5 kPa and 1.05 bar in millibars.

3. The pipe from a gas meter to a boiler is 18 m long and has elbows that cause a resistance equivalent to 25% of the measured length. Calculate the maximum allowable pressure loss rate for the pipeline.

4. If the maximum allowable pressure loss rate in a pipeline can be 2.3 Pa/m and the resistance of the pipe fittings amounts to 20% of straight pipe, what is the maximum length of pipe that can be used?

5. A gas boiler of 43 kW heat output and 75% efficiency is supplied from a meter by a pipe 23 m long. The resistance of the fittings amounts to 25% of the pipe length. Find the gas supply pipe size needed.

6. Calculate the actual gas pressure drop through a 22 mm pipe carrying 0.81 l/s when the pipe length is 12 m and the fittings resistance amounts to 20% of its length.

7. Sketch and describe the gas service entry and meter compartment arrangements for housing.

8. Explain how a gas meter measures gas flow rate and total quantity passed during a year.

9. List the methods of flueing gas appliances and compare them in relation to their application, complexity and expected cost.

10. Explain, with the aid of sketches, the sequence of operation of safety and efficiency controls on gas-fired appliances.

11. Around what pressures do natural gas and liquefied petroleum gas run at in pipe distribution systems in buildings? More than one correct answer.

    1. Up to 40 atmospheres.
    2. Usually above 200 kPa.
    3. Up to 3000 mm water gauge.
    4. 20–200 mm water gauge.
    5. Less than 100 kPa.

12. Natural gas metering for billing is carried out with:

    1. Rotary gas meter monitored by the BMS.
    2. Flow rate measured from the pressure drop through an orifice plate in the pipeline.
    3. Positive displacement gas consumption meter.
    4. Pitot-static tube in pipeline and a data logger.
    5. Rotary vane anemometer in pipeline and an integrating revolution-counter meter.

13. What happens to water vapour in flue gas?

    1. Condenses when flue gas cools to water vapour dew-point temperature.
    2. Remains as vapour at all times.
    3. Cools and appears as steam discharging into atmosphere.
    4. Becomes absorbed into flue system materials and drains.
    5. Combines with other flue gases.

13 Electrical installations

Learning objectives

Study of this chapter will enable the reader to:

1. understand how electricity is generated and distributed;
2. know the difference between single- and three-phase electricity;
3. distinguish line, neutral and earth conductors;
4. calculate the resistance of conductors;
5. understand the temperature effect of a current;
6. calculate current and power in electrical circuits;
7. know how to measure current and voltage;
8. use power factor;
9. calculate series and parallel circuit resistances;
10. find the current capacity of cables;
11. choose cable sizes;
12. calculate permissible cable lengths;
13. understand temporary electrical installations for construction sites;
14. calculate the total electrical loading in kilovolt-amperes and amperes for installation;
15. estimate the total cost of electricity likely to be consumed in an installation during normal use;
16. choose the correct fuse rating;
17. understand the operation of fuses and circuit-breakers;
18. know the distribution of electricity within buildings;
19. identify the use of isolating switches, distribution boards and meters;
20. understand earth bonding of services;
21. know the types of cable conduits and their applications;
22. understand the principles of ring circuits;
23. understand how electrical systems are tested;
24. be aware of telecommunications cables;
25. design lightning conductor systems;
26. identify the graphic symbols used on drawings.

Key terms and concepts

alternator 292; apparent power 296; balanced load 292; bonding 294; cable-sizing calculations 299; cables, trunking and conduits 307; capacitor 296; cartridge fuse 305; distribution board 307; electrical measurements 310; electric shock 304; fault current 305; fuses and circuit-breakers 304; graphic symbols 314; IEE Wiring Regulations 299; kilowatts (kW) and kilovolt-amperes (kVA) 296; lightning conductor systems 312; line current 292; line, neutral and earth conductor 293; methods of testing 310; miniature circuit-breaker 305; national grid 292; Ohm's law 295; power factor (*PF*) 296; real power 296; residual current device 305; ring circuit 308; series and parallel circuits 296; single- and three-phase 292; specific resistance 294; telecommunications 312; temperature coefficient of resistance 294; temporary electrical installations on sites 300; triple pole and neutral switch 308; voltage, current, power and resistance 295.

Introduction

The safe and economical use of electricity is of paramount importance to the building user and the world as it is the most highly refined form of energy available. Electricity production consumes up to three times its own energy value in fossil fuel, and electricity in its distributed form is potentially lethal.

In this chapter the handling methods and safety precautions for utilizing electricity are explained and a range of calculations, which can easily be performed by the services designer or constructor prior to employing specialist help, is introduced.

Electricity distribution

The electrical power-generating companies supply electrical power into the national 400 kV grid system of overhead bare wire conductors. This very high voltage is used to minimize the current carried by the cables over long distances. Step-down transformers reduce the voltage in steps down to 33 kV, when it can be supplied to industrial consumers and to other transformer stations on commercial and housing estates.

The electricity-generating alternator rotates at 50 Hz (3000 rev/mm) and has three coils in its stator. The output voltages and currents from each coil are identical but are spaced in time by one-third of a revolution, 120°. Each coil generates a sine wave or phase voltage that has the same heating effect as a 240 V continuous direct current supply. This is its root mean square (RMS) value. The RMS value of the three phases operating together is 415 V.

Figure 13.1 shows the connections to a three-wire three-phase 415 V, 50 Hz alternating current supply entering a non-domestic building. Various circuits of different voltages are supplied from the incoming mains. Equal amounts of power are fed into each phase, and so it is important that power consumption within a building is equally shared by each line. The neutral wire is a live conductor in that it is the return path to the alternator for the current which has been distributed.

Figure 13.2 shows a single-phase from one of the generator coils where the effective voltage is the RMS value:

$$\text{RMS} = \sqrt{\frac{\sum V_n^2}{n}}$$

where *V* is the voltage at each of *n* measurements, say at 15° intervals.

A balanced load, such as a three-phase electrical motor driving an air-conditioning fan, water pump or lift motor, does not produce a current in the neutral wire. This is because an alternating

13.1 Wiring circuits from a three-phase 415 V incoming supply.

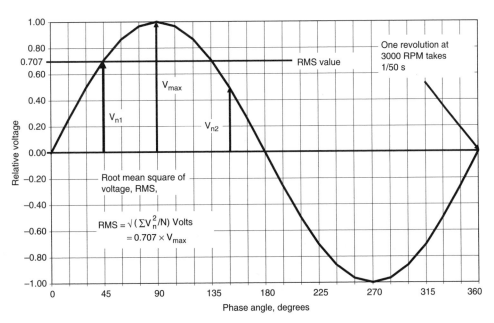

13.2 Single-phase RMS.

current flows alternately in the forward and backward directions along the line wire. The overall effect of three driving coils in the motor is a balance in the quantity and direction of the current taken from the line conductors. There is no net return current in the neutral wire from such a balanced load. Single-phase electrical loads, which are not in balance, produce a net current in the neutral conductor.

The casings of all electrical appliances are connected to earth by a protective conductor, the earth wire, connected to the earthed incoming service cable of the electricity supply authorities or an earth electrode in the ground outside the building. Gas and water service pipes are bonded to the earth by a protective conductor.

Circuit design

The resistance R ohms (Ω) of an electrical conductor depends on its specific resistance ρ Ωm, its length l m and its cross-sectional area A m^2. The specific resistance of annealed copper is 0.0172 $\mu\Omega$m (μ, micro stands for 10^{-6}) at 20°C.

$$R = \rho \frac{l}{A} \ \Omega$$

EXAMPLE 13.1

Calculate the electrical resistance per metre length at 20°C of a copper conductor of 2.5 mm^2 cross-sectional area.

$$R = \frac{0.0172}{10^6} \ \Omega\text{m} \times \frac{1 \text{ m}}{2.5 \text{ mm}^2} \times \frac{10^6 \text{ mm}^2}{1 \text{ m}^2}$$

$$= 0.0069 \ \Omega$$

The resistance of a cable increases with increase in temperature and the temperature coefficient of resistance (α) of copper is 0.00428 Ω/Ω°C at 0°C. If the resistance of the conductor is R_0 at 0°C, then its resistance at another temperature R_t can be found from:

$$R_t = R_0 (1 + \alpha t) \ \Omega$$

where t is the conductor temperature (°C).

EXAMPLE 13.2

Find the resistance of a 2.5 mm^2 copper conductor at 40°C.

R_0 is not known but the resistance of this conductor at 20°C was found in Example 13.1 and t can represent the increase in temperature above this value. A graph of resistance versus temperature would reveal a straight line of slope α.

$$R_{40} = R_{20} \, (1 + \alpha \times 20) \; \Omega$$
$$= 0.0069 \times (1 + 0.00428 \times 20) \; \Omega$$
$$= 0.0075 \; \Omega$$

The relation between applied voltage, electric current and resistance is given by Ohm's law:

$$I \text{ amps} = \frac{V \text{ volts}}{R \text{ ohms}}$$

Figure 13.3 shows how an ammeter and a voltmeter are connected into a circuit to measure power consumption. The load may be an electrical resistance heater or tungsten filament lamp, in which case the power consumption in watts is found from:

$$\text{power in watts} = V \text{ volts} \times A \text{ amps} \times \cos \phi$$

for single-phase and,

$$\text{power in watts} = V \text{ volts} \times A \text{ amps} \times \sqrt{3} \times \cos \phi$$

for three-phase, where $\cos \phi$ is the power factor (zero to unity). An electrical resistance heater and tungsten lamps are purely resistive loads whose power factor, $\cos \phi$, is unity, 1.

Loads such as electric motors and fluorescent lamps have a property known as inductance, which causes the current to lag behind the voltage that is producing that current. This is due to an electromotive force (emf), that is, a voltage which opposes the incoming emf. The densely packed electromagnetic windings of a motor have a high inductance and thus the 50 Hz cyclic variation of voltage and current is opposed by the 'inertia' of the equipment. The opposing emf comes from the expanding and collapsing magnetic fields of the input power around the conductors.

The lag angle $\phi°$ between peak voltage and peak current means that the instantaneous available power is less than the product of the two peaks. Figure 13.4 shows voltage applied across a circuit creating a current that lags in real time; peak current occurs after the peak voltage.

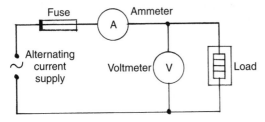

13.3 Measurements of power consumption in an electrical circuit.

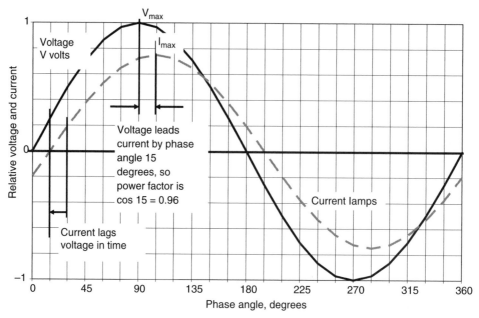

13.4 Phase angle lag.

Power factor is the term used to differentiate between useful output power in watts and the input instantaneous product of voltage and current to the load:

$$\text{power factor } PF = \frac{\text{useful power in watts}}{\text{input volts} \times \text{amps}} = \frac{\text{watts}}{\text{volt} \times \text{amps}}$$

$$= \frac{\text{real power (to do work)}}{\text{apparent power}}$$

$$PF = \frac{kW}{kVA}$$

$$= \frac{\text{kilowatts}}{\text{kilovolt-amps}} = \frac{\text{kW real}}{\text{kVA (apparent)}}$$

Figure 13.5 shows all three phases as they occur in real time, separated by 120° phase angle. Because of the three voltage sine waves within each cycle, the overall power generated is higher and smoother than with only a single-phase motor. Capacitors have an electrical storage capability, which is used to overcome the effects of inductance. Power factors of electrical equipment are commonly 0.85 and these can be improved to 0.95 by the addition of power-factor-correcting capacitors.

Several loads to a circuit may be connected either in series or in parallel with each other. For series-connected resistances:

$$R = R_1 + R_2 + R_3 + \cdots$$

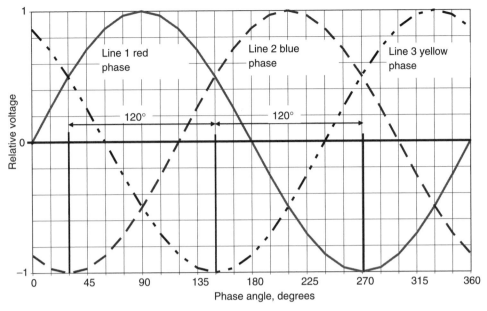

13.5 Three-phase.

For parallel-connected resistances:

$$\frac{1}{R} = \frac{1}{R_1} + \frac{1}{R_2} + \frac{1}{R_3} + \cdots$$

The resistance of electrical cables must be sufficiently low that the cables do not become significant sources of heat and run at temperatures that could be a fire hazard or damage their electrical insulation. Such heat generation would generally be wasted energy. The maximum voltage drop in a cable permitted in the Institution of Electrical Engineers (IEE) Wiring Regulations is 4% of the nominal supply voltage from the consumer's intake terminal to any point in the installation at full load current.

Cables that are grouped together, run in conduits or are covered with thermal insulation, say in a roof space, can operate at a temperature above the 30°C ambient condition assumed in the selection of their size and their electrical insulation. Where their temperature is likely to rise above this value, their current-carrying capacity is reduced by appropriate rating factors during design of the system. Care should be taken to allow natural cooling of all cable routes. The current-carrying capacity has to be 1.33 times the design current for cables partly surrounded by thermal insulation and twice the design current if they are wholly surrounded. This will generally mean an increase by one or two cable sizes.

EXAMPLE 13.3

Calculate the power consumption and resistance of a 240 V filament lamp if it has 1.5 A passing through it.

power in watts = volts × amps

$$= 240\,V \times 1.5\,A$$

$$= 360\,W$$

From Ohm's law:

$$I = \frac{V}{R}$$

and so,

$$R = \frac{V}{I}$$

$$= \frac{240}{1.5}$$

$$= 160\,\Omega$$

EXAMPLE 13.4

PVC insulation on a conductor carrying 415 V has an electrical resistance to earth of 500 MΩ. What leakage current could flow through the PVC when the cable is laid on an earthed metal support? (1 MΩ = 10^6 Ω.)

The difference between line and earth is 240 V. From Ohm's law

$$I = \frac{V}{R}$$

$$= \frac{240\,V}{500 \times 10^6\,\Omega}$$

$$= 0.48 \times 10^{-6}\,A$$

$$= 0.48\,\mu A$$

EXAMPLE 13.5

Compare the currents carried by an overhead line at 400 and 33 kV for the transmission of 10 MW of power for unity power factor, cos $\phi = 1$.

For a three-phase system, using:

watts = volts × amps × $\sqrt{3}$ × cos ϕ

current in amps = $\dfrac{\text{watts}}{\text{volts}} \times \dfrac{1}{\sqrt{3}} \times \dfrac{1}{\cos \phi}$

For 400 kV:

$$\text{current} = \frac{10 \times 10^6 \text{ W}}{400 \times 10^3 \text{ V}} \times \frac{1}{\sqrt{3}} = 14.4 \text{ A}$$

For 33 kV:

$$\text{current} = \frac{10 \times 10^6 \text{ W}}{33 \times 10^3 \text{ V}} \times \frac{1}{\sqrt{3}} = 175 \text{ A}$$

This demonstrates the advantages of high-voltage transmission of electrical power as smaller cable sizes can be used for the long distances involved.

Cable capacity and voltage drop

The maximum current-carrying capacities and actual voltage drops according to the LEE Regulations for Electrical Installations (16th edn, 1991) for unenclosed copper cables which are twin-sheathed in PVC, clipped to the surface of the building, are given in Table 13.1. Flexible connections to appliances may use 0.5 mm^2 conductors for 3 A and 0.75 mm^2 conductors for 6 A loads. The maximum voltage drop allowed is 4% of the 240 V nominal supply (Jenkins, 1991).

EXAMPLE 13.6

Find the maximum lengths of 1, 1.5 and 2.5 mm^2 copper cable which can be used on a 240 V circuit to a 3 kW immersion heater.

$$\text{current} = \frac{3000 \text{ W}}{240 \text{ V}} = 12.5 \text{ A}$$

$$\text{allowed voltage drop} = \frac{4}{100} \times 240 = 9.6 \text{ V}$$

$$\text{maximum length or run} = \frac{\text{maximum voltage drop allowed mV}}{\text{load current A} \times \text{voltage drop mV/A m}}$$

Table 13.1 Electrical cable capacities.

Nominal cross-sectional area of conductor (mm^2)	Maximum current rating (A)	Voltage drop in cable (mV/Am)
1	15	44
1.5	19.5	29
2.5	27	18
4	36	11
6	46	7.3
10	63	4.4
16	85	2.8

For 1 mm^2 cable:

$$I = \frac{9.6 \times 10^3}{12.5 \times 44}\, m = 17.5\ m$$

For 1.5 mm^2 cable:

$$I = \frac{9.6 \times 10^3}{12.5 \times 29}\, m = 26.5\ m$$

For 2.5 mm^2 cable:

$$I = \frac{9.6 \times 10^3}{12.8 \times 18}\, m = 42.7\ m$$

Construction site distribution

A list is made of all electrical equipment to be used on site in order to assess the maximum demand kilovolt-amperes and cable current rating required. An estimate of the cost of electricity for running the site may also be made.

EXAMPLE 13.7

A building site is to have the following electrical equipment available for use:

(a) tower crane, electric motors totalling 75 kW at 415 V;
(b) sump pump, 5 kW at 240 V;
(c) 60 tungsten lamps of 100 W each at 240 V;
(d) 12 flood lamps of 400 W each at 240 V;
(e) 20 hand tools of 400 W at 110 V.

1. Find the total kilovolt-amperes to be supplied to the site if the power factor of all rotary equipment is 0.8.
2. Find the electrical current rating for the incoming supply cable to the site.
3. Estimate the cost of electricity consumed on the site during a 12-month contract.

For rotating machinery:

$$\text{power, } VA = \frac{\text{useful power W}}{PF} = \frac{W}{0.8} \text{ and, } kVA = \frac{kW}{PF}$$

For single-phase current:

$$\text{line current} = \frac{VA}{V}\ A$$

$$= \frac{kVA}{kV}\ A$$

For three-phase current:

$$\text{line current} = \frac{VA}{V \times \sqrt{3}} \text{ A}$$

$$= \frac{kVA}{kV \times \sqrt{3}} \text{ A}$$

The results of the power calculations are given in Table 13.2.

The answers required are as follows.

1. The total input power kilovolt-amperes required for site is 120.8 kVA.
2. The incoming supply cable capacity at 415 volt, three-phase, 50 Hz required is:

$$\text{current} = \frac{120.8 \times 1000}{415 \times \sqrt{3}}$$

$$\text{current} = 168 \text{ A}$$

This is the input current to the site at the voltage of that cable. This is not the same as a total of the currents calculated in Table 13.2 as these larger numbers only appear at their reduced voltages in the relevant sub-circuits.

3. Assume that the crane, pump and tools are running for 25% of an 8-h working day, 5 days per week for 48 weeks, 20 of the tungsten lamps are for security lighting 16 h every night, and the remaining 40 tungsten lamps and the flood lamps are used for 3 h per day, 5 days per week for the winter period of 20 weeks. The crane, pump and tools are working for:

$$0.25 \times 8 \, \frac{h}{day} \times 5 \, \frac{days}{week} \times 48 \text{ weeks} = 480 \text{ h}$$

The security lamps are working for:

$$16 \, \frac{h}{day} \times 7 \, \frac{days}{week} \times 52 \text{ weeks} = 5840 \text{ h}$$

The other lamps are working for:

$$3 \, \frac{h}{day} \times 5 \, \frac{days}{week} \times 20 \text{ weeks} = 300 \text{ h}$$

Table 13.2 Building site plant schedule.

Equipment	Power (kW)	Number	kW	kVA	V	A
Tower crane	75	1	75	93.75	415	130.4
Sump pump	5	1	5	6.25	240	26
Lamps	0.1	60	6	6	240	25
Flood lamps	0.4	12	4.8	4.8	240	20
Hand tools	0.4	20	8	10	110	90.9
Total			98.8	120.8	n.a.	n.a.

Table 13.3 Building site energy use.

Equipment	Power (kW)	Number	Time (h)	Energy (kWh)
Tower crane	75	1	480	36 000
Sump pump	5	1	480	2400
Tungsten lamps, security	0.1	20	5824	11 648
Lamps	0.1	40	300	1200
Flood lamps	0.4	12	300	1440
Tools	0.4	20	480	3840
Total power used				56 528

The total energy used by the systems is found from,

kWh = number of appliances × kW per appliance × operation hours as shown in Table 13.3.

If electricity costs 8 p per unit (kWh) then the estimated cost for the 1-year contract will be:

$$\text{cost} = 8\frac{P}{kWh} \times 56\,528 \text{ kWh} \times \frac{£1}{100\,p}$$

$$= £4522$$

Site safety

Adequate safety in the use of electricity on site is essential and a legal obligation upon employers. The area electricity supply authority must be contacted before any site work, to establish the locations of overhead and underground power cables. Assume that all lines are live. Overhead lines are not insulated except at their suspension points. The Electricity At Work Regulations 1989 need to be consulted.

Roadways for site vehicles should be made underneath overhead cables by the erection of clearly marked goalposts, on each side of the cable route, through which traffic must pass. These goalposts form the entrance and exit from the danger area and are spaced at 1.25 jib lengths of the mobile crane to be used on site, or at a minimum of 6 m either side of the cables. Entry to the roadway other than through the goal posts is barred with wooden fencing or tensioned ropes with red and white bunting at high and low levels.

Underground cables that become exposed during excavations must remain untouched until the electricity authority has given advice. Safe working clearances will be ascertained at this time.

Hand lamps and tools are operated from a transformer at 110 V to reduce the damage caused by an electric shock. For work within tunnels, chimneys, tanks or drains, 25 V lamps, or battery lamps, are advised. Each portable appliance should be checked by the operator before use and also inspected and tested by a competent person at intervals not exceeding 7 days. Records of maintenance and safety checks should be kept.

A weatherproof cubicle is provided at the edge of the site by the main contractor for the electricity authority's temporary fuses and main switch. Site distribution cables are supported from hangers on an independent wire suspended between poles around the edge of the site, with spur branches to site accommodation and work areas. The minimum clear heights under cables should be 4.6 m in positions inaccessible to vehicles, 5.2 m in positions accessible to vehicles and 5.8 m across roads.

The site programme for the main contractor is as follows.

1. Arrange a pre-contract meeting between the executives responsible for the work.
2. List electrical requirements for all temporary plant.
3. Prepare layout drawings showing equipment siting and electrical loads, including site offices, stores, canteen, sanitary accommodation and illumination. Carefully site equipment to minimize interference with the construction work.
4. Apply to the electricity supply authority for a temporary supply to the site, stating maximum kilovolt-ampere demand and voltage and current requirements.
5. Provide the electrical distribution equipment: 415 V, three-phase, 50 Hz for fixed plant and movable plant fed by trailing cables; 240 V, single-phase, 50 Hz for site accommodation and site illumination; 110 V, single-phase, 50 Hz for portable lighting and tools; 50 or 25 V, single-phase, 50 Hz for portable lamps to be used in confined spaces and damp areas.
6. Once site accommodation is in place, ensure that a satisfactory semi-permanent electrical system is provided.
7. A competent electrician is to carry out all site work. His name, designation and location are to be prominently displayed on site, so that faults, accidents and alterations can be expedited. All plant and cables are regularly inspected and tested.
8. Display the electricity regulations placard and the first-aid instruction card.
9. Appraise the use of electrical equipment and distribution arrangements weekly to ensure that the most efficient use is made of the system. Idle equipment is returned to the supervised store.

Construction site electricity

Distribution equipment for site use is housed in weatherproof rugged steel boxes on skids. Built-in lifting lugs facilitate crane or manual transportation. The main items are as follows.

Supply incoming unit (SIU): a unit to house the electricity supply authority's incoming cable, service fuses, neutral link, current transformers and meters. Outgoing circuits of 100 A or more are controlled by triple pole and neutral (TPN) switches for three-phase and either cartridge fuses or residual current devices to break the circuits in the event of a current flowing to earth from a fault in a wire or item of plant.

Main distribution unit (MDU): a cubicle, which may be bolted to the SIU, providing a number of single- and three-phase outlets through weatherproof plugs and sockets. Each outlet is protected by a residual current device.

Transformer unit (TU): a unit providing 110 V single- and three-phase supplies with 65 V between any line and earth for safety in the use of hand tools. It may be coupled to an SIU. Each outlet circuit is protected by a residual current device.

Outlet unit (OU): easily portable distribution box fed from an MDU or TU for final connection of sub-circuits to tools, lighting or motors. Each circuit is protected by a residual current device and clearly labelled with its voltage, phases and maximum current capacity.

Extension outlet unit (EOU): similar to an OU but fed from a 16 A supply and circuit protection is by a cartridge fuse. It may have up to four 16 A outlets on a cubicle metal box.

Earth monitor unit (EMU): flexible cables supplying electricity to movable plant; may incorporate a separate pilot conductor in addition to the protective conductor to earth. A small current passes through the portable plant and the EMU via the pilot and protective conductors. If the earth conductor is broken, the EMU current is interrupted and the circuit is automatically isolated at its circuit-breaker.

13.6 Distribution of electricity during site construction.

Semi-permanent installations are run in metal-sheathed or armoured flexible cable. The metal sheath is permanently earthed as is the earth wire. An over-sheath of PVC or an oil-resisting and flame-retardant compound is provided.

Connections from outlet units to hand-operated tools and lighting systems are made in tough rubber-sheathed (TRS) flexible cable. Walkways and ladders must be kept clear of cables and the cables must be kept 150 mm from piped services. Cables under site roadways are installed in a temporary service duct, such as drain pipework, at a depth of 600 mm and with markers at each end of the crossing. Figure 13.6 shows an arrangement of a site's electrical distribution.

Safety cut-outs

An electric shock is sustained when part of the human body establishes contact between a current-carrying conductor and earth. It is also likely from contact with two conductors of different phase. Voltages of less than 100 V have proved fatal under certain circumstances. The size of the current depends on the applied voltage and the body's resistance to earth.

Rubber shoes or flooring greatly increase the resistance of the shock circuit. Body resistance with damp skin is around 1100 Ω at 240 V; thus a current of 218 mA could flow. At 55 V, body resistance is 1600 Ω and this could produce 34 mA. A current of 1–3 mA is generally not dangerous and can just be perceived. At 10–15 mA acute discomfort and muscle spasm occurs, making release from the conductor difficult. A current of 25–50 mA causes severe muscle spasm and heart fibrillation, and will probably be fatal.

Prolonged exposure to a shock current causes burns from the heating effect of the supply. If electric shock occurs, switch off the supply without contacting any metal component. If necessary, begin resuscitation and summon qualified medical assistance immediately.

During normal operation, current flows from the 240 V (or other nominal value) line, through the appliance and along the neutral conductor back to the power station alternator. The nominal drop of voltage across the appliance is 240 V. Should either the line or the neutral conductors come into contact with a conducting material that is earthed, owing to a wiring fault, the current will choose the lower-resistance path to earth on its return journey to the earthed alternator. Immediately, a higher current will flow and the appliance has become a shock hazard.

The increased heating effect of the fault current can be used to melt a rewirable or cartridge fuse at the appliance, its fault current being 60% above the stated continuous rating. High rupturing capacity (HRC) cartridge fuses have silver elements in a ceramic tube, which is packed with granulated silica. They allow for the high starting currents required for electric motors. The correct fuse rating must be used for each appliance to avoid damage to cables and buildings from overheating through the use of too high a fuse capacity. Fuse ratings are quoted in Table 13.4 and are found from

$$\text{fuse current rating} = 1.6 \times \frac{\text{appliance input VA}}{\text{circuit voltage}}$$

A faster-acting protection, with greater reliability, whose operation can be tested is provided by a miniature circuit-breaker (MCB), which opens switch contacts upon detection of an excess current. Circuit faults that cause a leakage to earth are detected by a residual current circuit-breaker (Fig. 13.7).

During normal operation, the line and neutral coils around the electromagnetic core generate equal and opposite magnetic fluxes, which cancel out. A current leakage to earth at the appliance reduces the current in the neutral conductor and a residual current is generated in the core by the line coil. This residual current generates an emf and current in the detector circuit, which in turn

Table 13.4 Fuse ratings for 240 V single-phase and unity power factor.

Power consumption (W)	Fuse required (A)
120	0.8
240	1.6
720	4.8
1200	8
3120	20.8
3600	24
7200	48

13.7 Residual current circuit-breaker.

energizes the trip solenoid, which opens the double-pole switch, or TPN switch in a three-phase circuit, and isolates the appliance.

Residual currents of 30 mA are set for sensitive applications and outdoor equipment. They are frequently used in addition to fuses. Pressing the test switch short-circuits the line coil and a residual current is generated in the core, tripping the residual current device for test purposes.

Fuses and circuit-breakers are selected for their time–current characteristics in relation to the risk that is being protected against. Figure 13.8 shows the performance curve for a cartridge fuse to British Standard 1361: 1971 type 1 having a 15 A continuous rating for domestic installations and a comparative miniature circuit-breaker (IEE Wiring Regulations for Electrical Installations, 16th edn, 1991). The horizontal and vertical scales of the graph are logarithmic. It can be seen that both devices will pass the design maximum 15 A current without opening the circuit. The cartridge fuse is designed to melt if a current of 46 A were to flow for a period of 5 s, or 97 A for 0.1 s. Other combinations of heating effect are in proportion. These points lie to the left of the fuse curve and do not cause it to break.

The MCB is designed to open the circuit when 60 A flows for between 0.1 and 5 s. A lower current value will take in excess of 20 s to open the contacts. A fault in the protected circuit that causes the current to rise above 60 A will open the MCB in less than 0.1 s. A miniature circuit-breaker and a residual current device (RCD) may be combined in one moulded casing to protect against excess current, short-circuit and a leakage to earth. The MCB has a bimetallic strip thermal and magnetic trip mechanism. The speed of operation of an RCD is typically 20 ms (Midland Electrical Manufacturing Company Limited).

13.8 Time–current characteristics of a cartridge fuse and a miniature circuit-breaker.

Electrical distribution within a building

The incoming cable, residual current device and meter are the property of the electricity supply authority. Underground cables are at a depth of 760 mm under roads, and enter the building through a large radius service duct of 100 mm internal diameter. A drainpipe can be used for this purpose, laid through the foundations and rising directly to the meter compartment. External meter compartments can be used. The meter should not be exposed to damp or hot conditions and the electricity supply authority's advice should be sought. Figure 13.9 shows a distribution system for a dwelling.

Each circuit has a fuse or circuit-breaker and the fused distribution board connects the neutral and earth protective conductors to the supply cable. Appliances have a cartridge fuse at their connecting plug. Three-phase distribution in a large building is shown in Fig. 13.10. Switches are used to enable separation of individual circuits as well as appliances. A fuse or circuit-breaker is always fitted on the live line so that the incoming current is disconnected.

Power socket outlets are fitted into a ring circuit as shown in Fig. 13.11. Care must be taken not to overload the circuit by connecting appliances whose total current consumption would exceed the 30 A limit, particularly in kitchens.

Types of cable used in distribution systems are as follows.

PVC-insulated and -sheathed

PVC-insulated and -sheathed cables consist of copper conductors of multi-stranded or solid wire having sizes from 1 to 16 mm^2 cross-sectional area. Single-, twin- or three-core cables, with or without earth wires, are used. They are among the cheapest cables available and can be pulled through conduits, trunking or holes bored in floor joists. Such holes are drilled 50 mm below the

13.9 Domestic electricity distribution.

13.10 Three-phase electricity distribution.

13.11 Ring main to socket outlets.

floorboards. Ambient temperature limits for the cable are 0–65°C and the cable must not exceed 70°C in use. Colour coding of the insulation ensures correct polarity at terminals. The earth, or protective conductor, is always green, neutral is always black and the line conductor is red on single-phase. Three-phase line conductors are one red, one yellow and one blue.

PVC is a flexible, non-hygroscopic, tough, durable, corrosion-resistant and chemically inert material, which is used for both electrical insulation and conduits. Installations are tested for earth leakage through their insulation at regular intervals and cables are replaced after about 20 years because of ageing of the PVC.

Mineral-insulated copper-sheathed

Solid copper conductors surrounded by compressed magnesium powder are factory-fitted inside a copper tube or sheath. A PVC over-sheath gives extra protection for cables that are to be buried in building materials. These cables are used for both internal and external wiring and withstand severe conditions, even continuing to operate during a fire. They can be operated continuously at temperatures of up to 250°C compared with only 70°C for PVC-covered cables.

The soft copper cable is supplied in rolls and is run continuously from the distribution board to the switch or power point. Screwed gland joints are designed to exclude dampness from the hygroscopic insulant. One sheath may encapsulate up to nineteen 1.5 mm^2 conductors. They are non-ageing and unlikely to require replacement during the building's period of service.

Particular applications are fire alarm systems, in petrol filling stations and within boiler plant rooms. The copper sheath is used as the earth protective conductor and also withstands severe abuse, flattening or twisting without short-circuiting the conductors. The cables can be bent by hand or machine, and conduit fittings are not needed. The overall diameters of mineral-insulated cables are much less than those of other types of comparably rated cable system. Only a thin plaster covering is needed if the cables are not to be surface mounted.

Armoured PVC-insulated, PVC-sheathed

Copper or aluminium conductors in PVC insulation, a PVC bedding, galvanized steel wire armour and a PVC sheath are used for heavy-current cables to large machinery and mobile plant on site. The cable can be run on the building surface, laid on the ground or put in a trench. Screwed gland nuts are used to bond the armour to the appliance casing.

Busbar

Bare copper or aluminium rectangular bar conductors are supported on insulators within a sheet metal duct or trunking. Vertical service shafts within buildings may have a rising busbar system with tap-off points at each floor level for the horizontal distribution of power with insulated cables. Small busbar distribution systems can be used in retail premises and within raised floors in office buildings. These provide flexibility in the siting of outlet boxes.

Overhead distribution in factories allows connections at any point along the busbar with plug-in boxes, allowing machinery to be moved at a later date. The trunking acts as the protective conductor. Three-phase 415 V supplies are distributed at current ratings of 100–600 A, branches being at 30, 60 or 100 A. An armoured PVC cable can supply the incoming end of the busbar. Where the system passes through a fire-resistant partition in the building, a fire barrier is fitted across the inside of the trunking.

Other cable insulants

Flexible external and special application cables are used as follows.

Butyl rubber: for up to 85°C continuous plus higher overload rating. Additional heat-resisting glass fibre wrapping increases the continuous temperature to 100°C. Flexible cord use.

Ethylene propylene rubber (EP rubber): an elastomer with similar properties to butyl rubber. Has improved resistance against the effects of water and long-term ageing. Retains its flexibility down to −70°C.

Silicone rubber: withstands −75°C and a wide variety of chemical and oxidizing agents, weak acids, salts and vegetable oils. It retains its insulation and elasticity at working temperatures of up to 150°C.

Polychloroprene (PCP): for general purpose heat-resisting, oil-resisting and flame-retardant (HOFR) use in the presence of oil and petrol.

Chlorosulphonated polyethylene (CSP): heavy duty use in aggressive atmospheres in laundries or in the presence of oil and petrol.

Conduit and trunking

Circular conduit systems are used to carry insulated cables and should last the service period of the building. The space occupied by the cable must not exceed 40% of the cross-sectional area of the inside of the conduit to allow for ventilation to remove the heat generated by cable resistance.

Materials used are light- or heavy-gauge steel, depending upon exposure to damp or explosive fumes. The external conduit will be galvanized. Lug grip connections are used for light-gauge pipework and screwed joints for heavy-gauge pipework. Pipe sizes are 16, 20, 25 and 32 mm. The conduits are used as the protective conductor.

PVC conduit, using solvent weld joints, is lighter and easier to handle and does not corrode but requires the cable to incorporate the protective conductor. Its upper temperature limit is 60°C.

Rectangular galvanized sheet steel or PVC trunking is used where large-cable carrying capacity is needed. These must not be filled to greater than 45% of their cross-sectional area with cables. Surface-mounted trunking can be incorporated into the interior decoration and up to three separate cable compartments are used for different services, including telecommunications, computer, power and lighting cables.

Trunking may be installed under raised timber flooring, within the concrete floor slab or screed, in a grid, branch duct or perimeter distribution arrangement. Outlets that are raised or flush with the floor are provided to suit either fixed or movable office layouts.

Testing

Inspection and testing of an electrical installation is carried out before it is put into service and at regular intervals during use. The main reasons for this are to ensure that its operation will be entirely safe, in accordance with the demands put upon it, and energy-efficient. The work entails the following tests.

Power measurement

The power consumed by a single of three-phase electrical system or item of plant, such as a refrigeration compressor, fan or pump motor, or a lighting or power sub-circuit, is measured with a portable instrument as shown in Fig. 13.12. Newly installed plant is checked for equal phase currents, voltage and power consumed during commissioning. The energy auditor measures these some years after installation, as commissioning data may no longer be relevant or plant may have been changed. The logger should also measure power factor so an assessment can be made of whether to install power factor capacitors to reduce incoming current. Each phase line conductor has one of the openable clamps fitted. Voltage difference across the circuit is measured

13.12 Electrical power measurement logger (reproduced by courtesy of Mobile Architecture and Built Environment Laboratory, Deakin University, Geelong).

with additional cables and attachment clips. Both useful power in kW and apparent power in kVA are measured. All work on live electrical wiring is only done by a registered electrician.

Verification of correct polarity

A visual inspection is carried out of all fuses and switches to check that they are fitted into a line conductor. The centre contact of each screw lamp holder is connected to the line conductor. Plugs and sockets must be correctly connected and wire rigidly held.

Tests of effective earthing

There are four separate tests:

Test of the protective conductor: A 40 V 50 Hz supply of up to 25 A is injected into the earth conductor. Its resistance is not to exceed 1 Ω. An impedance test meter is used.

Earth loop impedance test: A line-earth loop impedance test meter is attached to a 13 A three-pin plug. This is plugged into each power socket and the meter injects a current into the earth protective conductor. The current flows along the supply authority's cable sheath to the local transformer and back to the power socket along the line conductor.

Test of residual current devices: A test transformer providing 45 V is connected to a socket outlet. A short-circuit current is passed from the neutral to the protective conductor, causing the residual current device to trip instantaneously.

Measurement of consumer earth electrode resistance: Where this is used, a test electrode is put into the ground and a steady 50 Hz current is passed between the electrode and the consumer's earth electrode to determine its circuit resistance.

Insulation resistance tests

An insulation test meter is connected between the line and protective conductors. A 500 V direct current is applied to this circuit by the meter and an electrical resistance of 0.5 MΩ or more must be shown.

Test of ring circuit continuity

Each ring circuit is tested for resistance at the distribution board with an ohmmeter. Probes are connected to each side of the line conductor ring and a zero resistance proves a continuous circuit. The test is repeated on the neutral and protective conductor circuits and spur branches.

Tests on installations must be carried out in accordance with the IEE Wiring Regulations by a competent person, who should preferably be a professionally qualified electrical engineer having installation experience. IEE Completion and Inspection Certificates are issued by the engineer.

Telecommunications

Cables between the switchboard and socket for each telephone are accommodated within vertical and horizontal service ducts spaced 50 mm from alternating current cables to avoid speech interference. Alternatively, a partitioned chamber can be reserved throughout the cable trunking.

Lightning conductors

Rules are provided (BS Code of Practice 326: 1965) to determine whether a protection system is required. This depends on building construction, degree of isolation, height of the structure, topography, consequential effects and lightning prevalence. Recommendations on system types, including those for temporary structures, are given.

Copper and aluminium 10 mm rod, 25 mm × 3 mm strip, PVC-covered strip, copper strand and copper braid are used for conductors. The air terminal is sited above the highest point of the structure and a down conductor is bolted to the outside of the building so that side flashing between the lightning conductor and other metalwork will not occur.

Ground termination is with a series of earth rods driven to depths of up to 5 m, cast iron or copper plates 1 m square horizontally or vertically oriented 600 mm below ground, or a copper lattice of flat strips 3 m × 3 m at a depth of 600 mm. Where large floor areas containing earth rods are to be concreted, a precast concrete inspection pit is built over the rod location.

The electrical resistance to earth of the whole system is not to exceed 10 Ω (Butler, 1979a). Calculation of the ground earthing resistance R requires a knowledge of the earth type (BS Code of Practice 1013: 1965) and resistivity. Typical values of earth resistivity are 10 Ωm for clay, 50 Ωm for chalk, 100 Ωm for clay shale and 1000 Ω for slatey shales. The resistance of a rod electrode in earth is:

$$R = \frac{0.37\,\rho}{l}\log\left(\frac{4000\,l}{d}\right)\ \Omega$$

where l is the earth rod length (m), d is the earth rod diameter (mm) and ρ is the resistivity of the soil (Ωm). A number of rods are connected in parallel and spaced 3.5 m apart to provide the required resistance.

EXAMPLE 13.8

Design a lightning conductor system for a building 30 m high in an area where thunderstorms are expected. The ground has a high chalk content and rod electrodes 4 m long are to be used. The conductors are to be 25 mm × 3 mm copper strip. The specific resistance ρ of copper is 0.0172 $\mu\Omega$m.

length of conductor = air terminal + down conductor + ground lead

Take the length of the conductor as 40 m,

resistance of conductor $R = \rho \dfrac{l}{A} \, \Omega$

where A is the conductor cross-sectional area (m^2); hence:

$$R = 0.0172 \times 10^{-6} \, \Omega m \times \frac{40 \text{ m}}{(0.025 \times 0.003) \text{ m}^2}$$

$$= 0.092 \, \Omega$$

resistance R of earth electrode $= \dfrac{0.37 \, \rho}{l} \log \left(\dfrac{4000 \, l}{d} \right) \Omega$

where the earth resistivity ρ is 50 Ωm, the electrode length l is 4 m and the electrode diameter d is 10 mm; hence:

$$R = \frac{0.37 \times 50 \, \Omega m}{4 \text{ m}} \times \log \left(\frac{4000 \times 4}{10} \right) \Omega$$

$$= 4.625 \times \log 1600$$

$$= 14.819 \, \Omega$$

The resistance of one electrode in the ground plus the down conductor is greater than the 10 Ω allowed, and so we find the combined resistance of two electrodes connected in parallel:

$$\frac{1}{R} = \frac{1}{R_1} + \frac{1}{R_2} = \frac{1}{14.8} + \frac{1}{14.8} = 0.135$$

$$R = \frac{1}{0.135} = 7.4 \, \Omega$$

Two electrodes and the down conductor connected in series have a total resistance of:

$$R = (7.4 + 0.0092) \, \Omega = 7.4 \, \Omega$$

This is less than the 10 Ω allowed and is satisfactory. The resistance of the lightning conductor is negligible in relation to that of the earth electrodes. The calculations have been made on the assumption that the lightning discharges in a direct current. Lightning energy can produce a current to earth of 20 000 A for a few milliseconds. It causes physical damage to building structures, starts fires in combustible materials and injury to people, sometimes fatal.

Graphical symbols for installation diagrams

Some of the symbols used on drawings of electrical installations are listed in Fig. 13.13 as in accordance with BS 3939: 1985, Guide for Graphical Symbols.

Joint or junction box	
Lamp	
Single fluorescent lamp	
Time switch	
Switched socket outlet	
Intake and control point	
Main switch	
kWh meter	
Consumer's earthing terminal	
Electricity appliance	
Heater	
Telephone call point	
Earth	
Single-pole switch	

13.13 Drawing symbols for electrical installations.

Questions

1. Explain how electricity is generated and transmitted to the final user.
2. List the sources of energy used for the generation of electricity and state their immediate and long-term benefits.
3. Explain, with the aid of sketches, the meaning of the terms 'single-phase' and 'three-phase' electricity supplies and show how they are used within buildings.
4. What does 'balancing the phases' mean?
5. Calculate the electrical resistance per metre length at 20°C of a copper conductor of a 10 mm^2 cross-sectional area.
6. Find the electrical resistance of a copper conductor 1.5 mm^2 in cross-sectional area if its total length is 25 m and its temperature is 20°C.
7. A 28 m copper conductor 4 mm^2 in cross-sectional area is covered with thermal insulation, which causes the cable temperature to rise to 45°C. Calculate the percentage increase in electrical resistance compared with its value at a cable temperature of 20°C.
8. Sketch the methods of connection used for measurements of current, voltage drop and power consumption in an electrical resistance heater on an alternating current circuit.

9. State the function of power factor correction in alternating current circuits.
10. Calculate the apparent power, in kilovolt-amperes, of an electric motor which is connected to a 415 V AC three-phase supply and has a current flow of 17.5 A.
11. Calculate the resistance of a 3 kW immersion heater on a 240 V AC circuit.
12. What current, in milli-amperes, would flow to earth during an insulation resistance test when a 500 V DC emf is applied between the line and protective conductors and the resistance is found to be 1.75 MΩ?
13. Show by sample calculation why smaller cables can be used for long-distance power transmission when very high voltages are used.
14. A 33 kV supply to a factory carries 250 A per phase or line. Calculate the usable electrical power in the factory if the average power factor is 0.68.
15. Find the maximum length of 6 mm^2 cable that can be used if the maximum current-carrying capacity is to be utilized on a 240 V circuit.
16. A building site is to have the following electrical equipment in use each day:

    (a) concrete mixer, 5 kW, 4 h, 415 V;
    (b) sump pump, 1.5 kW, 6 h, 240 V;
    (c) 20 lamps, 150 W each, 4 h, 240 V;
    (d) 5 flood lamps, 300 W each, 4 h, 240 V;
    (e) 6 hand tools, 750 W each, 5 h, 110 V.

    The power factor of the machinery is 0.8. Site work takes place 5 days per week for 28 weeks. Electricity costs 6 p per unit. Find

    (i)   the total kilovolt-amperes and the line current of the required temporary incoming supply system, and
    (ii) the cost of the electricity used on site during the contract.

17. Sketch and describe the safety precautions taken to avoid contact with both overhead and underground electricity cables during site construction work.
18. Sketch a suitable arrangement of temporary wiring, control and safety equipment on a site where the following items are employed: tower crane, sump pump, five flood lamps, security lighting and circuits on each of three floors for hand lamps and tools.
19. List the site programme for the main contractor in the installation and operation of temporary site electrical services.
20. Sketch and describe the characteristics of rewirable and cartridge fuses and residual current devices.
21. Show how an underground electrical service cable enters a building. Sketch the arrangement of electricity distribution within a typical residence.
22. List the cable and conduit systems used for electricity distribution and state their applications.
23. Briefly describe the methods of testing electrical installations.
24. State the requirements of telecommunications installations.
25. Sketch and describe a lightning conductor installation for a city centre office block.
26. Design a lightning conductor system for a 60 m high building on clay shale using 4.5 m earth rods. The down conductor is to be a copper rod 10 mm in diameter.
27. Why is power factor in electrical systems an issue for concern? More than one correct answer.

    1. It is not of any concern.
    2. Low-power factor means electrical energy is used inefficiently.
    3. High-power factor means electrical energy is wasted.

4.   100% power factor is not usually attainable at a justifiable cost.
5.   Low-power factor means electrical supply system becomes oversized.

28. What is a low-power factor?

1.   Zero
2.   0.95
3.   0.85
4.   1.0
5.   0.8 and below.

29. Which is the correct formula for power factor?

1. $PF = \dfrac{kVA}{kVAh}$

2. $PF = \dfrac{kVAh}{kWh}$

3. $PF = \dfrac{kW}{kVA}$

4. $PF = \dfrac{\text{input energy}}{\text{kilovolt-ampere}}$

5. $kW = \dfrac{PF}{kVA}$

30. How does a 30 mA residual current circuit-breaker work? More than one correct answer.

1.   Line and neutral conductors wrap around an electromagnetic core and no current flows through the core as magnetic flux from each wire cancels the other.
2.   Line and neutral conductors wrap around an electromagnetic core; current flows through the core due to magnetic flux from each wire.
3.   Line and neutral conductors wrap around an electromagnetic core; current flows through the core due to magnetic flux induced from line conductor.
4.   Line and neutral conductors wrap around an electromagnetic core; magnetic flux circulates harmlessly through the core due to the alternating current line and neutral conductors.
5.   Current leakage to earth from the protected appliance loses neutral current, causing imbalance between line and neutral magnetic fluxes in core; imbalance flux generates 30 mA in trip solenoid circuit breaker.

31. How can cartridge fuses cope with high starting currents at electric motors?

1.   They cannot.
2.   Only micro-circuit-breakers are used to protect motors driving compressors.
3.   High rupturing capacity cartridge fuses regularly pass 500% of normal running current.
4.   High rupturing capacity cartridge fuses have silver elements and packed with silica to allow high starting currents for a known duration.
5.   High rupturing capacity cartridge fuses have bimetallic elements and packed with carbon granules to allow high starting currents.

32. Which are correct about mineral-insulated copper-sheathed electrical cables? More than one correct answer.

1.   Fragile and easily damaged.
2.   Withstand most fires and remain operational.
3.   Supplied in hard copper fixed lengths, like plumbing pipes.
4.   Screwed gland joints exclude water from the hygroscopic insulant.
5.   Cannot be installed outdoors.

33. What is meant by busbar? More than one correct answer.

    1. Communications bus, C-bus, in a computer.
    2. Ethernet communications cabling system around a large network.
    3. PVC-insulated circular bar conductors at high level through an industrial manufacturing plant.
    4. Bare copper or aluminium bar conductors carried on insulators within sheet metal trunking.
    5. Up to 600 A three-phase vertical or horizontal distribution allowing off-takes along length.

34. What is the meaning of inverter drive?

    1. An electric motor installed in an inverted position.
    2. Three-phase electric motor.
    3. Three-phase motor running in single-phase.
    4. Digitally driven motor.
    5. Motor driven at variable alternating current frequencies.

35. Which is correct about electrical systems?

    1. Watts = volts × amps.
    2. Three-phase current = 3 × single-phase current.
    3. Watts and volt-amperes are always the same value.
    4. Current = voltage × resistance.
    5. Current measured in mega-ohms.

36. How is power consumed by an electrical item measured?

    1. Wattmeter cut into the line conductor.
    2. Voltmeter and current meter both cut into line conductor to the load.
    3. Ammeter connected into the line conductor to the load and a voltmeter connected in parallel with the load.
    4. Voltmeter connected into line conductor to the load and an ammeter connected in parallel with the load.
    5. Wattmeter connected in parallel with the load.

37. Which system of public electricity is provided to small- to medium-sized non-domestic buildings in the UK and Australia?

    1. 50 Hz, 240 V, AC.
    2. 210 V, single-phase, 55 Hz, alternating current.
    3. Two-phase, 60 Hz, 200 V.
    4. 440 V, three-phase, 60 Hz.
    5. AC, 50 Hz, three-phase, 415 V.

38. How is three-phase electricity created at a power station?

    1. Three single-phase alternators have interconnected output power circuits.
    2. Single-phase generators with rectifiers creating three-phases.
    3. Multiple transformers create three-phases.
    4. Each alternator has three stator coils so one revolution of the rotor generates three separate sine wave outputs.
    5. Three single-phase alternators each power one line voltage; each building takes all three lines to have three-phase power.

39. Which of these does a single-phase electricity current look like on an oscilloscope screen? (Hint, draw them.)

    1. Zero phase angle zero current, 90° phase angle maximum positive current, 180° phase angle zero current.
    2. Zero phase angle maximum positive current, 90° phase angle zero current, 180° phase angle maximum negative current.
    3. Zero phase angle 50% maximum positive current, 90° phase angle 100% positive current, 180° phase angle zero current.
    4. 135° phase angle zero current, 225° phase angle maximum negative current, 315° phase angle zero current.
    5. Zero phase angle zero maximum negative current, 90° phase angle zero current, 180° phase angle maximum positive current.

40. Which correctly describes the relationship between voltage and current in an alternating current system?

    1. Voltage and amperes are synchronized.
    2. Voltage peak occurs behind peak value of current.
    3. Current always follows voltage producing it by exactly one phase.
    4. Inefficient generators produce a current flow lagging voltage.
    5. Current always follows fractionally behind the voltage that produces flow of electricity.

41. What is the advantage of a three-phase system?

    1. Widely variable phase current meets variable demands.
    2. Continuously stable power supply.
    3. More easily generated.
    4. Can generate at any desired frequency.
    5. Quieter than single-phase.

42. How should the services in a building take power from a three-phase supply?

    1. Each phase serves a different part of the building.
    2. The mechanical services distribution board always takes all its power from the yellow phase.
    3. Single-phase circuits take current from each phase.
    4. Equal current taken from each phase.
    5. 240 V circuits for lighting and small power equipment each connect to all three phases.

43. Where are cartridge fuses or MCBs always installed?

    1. Live phase cable.
    2. Neutral wire.
    3. Earth conductor.
    4. External to the building.
    5. Within a fire-resistant switchboard.

44. Which is true about three-phase motors?

    1. Run hot.
    2. Need built-in cooling fan.
    3. Generate more noise than an equivalent single-phase motor.

4. Quieter than single-phase motors.
5. Provides more power output for same line current as a single-phase motor.

45. What opposes flow of electrical current into an electric motor?

1. Resistance of the motor coils.
2. Inductance.
3. Temperature coefficient of resistance increases circuit electrical resistance.
4. Mechanical feedback from forces on motor output shaft.
5. Capacitance of motor control system.

46. How is electrical energy consumed by an item of plant, equipment or a whole building measured?

1. Magnetic field-sensing data logger is strapped to the single or three-phase cable.
2. Kilowatt meter measures instantaneous current flow and applied voltage.
3. Integrating data logger multiplies output signal from a magnetic current transducer with voltage applied at the same time.
4. Ammeter reading multiplied by voltmeter reading divided by the time in seconds of their duration is calculated and added to a running total of energy consumed.
5. Integrating meter multiplies instantaneous current and voltage with the time duration and records kWh consumed.

47. What does self-induced electromotive force do to an electrical circuit?

1. Nothing.
2. Speeds up current flow.
3. Opposes incoming current and causes it to lag behind voltage in real time.
4. Opposes incoming current and causes it to appear leading the cyclic pattern of the driving voltage frequency.
5. Supports the incoming current frequency and increases the current.

48. What does inductance do to an electrical system?

1. Speeds up current.
2. Multiplies available power by a percentage.
3. Reduces current.
4. Reduces available voltage.
5. Causes current to lag behind applied voltage.

49. What is the time difference between voltage and current in an AC system called?

1. Lead angle.
2. Microsecond gap.
3. Phase.
4. Peak difference.
5. Lag.

50. How is electrical power factor raised?

1. Cannot be improved after equipment installation.
2. Replace with more efficient specification motor.
3. Install digitally controlled AC/DC rectifiers at mechanical switchboard.
4. Install capacitor banks in parallel with each plant item.
5. Renegotiate electrical supply contract.

51. Which is a typical time interval for a residual current device to open a 60 A circuit-breaker double-pole switch when a fault current occurs?

    1. 6 min.
    2. 1 min.
    3. 1 s.
    4. 20 ms.
    5. Less than 0.001 s.

52. What does RCD stand for?

    1. Residual circuit device.
    2. Residual current device.
    3. Resistance circuit design.
    4. Radio carbon dating.
    5. Ratio circuit device.

53. Which is the most common form of mortality from electric shock?

    1. Burns.
    2. Ventricular fibrillation.
    3. Muscle spasm.
    4. Pain.
    5. Bleeding.

54. Which is not a normal application for use of MICC wiring?

    1. Fire alarms systems.
    2. Public buildings such as theatres.
    3. Tunnels.
    4. Temporary buildings.
    5. Power stations and heavy industrial buildings.

55. Why are cables installed within fixed conduit?

    1. Hides ugly cables.
    2. Conduit is a permanent fixture of the building while cables require replacement when aged.
    3. Conduit becomes earth continuity conductor.
    4. Reduce heat emission from cables.
    5. Protects PCV cables from heat gain from environment.

56. Which is correct about electrical systems?

    1. Watts = volts × amps.
    2. Three-phase current = 3 × single-phase current.
    3. Watts and volt-amperes are always the same value.
    4. Current = voltage × resistance.
    5. Current measured in mega-ohms.

14 Room acoustics

Learning objectives

Study of this chapter will enable the reader to:

1. know the potential sources of sound and vibration within buildings;
2. know what is meant by noise;
3. understand how sound travels through a building;
4. understand what is meant by sound pressure wave, sound power level and sound pressure level;
5. know how to calculate sound pressure levels for normal building services design examples;
6. use sound levels at the range of frequencies commonly used in building services engineering;
7. understand how sound and vibration are transmitted through buildings;
8. be able to identify the need for sound attenuation vibration isolation;
9. understand and use the decibel unit of measurement of sound energy;
10. know the meaning and use of direct and reverberant sound fields;
11. calculate the sound pressure level in a plant room, a space adjacent to the plant room, in the target occupied room and in the external environment outside the plant room;
12. use logarithms to base 10 in acoustic calculations;
13. understand the principle of sound absorption;
14. calculate the sound absorption constant for a room at different frequencies;
15. know the sound absorption coefficients for some common building materials and constructions;
16. understand and use reverberation time and attenuation;
17. calculate sound pressure levels at different frequencies within a plant room;
18. know what a reverberant room and an anechoic chamber are;
19. use directivity index sound absorption coefficients, mean absorption coefficient and room absorption constant;

20. understand the behaviour of equipment at resonant conditions and how to minimize or avoid its occurrence;
21. calculate and use the sound pressure level in a plant room;
22. calculate the sound pressure level experienced at an external location from a plant room;
23. calculate the sound pressure level generated in a room or space that is adjacent to a plant room;
24. calculate the sound pressure levels at different frequencies that are produced in the target occupied room;
25. understand, calculate and use noise rating data;
26. know how the acoustic design engineer relates the noise output from plant systems to the human response;
27. be able to calculate noise rating curves;
28. know the noise rating criteria used for building services design;
29. plot noise rating curves, plant and system sound pressure levels and find a suitable design solution;
30. know the formulae used in practical acoustic design work;
31. be able to carry out sound pressure level and noise rating design calculations, try different solutions to attenuate plant noise and be able to produce a practical design to meet a design brief.

Key terms and concepts

absorption 324; absorption coefficient 326; acoustic 324; acoustic barrier 326; acoustic energy 323; acoustic power 323; air-duct lining 324; air ducts 324; air pressure 323; anechoic chamber 327; atmospheric pressure 323; attenuation 332; audible frequencies 324; Bel 324; building 334; compressible medium 323; compressors 330; decibel 324; directivity 325; direct sound field 325; ear response 323; elastic medium 323; electric impulses 323; engines 330; fan blades 331; fan noise spectrum 331; fans 331; flexible connections 324; fluid 324; free field 325; frequency 324; hemispherical sound field 325; Hertz (Hz) 324; human ear 323; intermediate room 334; logarithm 325; mean absorption coefficient 326; mechanical service equipment 330; molecular vibration 324; multiple reflection 326; natural vibration 326; noise 324; noise rating 336; outdoor environment 333; pipes 324; plant room 325; plant vibrations 330; porous material 326; pressure wave 323; pump blades 331; pumps 331; resonance 332; reverberant sound field 326; reverberation 326; reverberation time 326; room absorption constant 326; rotation 330; rubber mountings 324; solid materials 324; sound 323; sound power 323; sound power level 324; sound pressure 323; sound pressure level 324; springs 324; structure 324; target room 335; total sound field 325; turbulent flow 324; vibration 324; Watts (W) 323; wave 323.

Introduction

This chapter uses the worksheet file DBPLANT.WKS to find the noise rating that will be produced within an occupied room by direct transmission through the building from the noise-producing plant. The plant noise source creates sound pressure levels within the plant room. The plant room noise can pass through an intermediate space, such as a corridor, and then into the target occupied space. Sound can be transmitted from the plant room to a recipient outdoors for an environmental impact noise rating.

Sufficient reference data is provided on the worksheet for examples within this chapter and for some real applications. Reference data from any source can easily be added. This chapter allows for most practical examples of mechanical plant to be assessed quickly and without having to deal with the equations themselves. Data is provided for frequencies from 125 to 4000 Hz as this range is likely to cover the important noise levels for comfort. The range of frequencies can be added to should the need arise. The reader may wish to study the principles of acoustics in the appropriate text books and the references made as it is not the intention of this chapter to teach the subject in its entirety. However, it is the purpose of this chapter to provide an easily understandable method of analysing practical noise applications. Consequently, the reader should not find it difficult to enter correct data and acquire suitable results for educational reasons and in practical design office cases. The worksheet DBDUCT.WKS is used to calculate the noise rating in the target room that is produced by noise being transmitted from the air-conditioning fan through the ductwork system. Further examples of spreadsheet applications and explanation of spreadsheet use are provided in *Building Services Engineering Spreadsheets* (Chadderton, 1997b).

Acoustic principles

The building services design engineer is primarily concerned with controlling the sound produced by items of plant such as boilers, supply air and exhaust air fans, refrigeration compressors, water pumps, diesel or gas engine-driven electrical generating sets and air compressors. An excess of sound that is produced by the plant, above that which is acceptable to the recipient, is termed noise. All the mechanical service equipment and distribution systems to be installed within an occupied building are capable of generating noise.

Sound travels through an elastic, compressible medium, such as air, in the form of waves of sound energy. These waves of energy are in the form of variations in the pressure of the air above and below the atmospheric air pressure. The human ear receives these air pressure fluctuations and converts the vibration generated at the eardrum into electric impulses to the brain. What we understand to be recognizable language, music and noise is the result of human brain activity. Animals and the mythical person from another planet do, or may, process what we determine as normal sounds and come to a different conclusion from those of us who are conditioned to life on earth. These variations in the pressure of the atmospheric air are very small when measured in the Pascal or millibar values that engineers use. A scale of measurement that relates to the subjective response of the human ear is used. Although absolute units of measurement are taken and normal calculation procedures are adopted, it is important to remember that the smallest unit of sound is that which can be detected by the human ear. The waves of air pressure which pass through the atmosphere are measured in relative pressure units. The acoustic energy of the source which caused the air pressure waves has an acoustic power, or rate of producing energy, in the same way that all thermodynamic devices have a power output. There are two ways of assessing the output and transmission of acoustic energy:

1. source sound power: Watts;
2. sound wave atmospheric pressure variation: Pascals.

Sound waves are generated at different frequencies measured in cycles per second, Hertz (Hz). The plant which produces the noise has components that rotate, move and vibrate at a range of different speeds, or frequencies of rotation. The flowing fluid is vibrated by the passage of fan or pump blades and it transfers the plant vibration through to downstream parts of the connected services systems. The fluid is either water, oil, air, gas, refrigerant or steam, and can either simply transmit the plant vibrations and noise or add to them by means of its own pulsations due to its turbulent flow. Turbulence means that a fluid flow contains recirculatory parcels of fluid in the form of eddy currents. These parcels of swirling eddy currents move in all directions, that is, along with the general direction of the main flow, but also in the reverse direction and transversely across the main flow. Viewing wave action on a beach or a fast river flow from a bridge or at a bend reveals the nature of turbulent flow. The turbulent eddy currents occur at a range of frequencies, parcels of recirculating fluid per second, depending upon the overall diameter of the eddy current. The physical movement of the swirling fluid can vibrate the containing water pipe or air duct, causing vibration and noise. Obstructions in the air or waterway occasioned by sharp edges, dampers, grilles, temperature sensors and changes in duct cross-section, can cause the turbulent fluid to shear into additional swirling eddy currents and produce more vibration and noise. Turbulent fluid can vibrate air ducts, pipes and terminal heat exchange units. The structure of the building transmits noise by the vibration of its solid material particles and continuous steel frame and reinforcing bars within concrete framework. Acoustic energy is transferred between pressure waves in the air and vibration through solid materials in either direction. The vibration of fans, compressors, engines and pumps is controlled by mounting them on coiled steel springs, rubber feet and rubber matt. Fluid pipes and air ducts are separated from fans, air-handling units and pumps with flexible connections. These minimize, or stop, plant vibration being transferred to the reticulation system. Fluid-borne noise is reduced by selective absorption with a porous lining to the air duct. Sound waves are absorbed into the thickness of the lining material through a perforated surface material which protects the absorber from fluid damage and erosion. Sound energy is dissipated within the absorber by multiple reflections among the fibrous material.

Sound power and pressure levels

Sound power and pressure levels are measured over a range of frequencies that are representative of the response of the human ear to sounds, Fig. 14.1. The unit of measurement of sound is the Bel (B). The smallest increment of sound that the human ear can detect is one-tenth of a Bel, one decibel (dB). This means that the smallest change in sound level that is perceptible by the human ear is 1 dB, so any decimal places that are produced from calculations using sound power or pressure level are not relevant. A calculated sound level of 84.86 can only be 84 dB as the 0.86 decimal portion is not detectable by the ear. The 'A' scale of measurement gives a weighting to each frequency in the range 20 Hz to 20 kHz in the same ratio as can be heard. For example, the human ear is more sensitive to sounds at 1000 Hz than at higher frequencies.

The acoustic output power of a machine is termed its sound power level, *SWL* dB. Think of *SWL* as the sound watts level of the acoustic output power of the machine. The value of acoustic power in watts from building services plant is very small, much less than 1 watt of power. The word level is used because it is not the actual value of the number of watts that is normally used; it is the sound level produced in acoustic units of measurement, dB, that are taken for practical use. The manufacturer of the plant provides the sound power levels produced by a particular machine from test results and predictions for known ranges of similar equipment. The sound power level of a machine at the range of frequencies from 125 to 8000 Hz is required by the building services design engineer in order to assess the acoustic affects upon the occupied spaces of the building. The overall sound power level for a range of frequencies is also quoted by the manufacturer of a machine.

14.1 Sound pressure level meter (reproduced by courtesy of Casella CEL Ltd).

Sound pressure level

A sound field is created by the sound power output from a machine within a plant room. It is made up of a direct sound field, that is, directly radiated sound, and a reverberant sound field, that is, general sound that reflects uniformly from the hard surfaces around the room. The direct sound field reduces with the inverse square of the distance from the sound source and is not normally of importance as it only applies to very short distances from the sound source. The reverberant sound field results from the average value of the sound pressure waves passing around the room. These waves try to escape from the plant room and find their way into the occupied spaces where the air-conditioning engineer is attempting to create a quiet and comfortable environment. The sound pressure level, SPL dB, of the total sound field, direct plus reverberant, that is generated within a room from a sound source of sound power level SWL dB, is found from

$$SPL = SWL + 10 \times \log \left(\frac{Q}{4 \times \pi \times r^2} + \frac{4}{R} \right) \text{ dB}$$

(CIBSE [1985] and Sound Research Laboratories Limited), where,

SPL = sound pressure level produced in room   dB
SWL = sound power level of acoustic source   dB
log = logarithm to base 10   dimensionless
Q = geometric directivity factor   dimensionless
r = distance from sound source to the receiver   m
R = room sound absorption constant m^2

Logarithms to base 10, $\log_{10}$, are used throughout the calculation of acoustic values. A sound source that radiates sound waves uniformly in all directions through unobstructed space will create an expanding spherical sound field and have a dimensionless geometric directionality factor Q of 1. A sound source that is on a plane surface radiates all its sound energy into a hemispherical sound field moving away from the surface. This has a directionality factor Q of 2, that is, twice the sound energy passes through a hemisphere. Similarly, if the sound source occurs at the junction of two adjacent surfaces that are at right angles to each other, such as the junction of a wall and ceiling, Q is 4. When there are three adjacent surfaces at the sound source, such as two walls and a ceiling, Q is 8. Distance r is that from the sound source to the receiving person, surface or measurement location, such as an air outlet duct from the plant room or outdoor air grille.

Absorption of sound

The room sound absorption constant, R m^2, is found from the total surface area of the enclosing room, S m^2, and the mean sound absorption coefficient of the room surfaces, α, at each of the relevant frequencies:

$$R = \frac{S \times \bar{\alpha}}{1 - \bar{\alpha}}$$

where

$\bar{\alpha} =$ mean absorption coefficient of room surfaces

$S =$ total room surface area m^2

Mean absorption coefficient, α, is found from the area and absorption coefficient for each surface of the enclosing space. All the absorbing surfaces within the space, such as seats and people in a theatre, are included in the overall sound absorbing ability of the room:

$$\bar{\alpha} = \frac{A_1 \times \alpha_1 + A_2 \times \alpha_2 + A_3 \times \alpha_3}{A_1 + A_2 + A_3}$$

where

$A_1 =$ surface area of surface number 1 m^2

$\alpha_1 =$ absorption coefficient of surface number 1

Materials absorb different amounts of sound energy at each frequency due to the frequency of natural vibration of their fibres and the method of their construction. Stiff, dense materials, such as brickwork walls, absorb sound by molecular vibration. Highly porous materials, such as glass wool, have large air passageways that allow the sound waves to penetrate the whole of the material thickness quickly. The strands of glass wool are vibrated by the sound waves and the sound energy is dissipated as heat. Dense materials are very efficient at absorbing acoustic energy. The reduction in sound level between the surfaces of a sound barrier is proportional to the mass of the barrier. The absorption coefficients of some common surface materials are given in Table 14.1. This data is repeated on the worksheet from line 201.

Reverberation time

Reverberation time is the time in seconds taken for a sound to decrease in value by 60 dB. This effectively means the time taken for the sound source to decay to an imperceptible level, as a sound pressure of 30 dB is very quiet to the human ear. An echo is produced by sound waves bouncing, or reverberating, from one or more hard surfaces and this may last for several seconds. A room that has a long reverberation time sounds noisy, lively and it allows echoes. A room having a short reverberation time, less than 1 s, sounds dull and there is no echo. The ultimate in short

Table 14.1 Absorption coefficients of common materials.

Material	Absorption coefficient at					
	125 Hz	250 Hz	500 Hz	1000 Hz	2000 Hz	4000 Hz
25 mm plaster, 18 mm plasterboard, 75 mm cavity	0.3	0.3	0.6	0.8	0.75	0.75
18 mm board floor on timber joists	0.15	0.2	0.1	0.1	0.1	0.1
Brickwork	0.05	0.04	0.04	0.03	0.03	0.02
Concrete	0.02	0.02	0.02	0.04	0.05	0.05
12 mm fibreboard, 25 mm cavity	0.35	0.35	0.2	0.2	0.25	0.3
Plastered wall	0.01	0.01	0.02	0.03	0.04	0.05
Pile carpet on thick underfelt	0.07	0.25	0.5	0.5	0.6	0.65
Fabric curtain hung in folds	0.05	0.15	0.35	0.55	0.65	0.65
15 mm acoustic ceiling tile, suspended 50 mm mineral fibre wool or glass fibre matt	0.5	0.6	0.65	0.75	0.8	0.75
50 mm polyester acoustic blanket, metallized film	0.25	0.55	0.75	1.05	0.8	0.7
50 mm glass fibre blanket, perforated surface finish	0.15	0.4	0.75	0.85	0.8	0.85

reverberation time is found in the anechoic chamber that is used for the acoustic testing of equipment. The walls and ceiling of the chamber are lined with thick acoustic absorbent wedges. The floor is a suspended wire mesh, and beneath the floor more absorbent wedges complete the coverage of all the room surfaces. The sound source radiates outward and upon reaching the surfaces is instantly absorbed, allowing no reverberation or echo. This is as close to a free field test method as can be achieved because there is no reverberant field caused by reflected sound waves.

An interesting example of a large semi-anechoic chamber is the car testing facility at Gaydon, England. The four walls and the ceiling are covered with acoustic wedges, while the floor is a plain concrete surface. This simulates an open road, hemispherical acoustic field under laboratory repeatable conditions (CIBSE, 1995).

Reverberation time of a room is found from

$$\text{reverberation time } T = \frac{0.161 \times V}{S \times \overline{\alpha}}$$

(CIBSE [1986] and Sound Research Laboratories Limited).

EXAMPLE 14.1

A plant room for an air-conditioning fan is 4 m × 3 m in plan and 2.5 m high. It has four brickwork walls, a concrete floor and a pitched sheet steel deck roof having 50 mm thickness of glass fibre and an aluminium foil finish to the underside. Ignore the effects of the metal plant, air ductwork and the door into the plant room. Calculate the room constant and the reverberation time for the plant room.

Table 14.2 Solution to Example 14.1.

Surface	Absorption data at frequency					
	125 Hz	250 Hz	500 Hz	1 kHz	2 kHz	4 kHz
Floor α	0.02	0.02	0.02	0.04	0.05	0.05
Ceiling α	0.15	0.4	0.75	0.85	0.8	0.85
Walls α	0.05	0.04	0.04	0.03	0.03	0.02
Floor ($S \times \alpha$)	0.24	0.24	0.24	0.48	0.6	0.6
Ceiling ($S \times \alpha$)	1.8	4.8	9.0	10.2	9.6	10.2
Walls ($S \times \alpha$)	1.75	1.4	1.4	1.05	1.05	0.7
$\bar{\alpha}$	0.064	0.109	0.18	0.199	0.191	0.195
Room constant R m^2	4.03	7.21	12.95	14.66	13.93	14.29
Reverberation T s	1.28	0.75	0.45	0.41	0.43	0.42

The surface absorption coefficients are selected from Table 14.1. It can be seen that there will be a different room constant and reverberation time for each frequency. The solution is presented in Table 14.2.

$$\text{Room volume } V = 4 \times 3 \times 2.5 \text{ m}^2$$
$$= 30 \text{ m}^3$$
$$\text{floor area} = 12 \text{ m}^2$$
$$\text{ceiling area} = 35 \text{ m}^2$$
$$\text{wall area} = 35 \text{ m}^2$$
$$\text{Room surface area } A = (2 \times 4 \times 3) + (4 + 4 + 3 + 3) \times 2.5 \text{ m}^2$$
$$= 59 \text{ m}^2$$

For 125 Hz, the mean absorption coefficient is,

$$\bar{\alpha} = \frac{12 \times 0.02 + 12 \times 0.15 + 35 \times 0.05}{12 + 12 + 35}$$
$$= 0.064$$
$$\text{room constant } R = \frac{S \times \bar{\alpha}}{1 - \bar{\alpha}} \text{ m}^2$$
$$= \frac{59 \times 0.064}{1 - 0.064} \text{ m}^2$$
$$= 4.03 \text{ m}^2$$
$$\text{reverberation time } T = \frac{0.161 \times V}{S \times \bar{\alpha}}$$
$$= \frac{0.161 \times 30}{59 \times 0.064} \text{ s}$$
$$= 1.28 \text{ s}$$

EXAMPLE 14.2

An air-conditioning centrifugal fan has an overall acoustic output power level *SWL* of 87 dB on the 'A' scale. The fan is to be installed centrally within the air-handling plant room described in Example 14.1. Calculate the sound pressure level that will be produced in the plant room at 1000 Hz when the fan is operating, close to the fan and also generally within the room.

Room absorption constant from Example 14.1 at 1000 Hz

$$R = 14.66 \text{ m}^2$$

The fan is on the centre of a concrete floor in the plant room. Sound pressure waves leaving the fan will radiate into a hemispherical field above floor level. The sound waves are concentrated into half of a completely free field. The directivity, Q, of the sound field is 2. A person within the plant room can stand in the range of 100 mm–2 m away from the fan. A typical distance between the fan and the recipient is 1 m. The room sound pressure level is calculated for 100 mm and 1 m distances from the sound source. When

$$r = 100 \text{ mm}$$

$$SPL = SWL + 10 \times \log_{10} \left(\frac{Q}{4 \times \pi \times r^2} + \frac{4}{R} \right) \text{ db}$$

$$= 87 + 10 \times \log_{10} \left(\frac{2}{4 \times \pi \times 0.1^2} + \frac{4}{14.66} \right) \text{ db}$$

$$= 87 + 10 \times \log_{10}(16.188) \text{ dB}$$

$$= 87 + 10 \times 1.2092 \text{ dB}$$

$$= 99 \text{ dB}$$

The smallest change in sound level that is perceptible by the human ear is 1 dB, so the decimal places are not relevant. The plant room sound pressure level at 100 mm radius from the fan is 99 dB. At 1 m from the fan, the recipient experiences a sound pressure level of

$$r = 1 \text{ m}$$

$$SPL = 87 + 10 \times \log_{10} \left(\frac{2}{4 \times \pi \times 1^2} + \frac{4}{14.66} \right) \text{ dB}$$

$$= 87 + 10 \times \log_{10}(0.432) \text{ dB}$$

$$= 87 + 10 \times -0.3645 \text{ dB}$$

$$= 83 \text{ dB}$$

The direct sound field diminishes with distance from the source. The reverberant sound field establishes the general room sound pressure level when the recipient is sufficiently far away from the source.

Table 14.3 Fan sound spectrum in Example 14.3.

Item	Data at frequency					
	125 Hz	250 Hz	500 Hz	1 kHz	2 kHz	4 kHz
Room constant R m^2	4.03	7.21	12.95	14.66	13.93	14.29
Reverberation T s	1.28	0.75	0.45	0.41	0.43	0.42
Fan SWL dB	78	82	86	87	70	60
Room SPL dB	78	79	81	82	65	55

EXAMPLE 14.3

The spectrum of sound power levels produced by the centrifugal fan being installed in the 4 m × 3 m × 2.5 m high plant room in Example 14.1 is 78 dB at 125 Hz, 82 dB at 250 Hz, 86 dB at 500 Hz, 87 dB at 1 kHz, 70 dB at 2 kHz, and 60 dB at 4 kHz. Use the surface absorption data from Example 14.1 and calculate the room sound pressure level at a radius of 1.5 m from the fan for each frequency from 125 Hz to 4 kHz.

At 125 Hz,

SWL is 78 dB.

$$r = 1.5 \text{ m}$$

$$Q = 2$$

$$R = 4.03 \text{ m}^2$$

$$SPL = SWL + 10 \times \log_{10}\left(\frac{Q}{4 \times \pi \times r^2} + \frac{4}{R}\right) \text{ dB}$$

$$= 78 + 10 \times \log_{10}\left(\frac{2}{4 \times \pi \times 1.5^2} + \frac{4}{4.03}\right) \text{ dB}$$

$$= 78 \text{ dB}$$

The results are shown in Table 14.3.

Plant sound power level

The design engineer requires to know the sound power level of the, potentially, noise-producing items of plant. These plant items will be the supply air fan, extract air fan, exhaust fans from toilets, kitchens and some store rooms, fan coil units in ceiling spaces above occupied rooms, packaged air-handling units incorporating fans, direct refrigerant expansion outdoor condensing units, direct refrigerant expansion packaged air-conditioning roof-mounted units, gas- and oil-fired boilers, packaged air conditioners and heat pumps within rooms, external cooling towers and dry-air-cooled heat exchangers, refrigeration compressors and water chilling refrigeration plant. In addition to these major items of plant, supply air grilles, extract air grilles, room termi-nal air-handling units, dampers, air volume control boxes and fan-powered variable air volume control boxes can also generate noise. The manufacturer of these items will provide the results of acoustic test data for the building services design engineer. Current acoustic data, rather than catalogue information, is acquired and the manufacturer then becomes responsible for the numbers used. The designer needs the sound power level at each frequency that is to be analysed. These are normally 125–4000 Hz. Often the critical frequency for design will be 1000 Hz and this corresponds to a sensitive band in the human ear response.

For the worked examples and questions within this book, sound power levels are provided, either in the form of a discrete value for each frequency, or a single value for all frequencies for the plant item. Figure 14.2 gives an indication of the variations in sound power level from a single value for centrifugal fans, axial fans, refrigeration compressors, cooling towers, fan coil units and boilers. The reader will find the spectral sound power level by subtracting the variances from the single value quoted in the example or question. This data is not to be used in real design work as it is provided for illustration purposes only. The numbers that were used to produce Fig. 14.2 are listed in Table 14.4. Figure 14.2 is also provided as a chart on the worksheet file.

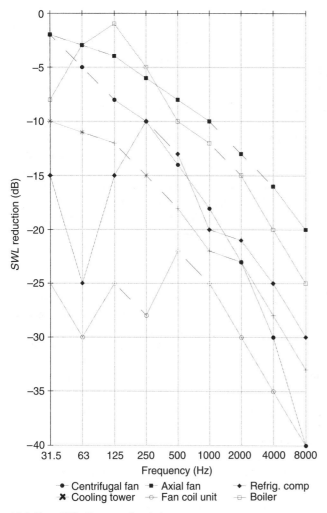

14.2 Plant *SWL* dB, spectral variation.

Table 14.4 Illustrative sound power level variances from Figure 14.1.

Plant item	Sound power level dB variance at frequency								
	31.5 Hz	63 Hz	125 Hz	250 Hz	500 Hz	1 kHz	2 kHz	4 kHz	8 kHz
Centrifugal fan	−2	−5	−8	−10	−14	−18	−23	−30	−40
Axial fan	−2	−3	−4	−6	−8	−10	−13	−16	−20
Refrigeration compressor	−15	−25	−15	−10	−13	−20	−21	−25	−30
Cooling tower	−10	−11	−12	−15	−18	−22	−23	−28	−33
Fan coil unit	−25	−30	−25	−28	−22	−25	−30	−35	−40
Boiler	−8	−3	−1	−5	−10	−12	−15	−20	−25

Transmission of sound

The sound pressure within a space will cause the flow of acoustic energy to an area that has a lower acoustic pressure. Sound energy converts into structural vibration and passes through solid barriers. A reduced level of sound pressure is established in the adjacent space due to the attenuation of the separating partition, wall, floor or ceiling. Air passageways through the separating partition act as sound channels that have little, or no, sound-reducing property, or attenuation. The reader can validate this effect by partially opening a window when the outdoor sound level is substantial. Compare the open and closed window performance when a train, lorry or high traffic volumes are present. A well-air-sealed single-glazed window imposes a sound reduction of 30 dB on external noise but a poorly sealed or open window has little attenuation.

Sound reduction by a surface is from the reflection of sound waves striking the surface and by the absorption of sound energy into a porous material. Absorbed acoustic energy is dissipated as heat within the solid components of the absorber. Dense materials are often efficient sound attenuators. The exception is metal. Sound travels easily through metals for great distances due to their molecular vibration. When the imposed sound frequency coincides with a natural frequency of vibration of the metal, resonance occurs and an increased sound level may be generated. This happens in particular when the shape of the metal creates an air space for the sound waves to resonate within, such as in a bell, an empty tank or a pipe. The structural steel within a building, service pipework, air ducts and railways lines can all transfer noise and vibration over long distances.

The sound pressure level generated within a room by mechanical plant, or sound systems for entertainment, will be passed through sound barrier materials and constructions such as walls and the ceiling, to adjacent spaces, occupied rooms and to the external environment around the building. The sound pressure levels received at each frequency depend upon the barrier attenuation, distance between the sound source and the recipient and the acoustic properties of the receiving space. The low-frequency sound waves, below 1000 Hz, are more difficult to attenuate than those above 1000 Hz. This is because the commonly used building and sound absorbing materials and vibration-isolating rubber all have a low natural frequency of vibration. They will resonate at a frequency often as low as 100 Hz. A material loses its attenuation property at the resonant frequency. Worse still, of course, is that when a rotary machine passes through or runs at its natural frequency of vibration, during start-up procedures, additional noise can be generated and the amplitude of its vibration may escalate to the point of physical destruction. It is vital that variable-speed controllers run the rotary machine speed through its resonant frequency band as quickly as possible to minimize noise and vibration. Attenuation materials such as brick, concrete, timber and acoustic fabric are good at absorbing sounds at the higher frequencies.

The human ear is most sensitive to sounds around 1000 Hz, making this the critical frequency for the acoustic design engineer.

Sound pressure level in a plant room

The sound source space is normally the mechanical services plant room. The reverberant sound pressure level in a plant room can be taken as

$$SPL_1 = SWL + 10 \times \log(T_1) - 10 \times \log(V_1) + 14 \text{ dB}$$

(Sound Research Laboratories Limited; see also, Smith *et al.* (1985)), where

$SPL_1 =$ sound pressure level in plant room   dB
$SWL =$ sound power level of source mechanical plant   dB
$T_1 =$ reverberation time of plant room   s
$V_1 =$ volume of plant room m^3

The reverberant sound pressure level is independent of the measurement location within the room. When a sound pressure level is required at a known location, the earlier equation is used with the radius from the source, r m,

$$SPL = SWL + 10 \times \log_{10} \left(\frac{Q}{4 \times \pi \times r^2} + \frac{4}{R} \right) \text{ dB}$$

EXAMPLE 14.4

A refrigeration compressor has an overall sound power level of 86 dB on the 'A' scale. The plant room has a reverberation time of 2 s and a volume of 70 m^3. Calculate the plant room reverberant sound pressure level.

$SWL = 86$ dBA

$T_1 = 2$ s

$V_1 = 70 \text{ m}^3$

$SPL_1 = SWL + 10 \times \log(T_1) - 10 \times \log(V_1) + 14$ dB

$= 86 + 10 \times \log(2) - 10 \times \log(70) + 14$ dB

$= 83 + 3 - 18 + 14$ dBA ignoring decimal places

$= 85$ dBA

Outdoor sound pressure level

The sound pressure level in the outdoor environment immediately external to the plant room can be taken as:

$$SPL_2 = SPL_1 - B + 10 \times \log(S_2) - 20 \times \log(d) + DI - 17 \text{ dB}$$

(Sound Research Laboratories Limited), where

SPL_2 = outdoor air sound pressure level   dB
SPL_1 = sound pressure level in source room   dB
B = sound reduction index of exterior wall or roof   dB
S_2 = surface area of external wall or roof   m^2
d = distance between plant room surface and recipient   m
DI = directivity index   dB

EXAMPLE 14.5

A refrigeration compressor generates an overall sound pressure level of 85 dBA within a plant room. The plant room has an external wall of 12 m^2 that has an acoustic attenuation of 30 dB. Sound radiates from the plant room wall into a hemispherical field that has a directivity index of 2 dB. Bedroom windows of an hotel are at a distance of 4 m from the plant room wall. Calculate the external sound pressure level at the hotel windows.

SPL_1 = 85 dBA

B = 30 dBA

S_2 = 12 m^2

d = 4 m

DI = 2 dB

$SPL_2 = SPL_1 - B + 10 \times \log(S_2) - 20 \times \log(d) + DI - 17$ db

$\quad = 85 - 30 + 10 \times \log(12) - 20 \times \log(4) + 2 - 17$ dB

$\quad = 85 - 30 + 10 - 12 + 2 - 17$ dB

$\quad = 38$ dBA

Sound pressure level in an intermediate space

The sound which is generated within a plant room may be transferred into an intermediate space within a building before being received in the target occupied room. Such intermediate spaces are corridors, store rooms, service ducts or roof voids. While it may not be important what the sound pressure level is within the intermediate space, the acoustic performance of this space affects the overall transfer of sound to the target occupied area. When the intermediate space is very large and has thermally insulated surfaces, for example, in a roof space, a considerable attenuation is possible. The sound pressure level in such an intermediate room or space can be taken as

$SPL_3 = SPL_1 - SRI + 10 \times \log(S_4) + 10 \times \log(T_2) - 10 \log(0.16 \times V_2)$ dB

(Sound Research Laboratories Limited), where

SPL_3 = sound pressure level in intermediate space   dB
SPL_1 = sound pressure level in plant room   dB

SRI = sound reduction index of common surface   dB
S_4 = area of surface common to both rooms m^2
T_2 = reverberation time of intermediate space   s
V_2 = volume of intermediate space m^3

EXAMPLE 14.6

A showroom has floor dimensions of 25 m × 10 m and a height of 3.6 m to a suspended tile ceiling. The average height of the ceiling void is 1.8 m. An air-conditioning system has distribution ductwork in the roof void above the suspended acoustic ceiling tiles. The air-handling plant room is adjacent to the roof void and there is a common plant room wall of 5 m × 2.5 m high in the roof void. The sound pressure level in the plant room is expected to be 50 dB. The reverberation time of the roof void is 0.8 s. The plant room wall adjoining the roof void has a sound reduction index of 10 dB. Calculate the sound pressure level that is produced within the roof void as the result of the air-handling plant room noise.

$SPL_3 = 50$ dB

$SRI = 10$ dB

$S_4 = 12.5$ m^2

$T_2 = 0.8$ s

$V_2 = 25 \times 10 \times 1.8$ m^3

$= 450$ m^3

$SPL_3 = SPL_1 - SRI + 10 \times \log(T_2) - 10 \times \log(0.16 \times V_2)$ dB

$= 50 - 10 + 10 \times \log(12.5) + 10 \times \log(0.8) - 10 \times \log(0.16 \times 450)$ dB

$= 50 - 10 + 10 + 0 - 11$ dB

$= 39$ dB

Sound pressure level in the target room

The sound pressure level in the target occupied room or space can be taken as:

$$SPL_4 = SPL_3 - SRI + 10 \times \log(S_5) + 10 \times \log(T_3) - 10 \times \log(0.16 \times V_3) \text{ dB}$$

(Sound Research Laboratories Limited), where

SPL_4 = sound pressure level in target room   dB
SPL_3 = sound pressure in adjacent room   dB
SRI = sound reduction index of common surface   dB
S_5 = area of surface common to both rooms m^2
T_3 = reverberation time of target room   s
V_3 = volume of target room m^3

The target room may be adjacent to, or close to, the plant room, or it may not be influenced by the plant room other than by the transfer of noise through the interconnected air-ductwork system. Analysis of the ductwork route for noise transfer is calculated separately and is not covered in this book.

EXAMPLE 14.7

The showroom in Example 14.6 has floor dimensions of 25 m × 10 m and a height of 3.6 m to a suspended tile ceiling. The reverberation time of the showroom is 0.5 s. The air-conditioning plant room generates a sound pressure level of 39 dB in the roof space. The acoustic tile ceiling has a sound reduction index of 12 dB. Calculate the sound pressure level that is produced within the showroom by the air-conditioning plant.

$SPL_3 = 39$ dB

$SRI = 12$ dB

$S_5 = 250$ m^2

$T_3 = 0.5$ s

$V_3 = 25 \times 10 \times 3.6$ m^2

$= 900$ m^3

$SPL_4 = SPL_3 - SRI + 10 \log(S_5) + 10 \times \log(T_3) - 10 \times \log(0.16 \times V_3)$ dB

$= 50 - 12 + 10 \times \log(250) + 10 \times \log(0.5) - 10 \times \log(0.16 \times 900)$ dB

$= 50 - 12 + 23 - 3 - 21$ dB

$= 37$ dB

Noise rating

The human ear has a different response to each frequency within the audible range of 20–20 000 Hz. It has been found that a low-frequency noise can be tolerated at a greater sound pressure level than a high-frequency noise. Noise rating (NR) curves are used to specify the loudness of sounds. Each curve is a representation of the response of the human ear in the range of audible frequencies.

The design engineer makes a comparison between the sound pressure level produced in the room at each frequency and the noise rating curve data at the same frequency. When all the noise levels within the room fall on or below a noise rating curve, that noise rating is attributed to the room. Noise rating curves for NR 25 to NR 50 are shown on Fig. 14.3. The values are plotted from,

$SPL = NR_f \times B_f + A_f$ dB

$SPL =$ sound pressure level at frequency f and noise rating NR dB
$NR_f =$ noise rating at frequency f Hz   dimensionless
B_f and $A_f =$ physical constants   dB
$f =$ frequency   Hz

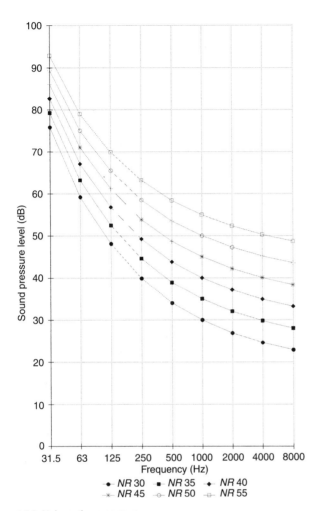

14.3 Noise rating curves.

Table 14.5 Physical constants for noise rating calculation.

Frequency f (Hz)	A_f (dB)	B_f (dB)
31.5	55.4	0.681
63	35.5	0.79
125	22.0	0.87
250	12.0	0.93
500	4.8	0.974
1000	0	1.0
2000	−3.5	1.015
4000	−6.1	1.025
8000	−8.0	1.03

The values of the physical constants to calculate noise rating are shown in Table 14.5 (Australian Standard AS 1469–1983).

The normal applications of noise rating are shown in Table 14.6.

Table 14.6 Noise rating applications.

Application	Noise rating	Comment
Acoustic laboratory	NR 15	Critical acoustics
Radio studio	NR 15	Critical acoustics
Concert hall	NR 15	Critical acoustics
TV studio	NR 20	Excellent listening
Large conference room	NR 25	Very good listening
Hospital, home, hotel	NR 30	Sleeping, relaxing
Library, private office	NR 35	Good listening
Office, restaurant, retail	NR 40	Fair listening
Cafeteria, corridor, workshop	NR 45	Moderate listening
Commercial garage, factory	NR 50	Minimum speech interference
Manufacturing	NR 55	Speech interference
Heavy engineering to industrial	NR 60 to NR 80	Sound levels judged on merits, leading to risk of hearing damage

---

EXAMPLE 14.8

An air-conditioning fan produces the sound spectrum shown in Table 14.7 within an occupied room. Calculate the sound pressure levels for noise ratings NR 35, NR 40, NR 45, NR 50 and NR 55, and plot the noise rating curves for the frequency range from 31.5 to 8000 Hz. Plot the room sound pressure levels on the same graph and find which noise rating is not exceeded.

Table 14.7 Noise spectrum in Example 14.8.

Frequency f (Hz)	31.5	63	125	250	500	1k	2k	4k	8k
Room SPL (dB)	39	44	48	52	55	49	36	33	28

A manually calculated example for one noise rating curve is shown. The reader should use the spreadsheet graph or chart facilities to plot the whole figure. Calculate the SPL values for NR 55.

$$NR_{fd} = 55$$

$$SPL = NR_f \times B_f + A_f \text{ dB}$$

Calculate the SPL at each frequency for the values of B_f and A_f from Table 14.5. For 31.5 Hz,

$$SPL = 55 \times 0.681 + 55.4 \text{ dB}$$

$$= 92 \text{ dB}$$

For 63 Hz,

$$SPL = 55 \times 0.79 + 35.5 \text{ dB}$$

$$= 78 \text{ dB}$$

For 125 Hz,

$$SPL = 55 \times 0.87 + 22 \text{ dB}$$
$$= 69 \text{ dB}$$

For 250 Hz,

$$SPL = 55 \times 0.93 + 12 \text{ dB}$$
$$= 63 \text{ dB}$$

For 500 Hz,

$$SPL = 55 \times 0.974 + 4.8 \text{ dB}$$
$$= 58 \text{ dB}$$

For 1000 Hz,

$$SPL = 55 \times 1.0 + 0 \text{ dB}$$
$$= 55 \text{ dB}$$

For 2000 Hz,

$$SPL = 55 \times 1.015 - 3.5 \text{ dB}$$
$$= 52 \text{ dB}$$

For 4000 Hz,

$$SPL = 55 \times 1.025 - 3.5 \text{ dB}$$
$$= 50 \text{ dB}$$

For 8000 Hz,

$$SPL = 55 \times 1.03 - 8.0 \text{ dB}$$
$$= 48 \text{ dB}$$

These sound pressure levels are compared to the room data in Table 14.8.

The closest approach to the *SPL* limit for *NR* 55 occurs at 500 Hz. Check that *NR* 50 is exceeded. For 500 Hz,

$$SPL = 50 \times 0.974 + 4.8 \text{ dB}$$
$$= 53 \text{ dB}$$

Table 14.8 Noise spectrum in Example 14.8.

Frequency f (Hz)	31.5	63	125	250	500	1k	2k	4k	8k
NR 55 *SPL* (dB)	92	78	69	63	58	55	52	50	48
Room *SPL* (dB)	39	44	48	52	55	49	36	33	28

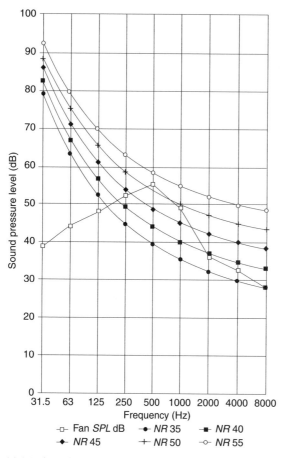

14.4 Noise ratings, solution to Example 14.8.

It should now be possible to check manually any sound pressure level against noise rating. Once the frequency that produces the greatest sound pressure level from the sound source is identified, other *SPL* values can be obtained for the peak frequency to check which *NR* is not exceeded.

The room does not exceed the *NR* 55 curve data but it does exceed *NR* 50. Plot the chart with the spreadsheet functions for *NR* 35 to *NR* 55 and with the room noise *SPL*. The spreadsheet will produce curves for six sets of data, so five *NR* curves and the one room curve can be displayed on one chart. The results are shown in Fig. 14.4.

Questions

Questions 1 to 31 do not need to be evaluated on the worksheet. The worksheet is to be used for Questions 32 onwards. The solution to Question 32 is shown on the original file DBPLANT.WKS. Solutions to descriptive questions are to be found within the chapter, except where specific answers are provided.

1. List the sources of noise that could be found within an air-conditioned building.
2. What is meant by noise?

3. State which items of mechanical services plant, equipment and systems within an occupied building are not likely to create noise.

4. Explain how sound travels from one location to another.

5. Explain what is meant by the term sound pressure wave.

6. Why is sound important?

7. Explain how we 'hear' sounds.

8. State what is meant by sound power and sound pressure level. State the units of measurement for sound power, sound pressure, sound power level and sound pressure level.

9. Explain why any decimal fraction of a decibel is not used in engineering design.

10. List the ways in which mechanical and electrical services plant, equipment and systems generate sound.

11. Explain, with the aid of sketches and examples, how sound is transferred, or can be, through a normally serviced multi-storey occupied building.

12. Discuss the statement: 'Turbulent flows in building services systems create a noise nuisance.'

13. State how the building structure transfers sound.

14. Explain, with the aid of sketches, ways in which the noise and vibration produced by the mechanical and electrical services of a building can be reduced before they become a nuisance for the building's users.

15. Explain how sound energy is dissipated into the environment.

16. State the range of frequencies that are detectable by the human ear and the frequencies that are used in acoustic design calculations. State the reasons for these two ranges being different, if they are.

17. Define the terms 'sound power level' and 'sound pressure level'.

18. Explain what is meant by direct and reverberant sound fields.

19. A plant room for a refrigeration compressor is 6 m × 4 m in plan and 3 m high. It has four brickwork walls, a concrete floor and a concrete roof. Select the surface absorption coefficients for the frequency range 125–4000 Hz from Table 14.1. Calculate the room absorption constant and the reverberation time for the plant room at each frequency. Do the calculations manually and then enter the same data onto the worksheet to validate the results.

20. An air-conditioning centrifugal fan has an overall sound power level SWL of 75 dBA. The fan is to be installed centrally within a plant room that has a room absorption constant R of 12 m². Calculate the sound pressure level that will be produced close to the fan, in the plant room at 1000 Hz when the fan is operating, and also generally within the room.

21. A 900 mm diameter axial fan is to be installed on the concrete floor of an 8 m × 4 m × 3 m high plant room. The fan sound power level at 1000 Hz is 89 dB. The room absorption constant R at 1000 Hz is 8 m² and the reverberation time is 0.4 s. Calculate the room sound pressure level at a radius of 300 mm from the fan, and the reverberant room sound pressure level.

22. A reciprocating water chilling refrigeration compressor has an overall sound power level of 92 dBA. It is to be located within a concrete-and-brick plant room that has a reverberation time of 1.8 s and a volume of 250 m³. Calculate the plant room reverberant sound pressure level.

23. An air-handling plant has an overall sound power level of 81 dB. The plant room has an external wall of 10 m² that has an acoustic attenuation of 35 dB and ventilation openings having a free area of 3 m². The windows of residential and office buildings are at a distance of 12 m from the plant room wall. Calculate the external sound pressure level at the windows and recommend what, if any, attenuation is needed at the plant room.

24. A forced-draught gas-fired boiler has an overall sound pressure level of 96 dB. The boiler plant room has an external wall of 60 m^2 that has an acoustic attenuation of 25 dB and two louvre doors to admit air for combustion. Calculate the external sound pressure level at a distance of 20 m from the plant room wall. State your recommendations for the attenuation of the boiler and the plant room.

25. A single-storey office building has floor dimensions of 40 m × 30 m and a height of 3 m to a suspended acoustic tile ceiling. The average height of the ceiling void is 1.5 m. A plant room is adjacent to the roof void. There is a common plant room wall of 10 m × 1.5 m high in the roof void. The sound pressure level in the plant room is expected to be 61 dB. The reverberation time of the roof void is 0.6 s. The plant room wall adjoining the roof void has a sound reduction index of 13 dB. Calculate the sound pressure level that is produced within the roof void as the result of the plant room noise. Comment on the resulting sound pressure level.

26. A hospital waiting area has floor dimensions of 8 m × 12 m and a height of 3 m to a plasterboard ceiling. A packaged air-conditioning unit is housed in an adjacent room. There is a common wall of 15 m^2 and sound reduction index of 35 dB to the two rooms. The sound pressure level in the plant room is expected to be 72 dB. The reverberation time of the waiting room is 1.3 s. Calculate the sound pressure level that will be produced in the waiting room.

27. A meeting room has floor dimensions of 8 m × 6 m and a height of 2.7 m to a suspended tile ceiling. The reverberation time of the room is 0.7 s. A fan coil heating and cooling unit creates a sound pressure level of 43 dB in the ceiling space. The acoustic tile ceiling has a sound reduction index of 8 dB. Calculate the sound pressure level in the meeting room.

28. An hotel bedroom is 6 m long, 5 m wide and 2.8 m high and it has a reverberation time of 0.4 s. The air-conditioning plant room generates a sound pressure level of 56 dB in the service space above the ceiling of the bedroom. The plasterboard ceiling has a sound reduction index of 16 dB. Calculate the sound pressure level in the bedroom.

29. Explain how noise rating curves relate to the response of the human ear and are used in the design of mechanical services plant and systems.

30. The centrifugal fan in an air-handling plant produces the noise spectrum shown in Table 14.9 within an office. Calculate the sound pressure levels for noise ratings NR 35, NR 40, NR 45, NR 50 and NR 55 and plot the noise rating curves for the frequency range 31.5 Hz–8 kHz. Plot the room sound pressure levels on the same graph and find which noise rating is not exceeded.

31. A model XT45 water chiller is to be located within a plant room on the roof of an hotel in a city centre. The plant room is 12 m long, 10 m wide and 3 m high. The room directivity index is 2. The plant operator will normally be 1 m from the noise source. The floor is concrete, the roof is lined internally with a 50 mm polyester acoustic blanket with a metallized film surface. The plant room walls are 115 mm brickwork. There are no windows. The water chiller manufacturer provided the sound power levels as 100 dB overall, 74 dB at 63 Hz, 89 dB at 125 Hz, 95dB at 250 Hz, 97 dB at 500 Hz, 99 Hz at 1 kHz, 97 dB at 2 kHz and 90 dB at 4 kHz.

    (a) Check that the correct data is entered onto the working copy of the original work-sheet file DBPLANT.WKS and find the noise rating that is not exceeded within the plant room.

Table 14.9 Noise spectrum in Question 30.

Frequency f (Hz)	31.5	63	125	250	500	1k	2k	4k	8k
Room SPL (dB)	30	35	32	40	42	31	28	20	10

(b)   The plant room has three external walls. The nearest openable window in nearby buildings is at a distance of 15 m from a plant room wall. There is no acoustic barrier between a plant room wall and the recipient's window. The directivity index for the outward projection of sound is taken as 3 dB. Find the noise rating at the recipient's window and state what the result means.

(c)   A corridor adjoins the plant room. The target sound space, an office, is on the opposite side of the corridor. The corridor is 10 m long, 1 m wide and 3 m high. It has a room directivity index of 2, a carpeted concrete floor, plastered brick walls and a plasterboard ceiling. The common wall between the plant room and the corridor is 10 m long, it is constructed with 115 mm plastered brickwork and it does not have a door. There are no windows. There is no other sound barrier. Find the noise rating which would be found at a distance of 0.5 m from the plant room wall while within the corridor.

(d)   The target office is 10 m long, 10 m wide and 3 m high. The room directivity index is 2. The nearest sedentary occupant of the office will be 1 m from the corridor wall. The floor has pile carpet, the walls are plastered brick and there is a suspended ceiling of 15 mm acoustic tile and 50 mm glass fibre matt. The office has four 2 m × 2 m single-glazed windows on two external walls. The office wall that adjoins the corridor is 115 mm plastered brickwork and it has one 2 m² door into the corridor. Find the noise rating, NR, and sound pressure levels, SPL dB, that are experienced in the target office. State what effect the office and plant room doors will have on the noise rating in the target room. Recommend appropriate action to be taken with these doors.

32.   A centrifugal fan is located within the basement plant room of an office building. The plant room is 8 m long, 6 m wide and 3 m high. The room directivity index is 2 and the plant operator will normally be 1 m from the noise source. The floor and ceiling are concrete, there are four 230 mm brick walls and one acoustically treated door. There are no windows in the plant room. The sound power levels of the fan are 86 dB overall, 64 dB at 63 Hz, 66 dB at 125 Hz, 72 dB at 250 Hz, 80 dB at 500 Hz, 86 Hz at 1 kHz, 82 dB at 2 kHz and 77 dB at 4 kHz.

(a)   Find the noise rating that is not exceeded within the plant room.

(b)   A corridor and staircase connect the plant room to the Reception area of the building. The corridor is 6 m long, 1 m wide and 3 m high. It has a room directivity index of 2. The corridor has a concrete floor, plastered brick walls and a plasterboard ceiling. The common wall between the plant room and corridor is 2 m long. The sound reduction index of the plant room door is 20 dB at each frequency from 125 Hz to 4 kHz. There is no other sound barrier. Find the noise rating that would be found at a distance of 1 m from the plant room in the corridor.

(c)   The Reception area is 12 m long, 8 m wide and 3 m high. The room directivity index is 2. There are 10 m² of single-glazed windows in the Reception. There is a door at the top of the staircase down to the plant room. The stairs door is 1 m wide, 2 m high and it has a sound reduction index of 20 dB at each frequency from 125 Hz to 4 kHz. The nearest occupant will be 1 m from the stairs door. The floor has thermoplastic tiles on concrete, the walls are plastered brick and there is a plasterboard ceiling. Find the noise rating which is not exceeded in the Reception.

33.   Oil-fired hot-water boilers are located in a plant room in the basement of an exhibition and trade centre building in a city centre. The plant room is 10 m long, 10 m wide and 5 m high.

The room directivity index is 2. The floor, walls and ceiling are concrete. There are no windows. The reference sound power level of the boiler plant is 88 dBA.

(a) Find the anticipated spectral variation in the sound power level for the frequency range from 63 Hz to 4 kHz from Table 14.4 and Fig. 14.1, enter the data onto the worksheet and find the noise rating that is not exceeded within the boiler plant room.

(b) The plant room has three 100 mm concrete external walls. The nearest recipient can be 1 m from the external surface of a boiler plant room wall. There is no acoustic barrier between a plant room wall and a recipient. The directivity index for the outward projection of sound is taken as 3 dB. Find the noise rating at the nearest recipient's position and state what the result means.

(c) A hot-water pipe and electrical cable service duct connects the boiler plant room to other parts of the building. The concrete-lined service duct is 30 m long, 2 m wide and 1 m high. Both ends of the service duct have a 100 mm concrete wall. Calculate the noise rating within the service duct at its opposite end from the boiler plant room.

(d) A conference room 115 mm brick wall adjoins the service duct at the furthest end from the boiler plant room. The conference room is 12 m long, 10 m wide and 4 m high. The room directivity index is 2. The nearest sedentary occupant will be 0.5 m from the service duct wall. The floor has pile carpet, the walls are plastered brick and there is a suspended ceiling of 15 mm acoustic tile and 50 mm glass fibre matt. There are no windows. Find the noise rating that is produced in the conference room by the boiler plant.

34. A four-pipe chilled- and hot-water fan coil unit is located within the false ceiling space above an office in an air-conditioned building. Conditioned outdoor air is passed to the fan coil unit through a duct system. The office is 5m long, 4 m wide and 3 m high. The room directivity index is 2. The office has a concrete floor with thermoplastic tiles and 115 mm plastered brick walls. The 700 mm deep suspended ceiling has 12 mm fibreboard acoustic tiles, recessed fluorescent luminaires, ducted supply and return air with a supply air diffuser, a return air grille and a concrete ribbed slab for the floor above. The office has a double-glazed window of 2 m x 2 m. The reference sound power level of the fan coil unit is 85 dBA. Enter the ceiling space as the plant room and bypass the intermediate space data as directed.

(a) Find the anticipated spectral variation in the sound power level of the fan coil unit for the frequency range from 63 Hz to 4 kHz from Table 14.4 and Fig. 14.2. Enter the data onto the worksheet and find the noise rating that is not exceeded within the ceiling space above the office.

(b) Find the noise rating that is not expected to be exceeded within the office at head height. Assume that the sound reduction of the acoustic tile ceiling is maintained across the whole ceiling area.

(c) Sketch a cross-section of the fan coil unit installation and identify all the possible noise paths into the office.

(d) List the ways in which the potential noise paths into the office can be, or may need to be, attenuated.

35. Presbycusis is:

1. Hearing loss due to long-term exposure to noise above 90 dBA.
2. Hearing loss due to ear disease.
3. Normal deterioration in hearing due to ageing.

4. A church presbytery committee.
5. Temporary shift in hearing ability from exposure to high industrial noise levels above 95 dBC.

36. Hearing range is:

1. 20 Hz to 20 kHz.
2. 2 Hz to 20 MHz.
3. 200 Hz to 200 MHz.
4. Infinitely wide.
5. 2 kHz to 20 MHz.

37. What is noise?

1. Sound.
2. Acoustic power.
3. Unwanted sound.
4. Age-related sound.
5. Traffic, aeroplanes, pneumatic drills, fans, refrigeration compressors.

38. How do we judge sound?

1. With absolute measurement.
2. Comparing a sound with absolute zero sound level.
3. Relatively.
4. Subjectively.
5. Qualitative judgement.

39. What is sound?

1. Electromagnetic radiation.
2. Molecular vibration of solid materials.
3. Radio frequency waves.
4. Anything that causes an ear response.
5. Pressure waves.

40. Sound travels through air because it is:

1. Incompressible.
2. Supporting molecular vibration.
3. Compressible.
4. Inelastic.
5. Plastic.

41. Reference point for sound level measurement is:

1. Absolute zero sound.
2. Lowest audible level by a domestic animal.
3. Smallest sound detectable by human ear.
4. Zero atmospheric pressure as found in space.
5. Inaudible level created in a test laboratory.

42. Sound waves repeat at a frequency due to:

1. Absorption by porous surfaces.
2. Wind forces.

3.   Multiple sources of sound.
4.   Passage of blades in a rotary machine such as a compressor, pump or turbine.
5.   Variations in air pressure.

43. An eight-cylinder formula one car engine peaks at 20 000 RPM. One of the sound frequencies it produces is:

1.   8 Hz.
2.   20 kHz.
3.   400 Hz.
4.   2000 Hz.
5.   2667 Hz.

44. A gas turbine rotates at 60 000 RPM and has 50 blades on its largest diameter. One of the sound frequencies it produces is around:

1.   50 kHz.
2.   50 Hz.
3.   5000 Hz.
4.   60 kHz.
5.   20 kHz.

45. How can the structure of a building transmit noise?

1.   Concrete-framed structures cannot as noise is dampened.
2.   Steel and concrete structures absorb all acoustic energy.
3.   Structures always absorb acoustic energy and dissipate it as heat.
4.   Molecular vibration.
5.   Physical movement.

46. How is noise transmission from plant reduced?

1.   Cannot be reduced, only contained within plant room.
2.   Select quieter plant.
3.   Seal plant room doors.
4.   Locate plant room away from occupied rooms.
5.   Flexible rubber and spring mountings.

47. Which is the smallest increment of sound pressure level detectable by the human ear?

1. 1 W/m^2.
2.   1 Bel.
3.   60 Bel.
4. 100 N/m^2.
5.   1 decibel.

48. Explain the meaning of *SWL*:

1.   Selective wind loading.
2.   Sound wind level.
3.   Sound watts level, meaning power.
4.   Sound pressure level, meaning energy.
5.   Sound watts loudness, meaning loudness power.

49. How much acoustic power is experienced within buildings?

    1. 10% of electric motor power becomes acoustic energy.
    2. Around 10 W/m^2 floor area.
    3. Above 1 kW.
    4. Always below 500 W.
    5. Less than 1 W.

50. The frequency range used for assessment of sound power level, SWL, from machines is:

    1. 0–200 MHz.
    2. 1 kHz to 2 MHz.
    3. 125 Hz to 8 kHz.
    4. 63–20 000 Hz.
    5. 125 kHz to 8 MHz.

51. By what mechanism do ears respond to sound power level, *SWL*?

    1. Ears have no mechanism.
    2. Sound power radiates to vibrate the eardrum.
    3. Acoustic vibration energy vibrates the body, which transfers through the body muscle and bone structure, to vibrate eardrums.
    4. Acoustic power raises air pressure on eardrums.
    5. Acoustic output power pulsates and vibrates air, raising air pressure waves; eardrum vibrates from air pressure waves.

51. What is a reverberant sound field?

    1. Sound transmitted over a large distance.
    2. Sound passing through a structure.
    3. What remains within an enclosure after source energy is absorbed by the building structure.
    4. Reflected sound.
    5. Sound pressure level measured in an anechoic laboratory chamber.

53. Direct sound field:

    1. Increases in intensity further from the source.
    2. Remains at a constant noise at any distance from the source while hearer remains in the source plant room.
    3. Decreases linearly with distance from source.
    4. Falls with the inverse square of the distance from the sound source.
    5. Doubles the value of the reverberant sound field.

54. Which of these is not correct about absorbing sound energy?

    1. Dense materials absorb acoustic energy efficiently.
    2. Highly porous materials are good sound absorbers.
    3. The denser the material mass, the greater the sound absorption.
    4. A 75 mm air cavity behind a sheet of plasterboard is a good sound absorber.
    5. A plastered brick wall has a low sound absorption coefficient.

55. Which is not correct about reverberation time:

    1.   When short, below a second, room seems lively.
    2.   Long reverberation time causes room to sound noisy and echoes.
    3.   A lecture theatre needs a short reverberation time.
    4.   A large volume car-manufacturing building has a long reverberation time.
    5.   When short, below a second, room seems dull.

56. Which is not correct about an anechoic chamber?

    1.   Used to measure sound power level from acoustic sources such as fans and compressors.
    2.   Must have no reverberant sound field.
    3.   Lined with fully absorbent foam wedges.
    4.   Sounds perfectly dull.
    5.   Used to measure reverberant field sound pressure level from acoustic sources such as fans and compressors.

57. What does natural frequency of vibration mean?

    1.   Damped vibration.
    2.   Strike a guitar string and it vibrates at up to four times its natural frequency depending on volume of sound box.
    3.   Bounce a coil spring and it vibrates at its natural frequency of vibration.
    4.   A frequency mechanically forced upon an item, such as by a motor.
    5.   A material never vibrates at this frequency.

58. Which does *NR* stand for?

    1.   Noise resonance.
    2.   Normal rating.
    3.   No resonance.
    4.   Noise ratification.
    5.   Noise rating.

59. How are noises related to human ear response?

    1.   Humans respond to sound power level within a range of audible frequencies.
    2.   Humans respond to loudness produced over a range of audible frequencies.
    3.   Sound pressure levels are added to create an overall relationship to ear response.
    4.   Sound power levels are added to create an overall relationship to ear response.
    5.   Loudest sound at any frequency is taken as ear response.

15 Fire protection

Learning objectives

Study of this chapter will enable the reader to:

1. classify fire hazards;
2. identify the necessary ingredients for a fire;
3. describe the development of a fire;
4. recognize the hazards of smoke;
5. apply the correct fire-fighting system, or combination of systems, to different fire classifications;
6. understand the principles of portable fire extinguishers;
7. know the criteria for the use of hose reel, dry riser, wet riser, foam, sprinkler, carbon dioxide (CO_2), vaporizing liquid, dry powder and deluge fixed fire-fighting systems;
8. know the sources of water used in fire-fighting;
9. identify how fire development can be detected;
10. recognize the importance of smoke ventilation;
11. identify the locations for fire dampers in air ductwork and know how they operate.

Key terms and concepts

break tank 353; carbon dioxide (CO_2) 351; dry hydrant rise 352; dry powder 351; fire damper 358; fire-fighting system classification 350; fire risk classification 350; foam 351; foam inlet 354; fuel, heat and oxygen 350; hazard, smoke and fire detectors 356; hose reels 352; hydrant valve 353; portable extinguishers 351; smoke 350; smoke ventilator 357; sprinkler systems 354; vaporizing liquid 351; water 351; water pumping 353; water sources 354; water storage 353; wet hydrant riser 353.

Introduction

The systems required to meet the needs of tackling small fires, evacuation and major fire-fighting both by the occupants and then the Fire Service are outlined. Building management systems under computer monitoring and control will incorporate such systems, together with security functions. Integration of such equipment with the architecture, decor and other services is planned from the earliest design stage.

Fire classification

A building's fire risk is classified according to its occupancy and use. Table 15.1 gives representative information (CIBSE, 1986).

A fire is supported by three essential ingredients: fuel, heat and oxygen. The absence of any one of these causes an established fire to be extinguished. The fire-fighting system must be appropriate to the location of the fire and preferably limited to that area in order to minimize damage to materials, plant and the building structure. Radiation from a fire may provoke damage or combustion of materials at a distance. Structural fire protection can include water sprays onto steelwork to avoid collapse, as used in the Concorde aircraft production hangar.

The system of fire-fighting employed depends upon the total combustible content of the building (fire load), the type of fire risk classification and the degree of involvement by the occupants. Fire escape design where children, the elderly or infirm are present needs particular care so that sufficient time is provided in the fire resistance of doors and partitions for the slower evacuation encountered.

Smoke contains hot and unpleasant fumes, which can be lethal when produced from certain chemicals and plastics. Visual obstruction makes escape hazardous and familiar routes become confused. Packaging materials, timber, plastics, liquefied petroleum gas cylinders and liquid chemicals must not be stacked in passageways or near fire exits in completed or partially completed buildings. Each working site or building needs a safety officer responsible for general oversight.

Fires are classified in Table 15.2.

Table 15.1 Classification of occupancies.

Category	Group	Hazard occupancy
Extra light	—	Public buildings
Ordinary	1	Restaurant
Ordinary	2	Motor garage
Ordinary	3	Warehouse
Ordinary	3 (special)	Woodwork
Extra high	—	Paint manufacture
Extra high (storage)	1	Electrical appliance
Extra high (storage)	2	Furniture
Extra high (storage)	3	Wood, plastic or rubber
Extra high (storage)	4	Foamed plastics or rubber

Table 15.2 Fire classifications.

Classification	Fire type	Fire-fighting system
A	Wood and textiles	Water, cools
B	Petroleum	Exclude oxygen
C	Gases	Exclude oxygen
D	Flammable metals	Exclude oxygen
E	Electrical	Exclude oxygen, non-conducting

Regular fire drills are conducted by the safety officer and employees are clearly notified of their responsibilities in an emergency. Staff duties will be to shepherd the public, patients or students out of the building to the rendezvous, while maintenance personnel may be required to operate fire-fighting equipment while awaiting the fire brigade.

Portable extinguishers

Portable extinguishers are manually operated first-aid appliances to stop or limit the growth of small fires. Staff are trained in their use and the appliances are regularly maintained by the suppliers. Table 15.3 summarizes their types and applications. Fire blankets are provided in kitchens where burning pans of oil or fat need to be covered or personnel need to be wrapped to smother ignited clothing.

Water

A 9 l water extinguisher is installed for each 210 m^2 floor area, with a minimum of two extinguishers per floor. A high-pressure CO_2 cartridge is punctured upon use and a 10 m jet of water is produced for 80 s. Water must not be used on petroleum, burning liquids or in kitchens as it could spread the fire.

Dry powder

Dry powder extinguishers contain from 1 to 11 kg of treated bicarbonate of soda powder pressurized with CO_2, nitrogen or dried air. A spray of 2–7 m is produced for 10–24 s depending on size. The powder interrupts the chemical reactions within the flame, producing rapid flame knockdown. The powder is non-conducting and does little damage to electric motors or appliances. A deposit of powder is left on the equipment.

Foam

Portable foam extinguishers may contain foaming chemicals that react upon mixing or a CO_2 pressure-driven foam. They cool the combustion, exclude oxygen and can be applied to wood, paper, textile or liquid fires. Garages are a particular application. Sizes range from 4.5 to 45 l. A 7 m jet is produced for 70 s with a 9 l capacity model.

Vaporizing liquid

Vaporizing liquid extinguishers use bromochlorodifluoromethane (BCF) or bromotrifluoromethane (BTM). These are 1–7 kg extinguishers containing a nitrogen-pressurized liquefied

Table 15.3 Type of portable fire extinguisher.

Group	Extinguishing agent	Fire type	Action	Colour
1	Water	Class A	Cools	Red
2	Dry powder	All	Flame interference	Blue
3	Foam	Class B	Excludes oxygen	Cream
4	Carbon dioxide (CO_2)	Classes B, E	Excludes oxygen	Black
5	Vaporizing liquid	Small fires, motor vehicles, class E	Flame interference	Green

halogen gas, which is highly efficient at interrupting the flames of chemical reactions and producing rapid knockdown without leaving any deposit. They are more powerful than CO_2 extinguishers and are used on electrical, electronic and liquid fires. Halogen is used for outdoor fires and motor vehicles, where the toxic vapour given off is adequately ventilated. They are not suitable for enclosed areas because of the danger to occupants. These are CFCs (p. 151) and are part of the international agreement to cease their use. A suitable replacement for fire-fighting is being sought.

Carbon dioxide

Pressurized CO_2 extinguishers leave no deposit and are used on small fires involving solids, liquids or electricity. They are recommended for use on delicate equipment such as electronic components and computers. The CO_2 vapour displaces air around the fire and combustion ceases. There is minimal cooling effect, and the fire may restart if high temperatures have become established. Water-cooling backup is used where appropriate.

Fixed fire-fighting installations

Various fire-fighting systems are employed in a building so that an appropriate response will minimize damage from the fire and the fire-fighting system itself. Backup support for portable extinguishers may be provided by a hose reel installation and this can be used by the staff while the fire brigade is called.

Some public buildings, shops and factories are protected by a sprinkler system, which only operates directly over the source of fire. This localizes the fire to allow evacuation. Where petroleum products are present, a mixture of foam and water is used. The Fire Officers' Committee (FOC) rules should be consulted for further information.

Hose reels

Hose reels are a rapid and easy to use first-aid method, complementary to other systems and used by the building's occupants. They are located in clearly visible recesses in corridors so that no part of the floor is further than 6 m from a nozzle when the 25 mm bore flexible hose is fully extended.

The protected floor area is an arc 18–30 m from the reel, depending on the length of the hose. A minimum water pressure of 200 kPa is available with the 6 mm diameter nozzle. This produces a jet 8 m horizontally or 5 m vertically. Minimum water flow rate at each nozzle is 0.4 l/s, and the installation should be designed to provide not less than three hose reels in simultaneous use: a flow rate of 1.2 l/s. Figure 15.1 shows a typical installation.

The local water supply authority might allow direct connection to the water main, and there may be sufficient main pressure to eliminate the need for pressure boosting. Pump flow capacity must be at least 2.5 l/s. The standby pump can be diesel-driven. Flow switches detect the operation of a hose and switch on the pump.

Dry hydrant riser

A dry hydrant riser is a hydrant installation for buildings 18–40 m high where prompt attendance by the fire brigade is guaranteed. A dry riser pipe 100 or 150 mm in diameter is sited within a staircase enclosure with a 65 mm instantaneous valved outlet terminal at each landing. All parts

15.1 Hose reel installation.

of the building floor are to be within 60 m of the hydrant, measured along the line on which a hose would be laid. A test hydrant is fitted at roof level, and also a 25 mm automatic air vent. A double inlet breeching piece with two 65 mm instantaneous terminals is located in a red-wired glass box in an external wall, 760 mm above ground level and not more than 12 m from the riser.

The inlet point is within 18 m of an access road suitable for the fire brigade pumping appliance. A brass blank cap and chain is fitted to each landing valve. The riser is electrically earthed. Landing valves are 1 m above floor level and are used by the fire brigade for their own hoses.

Wet hydrant riser

A permanently charged rising pipe 100 mm in diameter or greater supplies a 65 mm instantaneous valved outlet terminal at each floor at a pressure of between 410 and 520 kPa. The upper pressure limit is to protect the fire brigade hoses from bursting and is achieved by fitting an orifice plate restriction before the landing valve on the lower floors of a tall building. The maximum static pressure in the system when all the landing valves are shut is limited to 690 kPa by recirculating water to the supply tanks through a 75 mm return pipe.

Each hydrant valve is strapped and padlocked in the closed position. They are 1 m above floor level and are only used by the fire brigade for buildings over 60 m high which extend out of the reach of turntable ladders. The maximum normally permitted height is 60 m for a low-level break tank and booster set. Higher buildings have separate supply tank and pump sets for each 60 m height.

Pressure boosting of the water supply is provided by a duplicate pump installation capable of delivering at least 23 l/s. Pumps are started automatically on fall of water pressure or water flow commencement. Audible and visual alarms are triggered to indicate booster plant operation.

A break tank capacity of 11.4–45.5 m^3 is required and mains water make-up rate is 27 or 8 l/s for the larger tank. Additionally, four 65 mm instantaneous fire brigade inlet valved terminals are provided at a 150 mm breeching fitting in a red wired-glass box in an external wall, as described for the dry hydrant riser. The box is clearly labelled.

A nearby river, canal or lake may also be used as a water source with a permanently connected pipe from a jack well and duplicate pumps.

Pneumatic pressure boosting is used to maintain system pressure in a similar manner to that shown in Fig. 6.1. The standby pump may be driven by a diesel engine fed from a 3–6 h capacity fuel storage tank providing a gravity feed to the engine.

Foam inlets

Oil-fired boiler plant rooms and storage tank chambers in basements or parts of buildings have fixed foam inlet pipework from a red wired-glass foam inlet box in an outside wall as for the dry hydrant riser.

A 65 or 75 mm pipe runs for up to 18 m from the inlet box into the plant room. The fire brigade connect their foam-making branch pipe to the fixed inlet and pump high-expansion foam onto the fire. The foam inlet pipe terminates above the protected plant with a spreader plate. A short metal duct may be used as a foam inlet to a plant room close to the roadway. Vertical pipes cannot be used and the service is electrically bonded to earth.

On-site foam-generation equipment is available and may be used for oil-filled electrical transformer stations. In the event of a fire, the electricity supply is automatically shut off, a CO_2 cylinder pressurizes a foam and water solution and foam spreaders cover the protected equipment.

Automatic sprinkler

High-fire-risk public and manufacturing buildings are protected by automatic sprinklers. These may be a statutory requirement if the building exceeds a volume of 7000 m^3. Loss of life is very unlikely in a sprinkler-protected building. Sprinkler water outlets are located at about 3 m centres, usually at ceiling level, and spray water in a circular pattern. A deflector plate directs the water jet over the hazard or onto walls or the structure.

Each sprinkler has a frame containing a friable heat-sensing quartz bulb, containing a coloured liquid for leak detection, which seals the water inlet. Upon local overheating, the quartz expands and fractures, releasing the spray. Water flow is detected and starts an alarm, pressure boosting set and automatic link to the fire brigade monitoring station.

Acceptable sources of water for a sprinkler system are as follows:

1. a water main fed by a source of 1000 m^3 capacity where the correct pressure and flow rate can be guaranteed;
2. an elevated private reservoir of 500 m^3 or more depending on the fire risk category;
3. a gravity tank on site, which can be refilled in 6 h, with a capacity of 9–875 m^3 depending on the fire risk category;

4.   an automatic pump arranged to draw water from the main or a break tank of 9–875 m^3 capacity;
5.   a pressure tank: a pneumatic pressure tank source can be used for certain light fire risk categories or as a backup facility to some other system.

Sprinkler installations are classified under four principal types.

1.   Water-filled pipes are permanently charged with water.
2.   Dry pipework: pipes are filled with compressed air and used where pipework is exposed to air temperatures below 5°C or above 70°C.
3.   Alternate system: pipes are filled with water during the summer and air in the winter.
4.   A pre-action system is a dry pipe installation but has additional heat detectors which pre-empt the opening of sprinkler heads and admit water into the pipework, converting it to wet-pipe operation.

Different types of sprinkler head are used depending on the hazard protected, their object being to produce a uniform density of spray.

Fusible link: a soldered link in a system of levers holds the water outlet shut. At a predetermined temperature of 68°C or greater, the solder melts and water flow starts.
Chemical: similar to the fusible link but using a block of chemical, which melts at 71°C or greater, depending on the application.
Glass bulb: a quartz bulb containing a coloured fluid with a high coefficient of expansion, which fractures at 57°C or more.
Open sprinkler heads (deluge system): these are used to combat high-intensity fires and protect storage tanks or structural steelwork. They are controlled by a quick-opening valve actuated from a heat detector or a conventional sprinkler arrangement. A drencher system provides a discharge of water over the external openings of a building to prevent the spread of fire.

Each sprinkler installation must be provided with the following:

1.   main stop valve, which is strapped and padlocked in the open position to enable the water flow to be stopped after the fire is extinguished;
2.   alarm valve: differential pressure caused by water flow through the valve opens a branch pipe to the alarm gong motor;
3.   water motor alarm and gong: water flow through a turbine motor drives a rotary ball clapper within a domed gong to give audible warning of sprinkler operation and commence evacuation of the building.

A satisfactory pipework installation serving an automatic sprinkler distribution and hose reels is shown in Fig. 15.2. Two 65 mm instantaneous fire brigade inlet pipes are provided at a clearly marked access box.

Carbon dioxide

Carbon dioxide is used in fixed installations protecting electrical equipment such as computer rooms, transformers and switchgear. Heat or smoke detectors sound alarms and CO_2 gas floods the room from high-pressure storage cylinders. Pipework transfers the CO_2 to ceiling and

15.2 Water supply to hose reel and sprinkler installation.

under floor distributors. System initiation can be manual or automatic but complete personnel evacuation is essential before CO_2 flooding is allowed.

Fixed BCF, BTM and dry powder

Extinguishers are installed within rooms or false ceilings and are operated from a manual push-button or automatic fire detector. Personnel evacuation is followed by the release of halogen gas to flood the room with a 5% concentration in air, which is sufficient to inhibit fire.

Fire detectors and alarms

Detection of a potentially dangerous rise in air temperature or pressure or the presence of smoke is required at the earliest possible moment to start an alarm. Evacuation of the build-ing and manual or automatic contact with the fire brigade monitoring switchboard should take place before people are at risk. Means of detection can be combined with security surveillance. Fire detection takes the following forms.

Hazard detectors

Hazard detectors give an early warning of the risk of a fire or explosion.

Temperature rise: a local rise in temperature leads to the melting of a fusible link in a wire holding
    open a valve on a fuel pipe to a burner, thermal expansion of a fluid-filled bellows or capillary
    tube or movement of a bimetallic strip to make an alarm circuit.
Flammable vapour detector: gas, oil, petrol or chemical vapour presence is detected by a catalytic
    chemical reaction.
Diffusion: butane and propane vapour diffusion through membrane is detected.
Explosion: rise of local atmospheric pressure above a set value, or at a fast rate, is detected.

Ionization smoke detector

Ionization smoke detectors contain a radioactive source of around 1 micro curie, typically
americium-241, which bombards room air within the detector with alpha particles (ionization).
Electrical current consumption is 50 µA. The presence of smoke reduces the flow of alpha ions;
the electric current decreases and at a pre-set value an alarm is activated.

Visible smoke detector

A source of light is directed at a receiving photocell. Smoke obscures or scatters the light and an
alarm is triggered.

Laser beam

A laser beam is refracted by heat or smoke away from its target photocell and an alarm is
initiated. A continuous or pulsed infrared beam can be transmitted up to 100 m and can be
computer-controlled to scan the protected area. It can also serve as an intruder alarm.

Closed-circuit television

Manned security monitoring also acts as fire and smoke detection. Infrared imaging cameras
reveal overheating of buried pipes and cables and can detect heat sources unseen by visual
techniques.
    Fire alarms are a statutory requirement. Audible bells, sirens, klaxons, hooters and buzzers
are arranged so that they produce a distinctive warning. A visual alarm should also be provided
throughout a building. Breakable glass call points are located 1.4 m above floor level within 30 m
of any part of the premises.
    The electrical system for fire detectors comprises alarms, a central control panel, an incoming
supply and distribution board, emergency batteries, a battery charger and fire-resistant cable. A
permanent cable or telephone line connection is made to the fire brigade and computer-controlled
monitoring indicates any system faults.

Smoke ventilation

Positively removed smoke through automatically opened roof ventilators can greatly aid escape
and reduce smoke damage, often localizing a fire that would otherwise spread.
    The spread of smoke through ventilation ductwork is arrested by fire dampers where fire
compartments within the building are crossed. Fire dampers may be motorized or spring-loaded

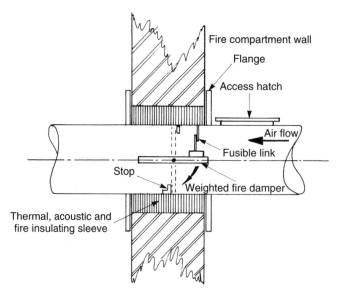

15.3 Hinged deadweight single-blade fire damper in a ventilation duct.

multi-leaf, eccentrically pivoted flaps, sliding plate or intumescent paint-coated honeycombs which swell and block on heating. A typical arrangement of a pivoted flap damper is shown in Fig. 15.3.

An air pressurization ductwork and fan system is switched on at the commencement of a fire to inject outdoor air into escape routes, corridors and staircases. The staircase static air pressure is maintained at 50 Pa above that of adjoining areas to overcome the adverse force caused by wind, mechanical ventilation and the fire-produced stack effect ventilation pressure. This ensures that clear air is provided in the escape route and smoke movement is controlled.

Questions

1. List the sources of fire within a building and describe how they may develop into a major conflagration. State how the spread of fire is expected to be limited by good building and services practice.
2. List the ways in which fire and smoke are detected and fire-fighting systems are brought into action.
3. Describe the methods and equipment used to fight fires within buildings in their likely order of use.
4. State the principal hazards faced by the occupants of a building during a fire. How are these hazards overcome? Give examples for housing, shops, cinemas, office blocks, single-storey factories and local government buildings.
5. Sketch and describe the fire-fighting provisions necessary in large industrial oil-fired boiler plant.
6. How are water and foam systems used to protect building structures from fire damage?
7. Compare a fixed sprinkler installation with other methods of fire-fighting. Give three applications for sprinklers.
8. Explain how sprinkler systems function, giving details of the alternative operating modes available. State the suitable sources of water for sprinklers.

9. Tabulate the combinations of fire classification and types of extinguisher to show the correct application for each. State the most appropriate fire-fighting system for each fire classification and show which combinations are not to be used.

10. Which are correct about fire-fighting services? More than one answer is correct.

    1. Primarily are to minimize damage to the building and its services systems.
    2. Primarily to prevent and minimize danger to people.
    3. Secondary purpose is to save the building and its continued use.
    4. Are only used by professional fire fighters.
    5. Have both fixed and portable fire extinguishers.

11. Hose reel fire-fighting systems:

    1. Are not suitable for office buildings.
    2. Never need to be tested.
    3. Should not be intrusively visible.
    4. Are for anyone to use to commence fire-fighting until professionals arrive.
    5. Only used in public buildings.

12. Which are true about fire-fighting systems in large buildings? More than one correct answer.

    1. Never need to be tested after being commissioned.
    2. Have a fire control panel identifying which parts of the system are activated.
    3. Have an indicator board in the entrance to the building advising fire location.
    4. Are all filled with water.
    5. Are always connected to the computer-based building management system.

13. Which are true about sprinkler fire-fighting systems? More than one correct answer.

    1. Have sprinklers located on a 3.0 m grid pattern covering the floors.
    2. Can have sprinklers mounted on a sidewall 150 mm from the ceiling.
    3. Have self-acting outlets heads that fracture on rise of air temperature.
    4. Are part of a fire-fighting strategy.
    5. Provide a high degree of security against fire damage.

14. Which are correct about aerosol chemical gas fire extinguishing? More than one correct answer.

    1. Is a solid aerosol-generating element.
    2. Developed from solid rocket fuel technology.
    3. Is an inert non-toxic solid that can be activated by heat to release a gas.
    4. Aerosol prevents fresh oxygen reaching the combustion zone.
    5. Aerosol is a fast-acting explosive response.

15. Sprinkler fire-fighting systems:

    1. Have sprinklers located on a 6.0 m grid pattern covering the floors.
    2. Must be manually turned on.
    3. Have self-acting outlets heads that fracture on rise of air temperature.
    4. Are the only fire-fighting system a building needs to have.
    5. Creates unnecessary water damage.

16. Which are correct about fire detection and alarm systems? More than one correct answer.

    1.   Thermal detectors in office ceilings.
    2.   Visible smoke laser light detectors in rooms.
    3.   Detailed drawings of detectors wired in series with each other.
    4.   Breakable glass alarm call points in all areas.
    5.   Automatic connection to the fire brigade call centre.

17. Smoke control during a building fire:

    1.   Is not a critical hazard for personnel.
    2.   Materials within buildings are sources of toxic chemicals.
    3.   Smoke control exhaust fans remove air from the building.
    4.   Escape routes are protected with water sprinkler systems.
    5.   Air pressure differentials across doorways control smoke movement to aid escape.

18. Identify the essential components of a fire:

    1.   Fuel and air.
    2.   Source of ignition and combustible material.
    3.   Paper, wood, solvents, air and warmth.
    4.   Fuel, oxygen and ignition temperature.
    5.   Fuel, air and high-temperature radiated heat.

19. How does a fire commence?

    1.   A small flame radiates combustibility over a long distance.
    2.   Adjacent buildings cannot be set on fire.
    3.   Combustible material, liquid or gas becomes raised to its ignition temperature in the presence of oxygen.
    4.   Electrical services often start fire.
    5.   Any spark from a light switch or plant switch can start a fire.

20. Which is not a means of extinguishing fire?

    1.   Deprive the fire of air.
    2.   Cool the burning material.
    3.   Cease the supply of more fuel.
    4.   Calling the fire brigade.
    5.   Closing doors and windows and evacuating the building.

21. Which is a means of extinguishing fire?

    1.   Disconnect electrical supply from electrical service or item on fire.
    2.   Switch lights off and leave that floor level.
    3.   Cover burning photocopier with wool blanket and leave that floor.
    4.   Spray water onto burning electrical heater and evacuate.
    5.   Throw a fireproof blanket over a burning computer.

16 Plant and service areas

Learning objectives

Study of this chapter will enable the reader to:

1. identify the actions necessary prior to commencement of a construction, in order to facilitate the correct provision of utility services;
2. coordinate utility services under public footpaths;
3. design suitable routes for utility services;
4. calculate the areas of buildings needed for services plant;
5. find the sizes of plant room needed for all services from preliminary building information at the design stage;
6. allocate routes for services through the building structure;
7. understand the need for fire barriers in service ducts and correctly locate them;
8. choose suitable sizes for pipe service ducts;
9. understand the need to allow space for pipes crossing over each other;
10. estimate sizes for service shafts carrying air ducts;
11. know the requirements for walkway arid crawl way ducts;
12. understand the need for expansion provision in pipework;
13. identify the ways in which pipes can be supported with allowance for thermal movement;
14. understand the ways in which thermal movement is accommodated;
15. know how thermal, fire, support and vibration measures are applied to pipes air ducts passing through fire barriers;
16. apply flexible connections to plant;
17. appreciate the need for coordinated drawings;
18. understand the use of services zones;
19. be able to design boiler house ventilation.

Key terms and concepts

air-duct support 376; air-handling plant 366; boiler room 365; cold- and hot-water storage 365; combustion air ventilation 378; cooling plant 366; coordinated drawings 377; coordinated service trench 362; crawl ways and walkways 370; electrical substation 367; fire compartment 368; flexible connection 377; foam seal 375; footpath 363; fuel handling 366; highway 363; lifts 367; noise and vibration 376; pipe anchor, guide, roller, loop expansion and insulation 375; pipe bellows, articulated and sliding joint 375; public utilities 362; service ducts 368; service plant area 363; services identification 364; services zones 377; standby generator 367; telephones 366; temporary works 362; under floor pipe ducts 369.

Introduction

The building design team needs information on the size and location of services plant spaces and their interconnecting ducts before the engineer has sufficient data on which to base calculations. There may be only general definitions of the spaces to be heated or air conditioned and preliminary design drawings. Past experience of similar constructions reveals the likely plant and service duct requirements.

The use of such data is explained and worked examples are used. The planning of external utility supplies is shown, together with typical arrangements for internal multi-service ducts.

Support and expansion provision for distribution services is demonstrated, as is the design of combustion air ventilation for boilers.

Mains and services

The planning and liaison with public utilities (National Joint Utilities Group, 1979) must be included in the initial application made by the developer for planning permission. Each utility requires detailed information at the estate design stage in order to facilitate the following:

1. siting of plant or governor houses, substations, service reservoirs, water towers and other large items of apparatus and also early completion of associated easements and acquisition of and early access to land in order to ensure service to the development by the programmed date;
2. design of mains and service layouts;
3. location of and requirements for road crossings;
4. provision and displacement of highway drainage;
5. programming of cut-offs from existing premises that are to be demolished;
6. arrangements for protecting and/or diverting existing plant and services;
7. provision of supplies to individual phases of the development, including temporary works services;
8. acquisition of materials and manpower resources;
9. siting of service termination and/or meter positions in premises and service entry details;
10. provision of meter-reading facilities;
11. provision of public lighting.

Developers must provide information on the following:

1. the intended position of public carriageways, verges, footpaths, amenity areas and open spaces;
2. existing and proposed ground levels;

3. the position and level of proposed foul and surface-water sewers and any underground structures.

The utility will inform the developer of the need to close or restrict roadways, and these matters will then be discussed with the local authorities.

All main services to more than one dwelling should be located on land adopted by the Highway Authority. The location on private property of a main designed to serve a number of dwellings can lead to friction between residents if excavation for repairs or maintenance is needed, and also makes it difficult for the utilities to gain ready access in an emergency.

With the exception of road crossings, mains and services other than sewers should not be placed in the carriageways. The routes chosen should be straight and on the side of the carriageway serving most properties. Any changes of slope should be gradual. The prior approval of the utilities must be sought if landscaping will alter the levels of underground services.

Public sewers must be laid to appropriate levels and gradients in straight lines between manholes, usually under the carriageway.

An underground clear width of 1.8 m is needed between a private boundary and the kerb foundations but extra allowance should be agreed for the following:

1. fire hydrants;
2. inspection covers and manholes;
3. large radius bends for pipes and cables;
4. fuel oil distribution pipes;
5. district heating pipework and manholes;
6. through-services not connected to the development;
7. cross-connections between services to form ring mains rather than dead ends;
8. imposed loads from adjacent buildings – medium-pressure gas mains must be 2 m from the building line.

Protective measures are taken where there is a risk to pipes or cables from vehicles that may park on soft ground. Where special paving is used, early consultation with the utilities will help to avoid subsequent defacement due to maintenance work.

Footpaths should be used for the utilities. Sewers need to be laid in conjunction with the early stages of road building. Utilities operate under statutory powers and will not carry out work as a subcontractor. On completion of site construction, a copy of the plans showing the installed routes and details of the mains and services is sent to each utility to enable permanent records to be established.

The minimum dimensions for the locations of mains and services under a pavement are shown in Fig. 16.1. Brick tiles, concrete covers or yellow marker tapes are put over 11 kV and 415 V electricity cables. The 11 kV cables have a red PVC over-sheath and 415 V cables have a black PVC over-sheath.

The mains and services are surrounded with selected backfill that is free of sharp or hard objects, and the trench is filled and compacted with earth that is free of rubble or site debris.

Plant room space requirements

General provision

Coordination of the services with architectural and structural design is required at the earliest possible moment during conception of the project. Plant rooms are places of physical work as

16.1 Positions of main services in straight routes under footpaths on residential estates.

well as equipment locations. Access to them is restricted to approved people under the control of the building manager. Safety, noise control, natural and mechanical ventilation, heating and cooling, ease of access from stairways and lifts, equipment and exit signage, adequate lighting and low-voltage power outlets all make for a successful building. Many plant rooms remain almost unchanged for 25 years or more, so a well thought-out facility that can be maintained is more than an after-thought. Keeping them clean, decorated and well lit is an important part of the workplace. Telephone communication with the building manager from each plant room is a life-safety issue as well as a vital means of communication when work is being conducted or faults occur; mobile phones may be out of contact due to the heavy presence of concrete, steel and noise. Plant rooms must not be used to store flammable materials, cleaning fluids, cleaning machines, furniture; they are not junk or store rooms, they are specialist and dangerous work places, often hot and noisy. Inappropriate access is not permitted.

New plant rooms must be designed to provide full access for maintenance and repair. Existing plant rooms often appear overcrowded to the maintenance engineer and energy auditor. Adequate head room for safe access and movement is necessary. Noise from refrigeration compressors, hospital medical air compressors and hospital medical suction vacuum pumps is always uncomfortable; hearing protection often advisable. Large fans generally do not create excessive noise within plant rooms; if they did, duct-borne noise would be troublesome downstream and possibly elsewhere. Fuel-fired water heaters and steam boilers usually do not make excessive noise that leaves the plant room. Large main plant rooms house the maintenance engineers office, record drawings and commissioning manuals of everything installed, equipment

Table 16.1 Services plant room space requirements.

Building type	Plant room floor area as a percentage of building floor area, excluding the plant room
Simple factory or warehouse	4
Most types	9
Small, well-serviced hospital	15

storage, duplicated water pumps, isolating valves, extensive pipework, water treatment tanks, anti-corrosion dosing tanks, mechanical switch boards and building management system field panels. This is where the building management system front end computer is often located as continuous access use is made of it by the maintenance team and contractors.

The design stage of a new building is too early for heating and cooling loads, plant sizes and system types to be known with any certainty, but reliable information is required to form the basis for decisions. Building Services Research and Information Association (BSRIA) surveys (Bowyer, 1979) of existing buildings have shown that their plant room requirements can be expressed as a percentage of floor area, as given in Table 16.1.

Some of the outline requirements for services plant rooms quoted by Bowyer (1979) are as follows.

Cold-water storage

The volume of cold water to be stored to cover a 24-h period is calculated from the building's occupancy and type. An incoming break tank may be required at ground or basement level for pneumatically boosted systems. Fire-fighting services may need water storage at ground level.

Tanks can be 1, 2 or 3 m high, with 1 m clearance allowed around them for insulation, pipework and access.

Hot-water storage

The space needed for vertical indirect hot-water storage cylinders, secondary pumps, pipework, valves, controls and heater coil withdrawal is given by:

$$\text{plant room floor area m}^2 = 1.7 \times \text{cylinder volume m}^3 + 10$$

Room heights of 3–4.8 m are needed depending on cylinder height.

Boilers

For buildings constructed to the Building Regulations, an assessment of the boiler power in watts can be made by multiplying the heated volume in metres cubed by 30.

The required boiler plant room floor area is:

$$\text{plant room floor area m}^2 = 80.99 + 31.46 \times \ln (\text{boiler capacity MW})$$

Plant room heights are up to 5 m. Domestic and small commercial boilers are accommodated within normal ceiling heights and their floor areas are not predicted with this equation. The area calculated allows for two equally sized boilers, pipework and water treatment, pressurization and pumping equipment.

Fuel storage and metering

Electricity and gas meters are part of the incoming services accommodated within the plant room space calculated for other equipment. Partition walls and access for reading and removal are required.

Two equally sized oil storage tanks, supports, tanked catch pit and access are accommodated in:

oil storage tank room area $m^2 = 22.52 + 0.64$ (oil storage volume m^3)

Plant room height is up to 4.5 m. Tanks are frequently located externally and stood on three brick or concrete block piers so that oil will flow by gravity to the burners. They are of mild steel and protected with bitumen paint.

Air handling

The air-handling plant room size is assessed by assuming that the mechanically ventilated parts of the building have between 6 and 10 air changes/h. The expected supply air volume flow rate is:

$$Q = \frac{NV}{3600} \ m^3/s$$

where V is the volume of ventilated space (m^3) and N is the number of air changes per hour (6–10).

The plant room will be 2.5–5 m high depending on the sizes of fans, ducts, filters, heater and cooler coils, humidifiers and control equipment. The floor area is:

plant room floor area $m^2 = 6.27 + 7.8 \times Q \ m^3/s$

A fresh-air-only system, such as an induction system, is sized on 1 air change/h.

Cooling plant

Refrigeration plant capacity may be as high as 30 W/m^3 of building volume, and an early estimate of heat gains should be made. A Building Energy Estimating Programme (BEEP) is available through the electricity supply authority, and other computer packages are in use.

Plant room height is 3–4.3 m and the floor area is:

plant room floor area $m^2 = 80.49 + 35.46 \times \ln$(cooling load MW)

The area is for two refrigeration machines, pumps, pipework and controls. Additional space on the roof is needed for the cooling tower.

Communications

Digital telephone technology has generally taken over from mechanical switch rooms within buildings. Telephone server computers may be required. Roof-mounted microwave communication transmitters, receivers and satellite dishes need clearly defined manual exclusion space due to high-intensity radiation hazards. A secure fenced communication roof area separates regular daily access from that to mechanical plant rooms and cooling towers. There are no water or gas services within communications plant rooms.

Computer servers

Commercial buildings with workstation computers have a server room in continuous operation. An office of 20 or more workstations has a significant server requirement and external cable data modem link equipment to the internet. Intermittent work at the PC front end computer by the network operator is needed, as well as space for expansion of electrical equipment, more cabling, fault servicing and storage of documents relating to the network. Continuous server computers with backup data storage disk drives generate considerable heat gains, often within internal closed rooms having no natural ventilation or cooling. Dedicated 24-h operation mechanical ventilation, recovery of useful heat and mechanical cooling often does not coordinate easily with the general office air-conditioning systems and requires careful design and implementation. Electrical equipment is only rated for operation at up to 40°C, easily reached within an unventilated closed internal room full of computers and screens, making suitable temperature control vital for the functioning of the purpose of the building, reliability of computer systems and physical security of the server installation. There are no water or gas services within these rooms. Adequate lighting and low-voltage power outlets are provided.

Lifts

An early assessment of requirements is made in conjunction with the lift engineer. Bowyer (1979) gives further information. Each electrical lift has a lift motor room on top of the shaft that is several times the floor area of the shaft. Lifts are grouped close together and the room level lift motor room covers all shafts plus control equipment. Older electrical lifts have mechanical switch control panels of considerable size. A lifting beam, lifting gear and lift motor concrete bases along with one or more spare lift motors are provided for repair work. Digital control with frequency inverters is normal for modern lift installations, requiring a large control panel. Lighting and access for regular inspections, maintenance and large item replacement add to the lift motor room size. A lift motor room on top of a tall air-conditioned building has significant internal and external heat gains. Natural ventilation is always provided but mechanical air flow and cooling are provided where there is a significant risk of the lift motor room exceeding a safe environmental working temperature for the staff and the electrical equipment. Smaller capacity electric lifts have the driving motor and control gear installed on the support frame of the top of each car, so only wire ropes and the counterweight are within the lift shaft and there is no lift motor room.

Hydraulic lifts may be used to service up to around six floors. The lift car is raised from hydraulic rams and cables within the lift shaft. They are silent and slow speed, some allowing glass-sided cars as there are no visible cables. Motive power comes from a hydraulic pump within the oil storage tank alongside the base of the lift shaft. Lowering the car pushes oil back into the tank while car raising is pumped. The submerged pump is oil cooled and uses minimum energy.

Electrical substation

The incoming high-voltage supply is located in a substation, which may be external or on an external wall of the building. The floor area needed is 35 m^2 for a 200 kVA load and up to 48 m^2 for a 2000 kVA load.

Standby diesel electric generator

A standby diesel electric generator supplies emergency electrical power of up to 100% of the connected load from the mains. Often only having the capacity to maintain essential lighting, a minimum lift service, stairway pressurization fans, hospital operating theatres and any other essential service for the site. The plant room will be adjacent to the substation and will be up to 4 m high. The floor area required is 18 m² for 50 kVA up to 37 m² for 600 kVA, plus a diesel oil storage tank for 7 days of continuous running.

Service ducts

Service ducts are passageways that traverse vertically and horizontally throughout a building, or between buildings, large enough to permit the satisfactory installation of pipes, cables and ducts, together with their supports, thermal insulation, control valves, expansion allowance and access for maintenance. Each service duct might be constructed as a fire compartment, and BS Code of Practice 413: 1973 should be consulted. An example of current practice is given in Fig. 16.2.

Casings and chases of 100 mm diameter or less are fire stopped to the full thickness of the wall or floor. The passage of a service must not reduce the resistance of the fire barrier. Plastic pipes can soften and collapse during a fire and allow the passage of flames and smoke. A galvanized steel sleeve with an intumescent liner can be used to surround the plastic pipe where a fire stop is needed. When its temperature rises to 150°C, the intumescent liner expands inwards to close the softened pipe and seal the wall aperture.

Service duct sizes can be found from an estimate of ventilation air supply rate Q m³/s, doubled to allow for the recirculation duct, with an assumed air velocity of 10 m/s for vertical ducts within brick or concrete enclosures or 5 m/s in false ceilings where quiet operation is important. At least 150 mm clearance is allowed between ducts and other surfaces for thermal insulation, jointing, supports and access.

16.2 Service duct fire compartment.

Fan noise is contained within the air-handling plant room by acoustic attenuators, anti-vibration machine mountings and heavy concrete construction.

The total floor space taken by vertical pipe and cable routes will be up to 1% of the gross floor area. Horizontal service ducts and false ceilings 500 mm deep are used for air-conditioning ducts and other services. Recessed luminaires and structural beams encroach into the nominally available spaces.

Under floor service ducts should be constructed to allow access for jointing and maintenance. The minimum standard for an under floor duct is shown in Fig. 16.3. The duct route is accurately marked on installation drawings and access is gained by breaking the screed. Hot-water pipes are insulated with 50 mm thick rigid glass fibre and wrapped with polyethylene sheet, sealed with waterproof tape. The duct is filled with dry sand. Sufficient depth is allowed for branches to cross over the other pipes. Recommended sizes are given in Table 16.2 (Butler, 1979b).

Vertical and other service ducts can be sized in a similar manner by allowing a 50 mm gap between thermal insulation and other surfaces. Additional smaller pipes run in the same duct will require an increase in the width.

When builder's work holes for services are specified, the dimensions of the structural opening required should always be used, rather than the nominal pipe diameter.

16.3 Minimum standard for an under floor service duct for pipework.

Table 16.2 Minimum under floor duct sizes for pipework.

Hot-water flow and return nominal diameter (mm)	Under floor duct dimensions width and depth (mm)
15	294
22	304
28	348
35	364
42	376
54	400

An under floor, or underground, crawl way or walkway has the following features:

1. crawl way duct height 1.4 m;
2. walkway duct height 2 m;
3. 750 mm clear width between fixtures;
4. reinforced concrete construction;
5. floor laid to fall to a drainage channel along the length of the duct, with connections to the surface drainage system;
6. watertight access manhole covers at intervals with built-in galvanized steel stepladders;
7. watertight lighting fittings and power sockets;
8. services are painted to appropriate British Standards colours and clearly labelled, and control valves are numbered with an explanation list provided;
9. services branching into side ducts do not block through-access.

EXAMPLE 16.1

A six-storey office building of 30 m × 25 m is to be constructed on a suburban green field site. It is to be air conditioned with a low-energy system using gas and electricity as energy sources. The main air-handling unit, AHU, plant room is to be built on the roof. There will be 225 occupants. Calculate the plant room and service duct space requirements for the preliminary design stage. Only outside air is to be passed through the distribution ductwork system.

It is expected that the plant rooms will require 9% of the floor area of $(30 \times 25 \times 6)$ m^2, and so a first estimate is 405 m^2. This will be mainly on the roof as the heating system water heater, air-conditioning cooling plant of chiller and cooling tower, domestic hot-water heater and storage cylinder, pumps and main switchboards are normally located there also; thus an oblong room of dimensions l and $2l$ could be used to establish the overall space requirement. This space can then be distributed around the building for the various services. Cold-water storage tanks are often located on a mezzanine floor above the heating plant. Fire-fighting water storage tanks are used in some buildings where needed for the sprinkler system; these, and any oil storage tanks, are always located in basement or ground floor plant rooms; neither are needed in this case. The main electricity meters for the central building services and tenants floors, main electricity distribution board, water and gas meters are usually located in a basement or ground floor plant room. The building management system computer, a server computer with screen, keyboard, printer and data backup storage tape or disc, is sometimes located in a plant room, where only the BMS maintenance contractor ever uses it, but it should definitely be in the building manager's office where it can be used every day. The building manager should be using this computer for condition, fault and energy monitoring continuously.

$$\text{area} = l \times (2l) = 405 \text{ m}^2$$

$$2l^2 = 405 \text{ m}^2$$

$$l = \left(\frac{405}{2}\right)^{0.5} = 14.23 \text{ m}$$

A plant room of 14.23 m × 28.5 m could be accommodated on the roof. However, some of this space will be located lower down the building.

A further estimate of the requirements can be made through consideration of each service.

1. Cold-water storage of 45 l per person:

$$\text{volume} = 225 \text{ people} \times 45 \frac{l}{\text{person}} \times \frac{1 \text{ m}^3}{10^3 \text{ l}} = 10.1 \text{ m}^3$$

tank dimensions = 2.25 m × 4.5 m × 1 m

Add 1 m all round the tanks. Thus the plant room floor space is 27.6 m². This could be reduced by stacking the tanks if there is sufficient headroom. The water main pressure will be sufficient to reach the roof; ground-level break tanks and pumps will not be needed.

2. Hot-water storage of 5 l per person:

$$\text{volume} = 225 \text{ people} \times 5 \frac{l}{\text{person}} \times \frac{1 \text{ m}^3}{10^3 \text{ litre}} = 1.125 \text{ m}^3$$

$$\text{volume of an indirect cylinder} = \frac{\pi d^2}{4} \times l \text{ m}^3$$

For a cylinder 1 m in diameter, its length l is:

$$l = 1.125 \times \frac{4}{\pi} = 1.43 \text{ m}$$

The floor area required is 1 m².

3. The boiler power is given by:

$$\text{boiler power} = (30 \times 25 \times 6 \times 3) \text{ m}^3 \times 30 \frac{W}{\text{m}^3} \times \frac{1 \text{ MW}}{10^6 \text{ W}}$$

$$= 0.405 \text{ MW}$$

$$\text{plant room floor area} = 80.99 + 31.46 \times \ln 0.405 \text{ m}^2$$

$$= 52.6 \text{ m}^2$$

4. Gas and electricity meters will be housed either in the roof plant room or in cubicles at the rear of the building on the ground floor.

5. The low-energy system air-handling plant will pass only the fresh air supply. The volume flow rate of supply air will be a maximum of 10 l/s per person,

$$Q = \frac{10 \text{ l}}{s} \times 225 \text{ people} \times \frac{1 \text{ m}^3}{10^3 \text{ l}} \text{ m}^3/\text{s}$$

$$= 2.25 \text{ m}^3/\text{s}$$

$$\text{plant room area} = 6.27 + 7.8 \times 2.25 \text{ m}^2$$

$$= 23.8 \text{ m}^2$$

Air change rate from the provision of outdoor air is:

$$N = 3600 \times \frac{2.25 \text{ m}^3}{\text{s}} \times \frac{1 \text{ air change}}{30 \times 25 \times 6 \text{ m}^3}$$

$$= 1.8 \text{ air changes/h}$$

This is acceptable and reasonable. There may be additional recirculated air within the floors from distributed terminal heating/cooling air control units such as variable air volume, induction or fan coil units.

6.　The cooling plant capacity is 30 W/m^3 = 0.405 MW, same as heating power calculated.

$$\text{plant room floor area} = 80.49 + 35.46 \times \ln 0.405 \text{ m}^2$$

$$= 48.4 \text{ m}^2$$

Data provided by Bowyer (1979) do not cover plant smaller than 0.3 MW, so this is likely to be realistic. The floor space needed for chillers and cooling towers does not increase as fast as for heating plant with increasing capacity.

Thus the total plant room space requirements are estimated to be

$$(27.6 + 1 + 52.6 + 23.8 + 48.4) \text{ m}^2 = 153.4 \text{ m}^2$$

These two methods of estimation show that plant room space requirements are of the order of 153–405 m^2. This is a wide spread of answers provided for the design concept stage. It will be refined by detailed design and better knowledge of the plant loads and locations to be used. A low-energy building has the minimum of mechanical services plant for air handling and cooling. Heating systems have traditionally not required much plant room space and hot water radiators and convectors are very compact, mainly consuming wall space. The historical precedent for a 9% floor area requirement is likely to be too high for a modern system.

A vertical service duct is needed from the roof plant room to ground level carrying supply and exhaust air ducts, drainage and water pipework and cables. If the maximum air velocity in the air ducts is 6 m/s, their sizes will be

$$\text{duct cross-sectional area} = \frac{Q \text{ m}^3/\text{s}}{V \text{ m/s}}$$

$$= \frac{2.25}{6} \text{ m}^2$$

$$= 0.375 \text{ m}^2$$

If square ducts are used, they will be 612 mm × 612 mm or larger, such as with standard sizes of 700 mm × 600 mm. An estimated service duct arrangement is shown in Fig. 16.4. This allows for thermal insulation and access to all the services.

False ceilings provide space for the horizontal distribution of services. The induction units will be located along the external perimeter under the windows or within the false ceiling. Holes 150 mm in diameter are needed in the floor slab, one for each unit, for the air duct and, close by, two holes 50 mm in diameter for pipes.

Electrical power distribution cables, building management system control panels, computer data network LAN communication panels, telephone distribution panels, electrical sub-meters for BMS, tenant electrical meters, sub-circuit switchboards. Separate from water services

Metal lattice walkway does not obstruct natural ventilation of vertical service shaft

Doorway

150   750

1545

2500   Piped services

Supply air and exhaust air ducts

16.4 Layout of the vertical service duct in Example 16.1.

Pipe, duct and cable supports

Hot-water pipework can be supported with hardwood insulation rings clamped in mild steel brackets, as shown in Fig. 16.5. Saddle pipe and cable clips, shown in Fig. 16.6, are extensively used because of their low cost. They should be made of a material that is compatible with that of the service. A row of services may be bolted to a mild steel angle iron whose ends are built into the structure. The longitudinal spacing of supports depends on pipe size, material and whether the service is horizontal or vertical. Rollers, as shown in Fig. 16.7, allow pipes to move freely during thermal movement.

Expansion and contraction of short pipe runs is accommodated at frequent bends and branches, the pipes moving within their thermal insulation and non-rigid brackets. Spaces between pipes are sufficient to avoid contact. Long pipe runs need expansion devices, anchors and guides.

1. With a pipe anchor the pipe is rigidly bolted or welded to a steel bracket which is firmly built into a brick or concrete structure.
2. Pipes can be supported by tubular guides, as shown in Fig. 16.8, which allow longitudinal movement with minimum metal contact.
3. Several types of thermal expansion device are used, depending on application, space available and fluid pressure.

    (a) Pipe loops are least expensive in some cases and can be formed where external pipes pass over a roadway. They can be prefabricated with pipe fittings, welded, bent or factory formed, as shown in Fig. 16.9.
    (b) Bellows are made of thin copper or stainless steel and have hydraulic pressure limitations. A complete installation is shown in Fig. 16.10.
    (c) Articulated ball joints take up pipe movement at a change in direction, as shown in Fig. 16.11.
    (d) Sliding joints are packed with grease and the pipe slides inside a larger diameter sleeve.

16.5 Insulated pipe support ring.

16.6 Pipe and cable saddle clip.

16.7 Roller pipe support.

16.8 Tubular guide support.

Formed on site with
pipe fittings or a
bending machine

Prefabricated unit

16.9 Pipe expansion loop.

16.10 Complete pipework installation for thermal expansion provision.

16.11 Articulated expansion joint.

A fire stop unit is used where pipes pass through a fire compartment wall or floor and incorporates structural support, vibration insulation and fire resistance within a steel flanged sleeve which is in two halves, as shown in Fig. 16.12. Silicone fire stop foam is used to seal the space around pipes. When it is exposed to heat, the foam chars to form a hard flame-resistant clinker.

16.12 Insulated pipe sleeve (reproduced by courtesy of Stuart Forbes (Grips Units) Ltd, Working).

16.13 Insulated duct support.

Ventilation ducts are fixed to the building with galvanized mild steel saddle clips for up to 300 mm diameter light-gauge metal; larger ducts have flanged joints, which are suspended with rods from angle brackets. Figure 16.13 shows a typical fixing.

Cables are supported along their entire length by the conduit or a perforated metal tray.

Plant connections

Connections to plant are made in flexible materials to reduce the transmission of vibration from fans, pumps and refrigeration compressors to the distribution services, or to allow greater flexibility in siting the equipment. A fan installation is shown in Fig. 16.14. The discharge air duct has a sound attenuator to absorb excess fan noise. Polyurethane foam held in place by perforated metal sheet is used in the attenuator.

16.14 Flexible connections to an air-conditioning fan.

16.15 Service zones in a false ceiling over a corridor.

Coordinated service drawings

A master set of all service drawings is maintained under the overall charge of a coordinating engineer for the project (Crawshaw, 1976). Drawings and a schedule of builder's work associated with the services are circulated round the construction team. The structural engineering implications of holes through floors and beams are checked at an early stage.

Service space allocation is made on the basis of zones for particular equipment within structural shafts and false ceilings. Each engineering service is restricted to its own zone. Common areas are provided for branches, as shown in Fig. 16.15.

Table 16.3 Minimum free areas of ventilation openings for combustion appliances for outdoor air per kilowatt of heat output.

Position of opening	Minimum free area (mm²)	
	Conventionally flued	Balanced flue
High level	550	550
Low level	1100	550

Boiler room ventilation

Combustion appliances must have an adequate supply of outdoor air, otherwise the fuel will not burn properly and carbon monoxide will be produced. Under down-draught conditions, this will be a danger to occupants. Fatalities have occurred through improper appliance operation.

Good installation practice is to introduce combustion air so that it does not cause a nuisance through draught, noise or poor appearance. Heat and fumes produced by the appliance are ventilated to outdoors through high-level openings.

Any room containing a fuel-burning appliance may be positively supplied with ventilation air by a fan if this is needed for some other purpose. An extract fan must not be used, as combustion products can be drawn into the occupied room. Natural inlet and outlet ventilation is predominant.

The combustion air inlet for a domestic kitchen or living area can be (CIBSE, 1973):

1. through an external wall, just below ceiling level, to enable the incoming cold air to diffuse with the room air above head height; this avoids most draughts and occasional blockage by snow or debris;
2. through an external wall at low level behind a hot-water radiator or other heat emitter; a frost thermostat switches on the heating system at an internal air temperature of 5°C;
3. by direct connection of the combustion air from outside to the appliance casing, locality or enclosing cupboard with an under floor duct. Two suitably sized air bricks are fitted into the external walls of a suspended timber floor on opposite sides of the building. Either a duct connection between the appliance casing and the floor space or a ventilation grille is put into the floor by the heater. A drain pipe or galvanized steel duct can be cast into a concrete floor slab for this purpose.

An appliance of up to 30 kW heat output may be fitted within a compartment, which is ventilated to an adjoining room, which in turn is ventilated to outdoors by the stated openings. Recommended ventilation openings are given in Table 16.3.

Questions

1. List the principal information and activities involved in the provision of main services throughout a housing estate.
2. Sketch a suitable arrangement for the services beneath the public highway and leading into a dwelling. Show the recommended dimensions and explain how the ground is to be reinstated.

3. Estimate the plant room and service duct space requirements of the following buildings, using the preliminary design information given.

(a) A naturally ventilated hotel with a hot-water radiator heating system. Roof and base-ment plant rooms are available. The hotel dimensions are 50 m × 30 m, with 10 storeys 3 m high. Total occupancy is 750. An oil-fired boiler plant is to be used.

(b) A single-storey engineering factory of dimensions 100 m × 40 m, using overhead gas-fired radiant heating. The roof height slopes from 3 m to 5 m at the central ridge. There are 300 occupants. Mechanical ventilators and smoke extractors will be fit-ted in the roof. A standby diesel electricity generator and an electrical substation are required.

(c) A 12-storey city centre educational building of 40 m × 20 m, 3 m ceiling height, with a single-storey workshop block and laboratory area 40 m × 60 m × 4 m. The whole complex is to be mechanically ventilated with 4 air changes/h. Hot-water radiators and fan convectors provide additional heating. Gas and electricity are to be used.

The tower building has a basement with ramp access to ground level. A refectory is located at ground level. The total building occupancy is 2000.

4. Draw the installation of services in a vertical duct through a three-storey office building. The duct is 2.5 m × 1.2 m. Boiler and ventilation plant are in the basement. There are false ceilings on all floors.

5. Sketch and describe how the spread of fire through a building is limited by the services installation.

6. A false ceiling over a supermarket contains recessed luminaires, sprinkler pipework and a single-duct air-conditioning system. The false ceiling is 400 mm and has structural steel beams 250 mm deep. Extract air from the shop passes through the luminaires. Draw the installation to scale.

7. A concrete floor with a wood block finish houses a service duct carrying two 35 mm heat-ing services, two 28 mm hot-water services, a 42 mm cold-water service and 54 mm gas pipework. Side branches are required to carry a maximum of three 22 mm pipes. Continuous access covers are to be provided. The hot-water pipes are to have 50 mm thick thermal insu-lation, and at least 25 mm clearance is needed around the pipes. Draw a suitably detailed design showing dimensions, materials, pipe support, cover construction and pipe routes at the branch.

8. Describe, with the aid of sketches, how successful coordination between all the services can be achieved within builders' work ducts.

9. Explain how fuel-burning appliances fitted in kitchens, living rooms, cupboards and domes-tic garages can be adequately ventilated. Illustrate an example of each location and state the areas of ventilation openings required for appliances of 3, 18 and 40 kW heat output.

10. Who are allowed into plant rooms?

    1.   Everyone in the building.
    2.   All contractors.
    3.   Members of the public.
    4.   Only employed maintenance engineers.
    5.   Those approved by the building manager.

11. What are plant rooms designed to accommodate?

    1.   Only mechanical plant.
    2.   Only electrical plant.
    3.   Plant and people.
    4.   Wiring and pipes.
    5.   Pumps and boilers.

12. What are plant room conditions supposed to be?

    1.   Safe work places.
    2.   Minimum size possible.
    3.   Maximum size needed.
    4.   Showcase for plant items.
    5.   Always out of site.

13. What do we know about computer server spaces?

    1.   None needed, located beneath a desk.
    2.   Located in roof plant space, out of sight.
    3.   Any corner of a room will do.
    4.   Secure accessible and safe room to work in.
    5.   Always very hot places.

14. What do we know about computer server spaces?

    1.   An inconvenient collection of electrical boxes.
    2.   Vital hub of every business and office.
    3.   Nobody ever works in there.
    4.   Do not need ventilation.
    5.   Provide useful heat into the building during winter.

15. What is essential for a computer server facility?

    1.   Basement store room location.
    2.   Empty internal office with lockable door.
    3.   Interior secure work room with mechanical ventilation and temperature-controlled cooling 24 h a day, 365 days a year.
    4.   Partitioned space in basement car park as it is cool there.
    5.   Any office or store room where enough space can be made available.

16. Where is the building maintenance manager's office likely to be located?

    1.   Alongside reception area.
    2.   By the main plant room, often in the basement.
    3.   In the executive office suite on a high-level floor.
    4.   Basement car park.
    5.   In entrance foyer.

17. Which are the most problematic noise sources in plant rooms for maintenance workers?

    1.   None of them are as plant is switched off when work is undertaken.
    2.   Toilet exhaust fans.
    3.   Air-handling units.

    4.   Rotary and reciprocating compressors.

    5.   Gas-fired water heaters.

18. How often do technical workers enter large plant rooms?

    1.   Once a year.

    2.   Daily, several times.

    3.   Monthly service checks and fan belt changes.

    4.   Hourly logging of energy and operational data.

    5.   Once a week.

19. What is the space temperature control requirement for plant rooms?

    1.   None, they always remain cool.

    2.   Does not matter as air temperature never gets too hot for the mechanical plant.

    3.   Wall and roof ventilation openings.

    4.   Nobody works in there so it does not matter.

    5.   Natural and mechanical ventilation for workers and where necessary, cooling to limit room temperature for workers and electrical systems.

20. What is involved with lift motor rooms?

    1.   Driving motor and winding gear are located in basement plant areas.

    2.   Each lift has its driving motor and winding gear mounted above the shaft within a ventilated and cooled roof-level plant room.

    3.   All motors are located on top of each passenger car.

    4.   One electric motor drives all lift cars in a group from a roof-level plant room.

    5.   Sealed concrete plant room above each lift shaft.

17 Mechanical transportation

Learning objectives

Study of this chapter will enable the reader to:

1. understand the principles of passenger and goods transportation within and between buildings;
2. discuss the applications of passenger lift systems;
3. know the speeds and carrying capacities of lifts;
4. know how lift systems are controlled and used during the outbreak of fire in a building;
5. know the principles of electric motor and hydraulic lifts;
6. know the principles and carrying capacities of escalators and passenger conveyors;
7. understand the importance of lift shaft and motor room ventilation;
8. recognize the builder's work required for a lift installation.

Key terms and concepts

builder's work 388; bypass holes 386; carrying capacity 383; car speed 384; collecting mode 383; computer control 383; counterweight 387; driving motor 386; driving pulley 387; escalator 385; fireman's lift 384; geared drive 386; goods lift 384; hydraulic lift 387; lift motor room 389; lift shaft 389; noise 386; passenger conveyor 386; passenger lift 383; paternoster 385; peak demand 383; roping 387; service conveyor 384; service lift 384; travel distance 386; variable-voltage control 386; ventilation 386

Introduction

The mechanical transportation of people and goods is an energy-using service which needs the designer's attention at the earliest stages of building design. Standards of service rise with expectations of quality by the final user and with the provision of access for disabled people.

The principles of transportation systems are outlined and reference is made to movement between buildings. Their energy consumption may be low, but the electrical power requirement is significant for short periods. Integration with other services, fire protection, means of escape and correct maintenance are of the highest importance.

Transportation systems

The mechanical transportation of people and equipment around and between buildings is of considerable importance in relation to the degree of satisfactory service provided. Increasing usage of computer networks, internet, digital communications will gradually reduce travel requirements. Cost-effective and energy-efficient transportation will always be in demand. Walking and cycling are supreme of personal low-cost mobility for the majority of the population. City express cycle lanes for aerodynamically enclosed human-powered vehicles can be provided with present-day technology for cruising speeds of 45.0 km/h. Tunnels and covered above-ground routes could be used for conventional bicycle traffic on large building developments or around towns. When the total concept of global sustainability is thought about, alternatives to consuming highly refined forms of primary energy, as electricity, must be considered.

Permanently installed energy-consuming systems in use are as follows:

Lifts

Passenger lifts are provided for buildings of over three storeys, or less when wheelchair movement is needed. The minimum standard of service is one lift for each four storeys, with a maximum walking distance of 45.0 m between workstation and lobby. Higher standards of service are provided in direct proportion to rent earning potential of the building and prestige requirements.

The peak demand for lift service is assessed from the building size, shape, height and population. Up to 25% of the population will require transportation during a 5-min peak period. Congestion at peak travel times is minimized by arranging the lift lobbies in a cul-de-sac of, say, two lift doors on either side of a walkway, rather than in a line of four doors along one wall.

Computer-controlled installations can be programmed to maximize their performance in a particular direction at different times of the day. Each lift car can be parked at an appropriate level to minimize waiting time. Two lifts of 680 kg carrying capacity, 10 people, provide a better service than one 1360 kg, 20-person lift. The large single lift would run only partly loaded during the major part of the day, with a resulting decrease in efficiency and increased running cost. One of the smaller lifts could be parked for long periods to reduce costs. The advantages of using two smaller lifts may be considered partly to outweigh the additional capital and maintenance costs.

Car speed is determined by travel distance and standard of service. Buildings of more than 15 storeys may have some high-speed lifts not stopping for the lower 10 storeys. A 49-storey tower in the City of London has double-deck cars serving two reception floor levels simultaneously. Sky lobbies are halfway up and near the top of the tower; non-stop service is provided in both lower and upper sectors between the main lobbies, at a speed of 7 m/s, intermediate floors being served with lower-speed stopping lifts. Travel from basement car parking to the main street lobby is usually a dedicated low-speed service.

Car speeds for various travel distances are shown in Table 17.1.

Car speed is chosen so that the driving motor can be run at full speed for much of the running time to maximize the efficiency of power consumption. Starting and accelerating power is greater than steady speed energy use, as it is with a road motor vehicle. Deceleration during braking dissipates the momentum gained by the car and counterweight in friction-generated heat, lost to the atmosphere and into the lift motor plant room at the top of the shaft. The overall speed of operation is determined by the acceleration time, braking time, contract speed (maximum

Table 17.1 Design lift car speeds.

Floors	Car speed (m/s)
4	0.75
9	2
15	3
Over 15	5–7

car speed), speed of door opening, degree of advance door opening, floor levelling accuracy required, switch timing and variation of car performance with car load.

The automatic control system should function in an upward collecting and downward collecting mode. Requests for service made sufficiently early at a lobby cause the car in that shaft to break its original journey instruction and stop. Computer controls are used to optimize the overall performance of the installation, by causing the nearest car to stop and to minimize electricity consumption.

In the event of a fire within the building, the central fire detector and alarm system signal causes all the lifts to run to the ground floor. Where the building extends out of the reach of conventional fire-fighting turntable ladders on vehicles, at least one of the lifts is designated as the fireman's lift. Its main features are the following:

1. platform area and contract load of at least 1.45 m^2 and 550 kg;
2. reaches the top of the building within 1 min;
3. has power-operated doors of not less than 0.80 m clear opening that are arranged to remain open at any floor;
4. has an overriding 'Fire Control' switch at the fire control floor level, to bring the lift under manual control of the fire officer. This switch brings the lift immediately to the fire control floor, which is the fire officers' entrance level; all other controls are made inoperative.

Goods lifts travel at a maximum of 1.0 m/s and have full width doors, sometimes at each end of the car. Accurate floor levelling, to within 5.0 mm, may be provided to facilitate smooth passage of trolleys carrying fragile goods, fluids or patients in a hospital. Passengers can use goods lifts but service is slow. A variable-voltage electrical or hydraulic power supply is used. Hydraulically operated lifts have the advantage of very quiet operation and low running costs. The only power-consuming plant item is a small hydraulic pump immersed in an oil tank. Goods lifts complete each journey instruction before accepting another. Door operation can be manual or automatic. Additional structural supports are needed for lifts with high carrying capacity and well-designed brakes. Non-metallic serrated inserts may be fitted into the grooves of the driving pulley (sheave) to reduce wear and increase traction.

Service lifts are small goods lifts with car floor areas of up to 1.20 m^2 and heights of less than 1.40 m. The serving level may be 0.850 m above floor level to coincide with the working plane. Documents, goods or food are carried at up to 0.50 m/s with a maximum contract load of 260 kg. Control can be manual or semi-automatic. Prefabricated service lifts can be installed in a day and require minimal builder's work. Controls and the electric driving motor can be at the base of the shaft and a chain drive used.

Service conveyors for document transportation combine horizontal and vertical movement. An installation in the City of London transfers documents throughout a bank complex using briefcase-sized carriages on a continuously moving railway track with sidings for loading and unloading.

A tunnel connects buildings on either side of a street. The Post Office unmanned underground electric mail trains are another example.

Paternoster

A number of doorless two-person lift cars travel continuously in a clockwise direction within a common vertical shaft. The principle of operation is shown in Fig. 17.1. This system is used in an office or an adult education building which requires regular inter-floor travel over short distances for normally agile people. A conventional passenger lift is installed alongside the paternoster for the carriage of more passengers, goods, children, the elderly or the infirm. Paternoster performance over more than about six storeys is limited by its low speed, 0.40 m/s. Its carrying capacity is 720 passengers per hour. A degree of acclimatization is needed for its use. Car spacing is 4.0 m. Chain drive from electric motors is employed. The driving motor, brake, gearing and control equipment is fitted in a machine room at the top of the shaft. Emergency stop buttons are at each floor level and are linked to an audible alarm.

Escalator

Escalators and passenger conveyors are primarily used where large numbers of passengers form surges at discharge times from offices, railway underground stations and airport terminals.

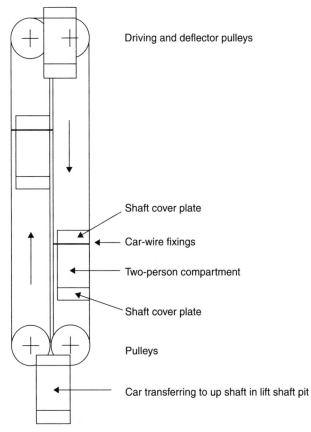

Driving and deflector pulleys

Shaft cover plate

Car-wire fixings

Two-person compartment

Shaft cover plate

Pulleys

Car transferring to up shaft in lift shaft pit

17.1 Paternoster lift.

Crowd flow, in plan, is similar to two-dimensional turbulent fluid behaviour and design for passenger routes can be regarded in a similar manner. Escalators provide suitable transport for all ages, laden or unladen. Their operating direction is reversible to correspond to peak travel times. Tread widths are from 0.60 to 1.050 m. For a given quality of service, they require less horizontal floor space than a lift. The angle of inclination is normally 30°, but 35° can be used for a vertical rise of less than 6.0 m and a speed of less than 0.5 m/s. Speeds of up to 0.75 m/s are permissible as this is the maximum safe entry and exit velocity.

Passenger conveyors

Passenger conveyors are moving pavements which can carry wheeled vehicles such as shopping or luggage trolleys, prams or wheelchairs. Distances of up to 300 m can be travelled at speeds of up to 0.90 m/s with an 8° slope. Combinations of grade, horizontal and up grade can be included, as in a road underpass. An S-shaped track overcame the limitation of entry and exit speed restriction on journey time by having its floor constructed from metal plates which slide relative to each other. The plates bunch together at the entry and exit curves of the track but spread out along the central straight which may be up to 1000 m length. Travel speed along the straight can be five times the entry speed. An electric motor of 19 kW drives the conveyor through a reduction gearbox and chain and 7200 passengers per hour can be carried. A concrete ramp forms the structural base for the entire conveyor. Emergency stop buttons are provided.

Driving machinery

The lift car and its load are partly balanced with a counterweight and this reduces motor power consumption. Motor power is used to overcome friction, acceleration, inertia and the unbalanced load during lifting. Power is transmitted to the traction sheave through a gearbox, two-speed or variable-speed motor driving sets, sometimes using direct current motors. The motor and driving sheave are mounted on a load-bearing concrete base at the top of the lift shaft. Considerable heat output is created and lift motor room natural or mechanical ventilation is essential plus in some cases, mechanical cooling from an air-handling unit, in hot climates.

The wire rope configuration for higher-speed lifts is shown in Fig. 17.2.

The lift motor room must be maintained at between 10°C and 40°C by natural or mechanical heating and ventilation if necessary. This ensures a condensation-free atmosphere in winter and adequate motor cooling in summer. A smoke extract ventilation grille of 0.10 m² free area must be provided at the top of each lift shaft. Some noise is produced in the motor room and its escape from this area is limited by a vibration-isolating concrete machine base and the concrete construction of the lift motor room. The external noise level produced is considered in conjunction with nearby room usage to assess whether additional attenuation is required.

Air bypass holes, as large and as frequent as possible, cross-connect adjacent lift shafts to allow car-induced draughts to circulate with minimum restriction. An emergency electrical power generator is often installed so that, in the event of mains failure, one lift at a time can be run to the ground floor and the doors opened. One lift can then be made available for the emergency services and emergency lighting is provided.

Hydraulic drive is often used for lift speeds up to 1.75 m/s for passenger travel and goods lifts of up to six storeys. Alternative bore hole and rope drum drive operation principles are shown in Figs 17.3 and 17.4.

Service lifts can use an electric-motor-driven winding drum with a deflector pulley at the top of the shaft, a pulley on top of the car and the motor at the bottom of the shaft. This is less efficient than counterweighted designs but saves space and complexity.

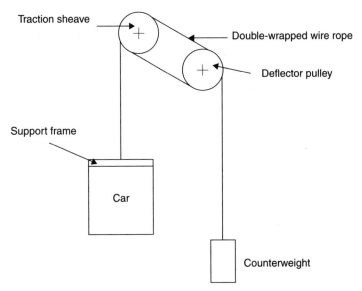

17.2 Traction arrangement for a high-speed passenger lift.

17.3 Hydraulic lift drive.

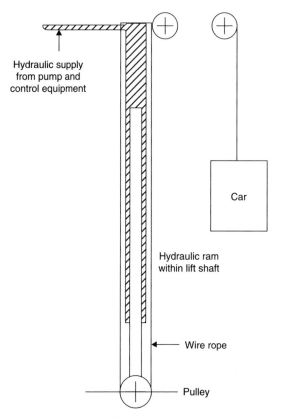

Hydraulic supply
from pump and
control equipment

Car

Hydraulic ram
within lift shaft

Wire rope

Pulley

17.4 In-shaft hydraulic lift drive.

Each passenger lift car has a service hatch in its ceiling. When this is opened, an electrical interlock switch opens to prevent the lift from being operated while escapees or a maintenance engineer may be in the shaft. A mechanical extract fan is often fitted in the roof of the car.

A lift motor room has the following features:

1. a concrete machine base incorporating a vibration-isolating cork slab to separate its upper and lower parts;
2. motor and brake equipment bolted to the upper, vibration-isolated, concrete slab;
3. flexible armoured electrical cable connections to the motor;
4. lift motor main isolating switch close to the plant room door;
5. access hatch into the lift shaft;
6. electrical control panel, switches or digital;
7. lifting beam built into the structure;
8. adequate artificial illumination;
9. natural or mechanical ventilation;
10. 13 A power point;
11. locked door;
12. light-coloured walls and ceiling;
13. emergency telephone.

The main constructional features of a lift shaft and motor room are shown in Fig. 17.5.

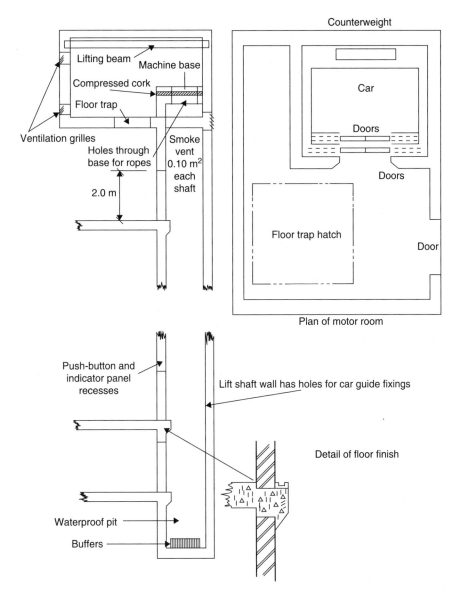

17.5 Constructional features of a passenger lift shaft.

Questions

1. Explain how transportation systems are employed within and between buildings to assist the movement of people and goods. Include details of the main characteristics of the systems, their performance and costs.

2. Explain with the aid of sketches, drawings or illustrations from the internet how high-speed electrically driven lifts operate and where they are used.

3. Give examples of mechanical transportation systems used in buildings. Take photographs of installations available to you if permission is granted. Use the internet and manufacturers resources to illustrate current practices.

4. Comment on improvements which might be made in the current designs mechanical transportation systems used in buildings.

5. Explain with the aid of sketches, drawings or illustrations from the internet how goods and hospital lifts operate and where they are suitable.

6. Suggest some probable future developments that may occur for mechanical transportation systems used in buildings, giving reasons.

7. Discuss the statement, 'Having machines to move people through buildings contributes significantly to obesity and lack of physical fitness.'

8. Compare the use of primary energy to operate mechanical transportation systems used in buildings with alternatives and comment on the differences.

9. Discuss the statement, 'Having tall buildings where vertical travel can only be accomplished by consuming primary energy is not sustainable.'

10. Explain with the aid of sketches, drawings or illustrations from the internet how hydraulic lifts operate and where they are suitable.

11. Visit a lift motor room in a large building, with the permission of the building manager. Photograph, report and draw a detailed description of the installation, room space dimensions, control cabinets, motor and drive arrangement, ventilation and temperature control systems.

12. Use a lift system in a building where you have permission to enter, and measure the typical round trip time for a journey from the entrance foyer to the top of the building and back to the foyer. Record the number of stops, passengers carried, type of lift system and time taken by the door systems. Comment on the quality of service provided.

13. Visit an escalator system, with the permission of the building manager if it is not a public facility. Photograph (if possible), report and draw a detailed description of the installation, space dimensions and suitability for the application. Comment on the quality of service provided.

14. Sketch, draw or illustrate from the internet and publications how drive systems for mechanical transportation systems work. Clearly show their principles of operation, how they use primary energy and where they impose loads on the building structure.

15. List the builder's work associated with lift installations. Show by means of sketches or drawings where these items will be located and how they will be performed by the construction team. At what stage of the building will the lift installation be constructed and how will it be achieved?

16. Which are correct about escalators? More than one correct answer.

    1. Run continuously.
    2. Reversible in flow direction.
    3. Stop and start frequently to save energy.
    4. Require electrical drive.
    5. Provide sloping and horizontal travel.

17. Which are correct about lifts? More than one correct answer.

    1. Lift capacity is designed to move the whole population of the building in or out within a 10-min period.
    2. Number and speed of lifts in an office building are designed to match the likely inflow of people into their building.
    3. Need negligible maintenance.
    4. Lift cars have to be air-tight.
    5. Lift cars have an access hatch.

18. Quality of a lift service is determined from:

    1. Number of passengers carried in each car.
    2. Highest car speed attainable.
    3. Round trip time for one lift divided by the number of lifts available.
    4. Passenger comfort.
    5. Quietness of the lift system.

19. Which are correct about lift motor rooms? More than one correct answer.

    1. There aren't any, all components are within the lift shafts.
    2. The motor sits on top of each car and has a pulley at the top of the shaft.
    3. Electrical lift motor room is situated at the top of each lift shaft.
    4. Hydraulic lift motor room is at the base of the lift shaft.
    5. Lift motor rooms are always halfway up the lift shaft.

20. Lift and escalator installations:

    1. Lift systems can be added to the building design after important design decisions have been agreed.
    2. Escalators are just motorized stairways and do not need much design work.
    3. Lift systems are an after-thought in the overall concept of the building.
    4. Transportation systems do not cost much.
    5. Mechanical movement systems for the building's occupants are critical to the success of the project.

21. Transportation systems:

    1. Allow people to become unfit; we should walk.
    2. Are not essential for several categories of buildings and building user.
    3. Allow rapid movement around large sites.
    4. Do not need much energy to operate.
    5. Need no maintenance.

22. Where could a double-deck lift system apply?

    1. Combined goods and passenger lift cars.
    2. Commercial buildings of around 20 stories.
    3. High-speed long-travel lifts.
    4. Low-speed short-travel lifts.
    5. Hospitals.

18 Question bank

Key terms and concepts

acronyms 393; data transmission 394; greenhouse 393; heat transfer 394; humidity 394; hydrocarbons 394; insulation 395; inverter 392; low-energy building 393; obesity 393; odours 395; refrigeration 396; sustainability 393; temperature 394; thermal comfort 395.

Introduction

This is a general knowledge section of questions covering topics within the fifth edition. Some questions may require the reader to look up answers in additional resources or use the internet with a search engine. There is only one correct answer to each question unless specified as having more. Incorrect answers may be partially true but not considered by the author to be the entirely correct response for the purpose of this book; these may stimulate additional study, discussion, questioning with peers or the instructor.

Question bank

1. What does inverter mean?

    1. Alternating current phases are reversed.
    2. It is an electronic soft starter for a three-phase motor.

3. Incoming 50 Hz alternating current is digitally reformed into an output frequency to a motor in the range from 0 to 20 000 Hz.

4. Incoming 50 Hz alternating current is digitally reformed into an output frequency to a motor in the range from 0 to 50 Hz.

5. Alternating current is electronically converted into direct current to drive a motor.

2. Which of these acronyms is correct?

1. ASHRAE means Australian Society for Heating, Refrigerating and Air Engineering.
2. AIRAH means American Institute for Refrigeration and Air Heating.
3. CIBSE stands for The Chartered Institution of Building Services Engineers.
4. BSRIA stands for British Services Refrigeration Institute for Air Conditioning.
5. CIC is the Council for Industry and Construction.

3. Does air conditioning lead to obesity? Which of these may not be a valid argument?

1. Obesity is due to overeating.
2. Obesity is due to working at computers too long each day.
3. Humans need outdoor environment exposure to regulate body temperature.
4. Metabolism slows and we eat more in air-conditioned environments.
5. People eat less when the air around us is very warm.

4. CFD stands for:

1. Comfort for disabled people.
2. Computational fluid dynamics.
3. Computer fluid dynamics.
4. Comfort frequency diagram.
5. Computer flow diagnostics.

5. What does BREEAM stand for?

1. Building Rehabilitation Electrical Energy Alternative Methodology.
2. Building Research Establishment Energy Audit Methodology.
3. Building Recycling Energy Effectiveness Association Member.
4. Brick Recycling Energy and Environment Assessment Method.
5. Building Research Establishment Environmental Assessment Method.

6. What does greenhouse rating of a building stand for?

1. The higher the greenhouse gas production due to the building, the higher the greenhouse rating.
2. A 10-star building produces no greenhouse gases.
3. Assessed greenhouse gas emission standard of a building.
4. Greenhouse rating stars awarded are inversely proportional to the tonnes of carbon dioxide created by the building.
5. An emission standard applied to all types of buildings.

7. What does sustainability mean for low-energy buildings?

1. No such thing as a modern sustainable building.
2. Everything used in the building's service life comes from globally sustainable resources.
3. This building is an example of good modern design practice.
4. All waste output from this building is recycled.
5. The building has been constructed from organically grown materials.

8. Which correctly describes heat transfer?

   1. Sensible heat transfer comprises all types.
   2. Latent heat transfer raises temperature.
   3. Sensible heat transfer is logged by a thermocouple and thermistor.
   4. Latent heat transfer is hidden from view.
   5. Sensible heat transfer only takes place through conduction and convection.

9. What are fluorinated hydrocarbons used for?

   1. Swimming pool water treatment.
   2. Biocide decontamination of cooling towers.
   3. Ozone-depleting refrigerants.
   4. Non-CFC foam insulation and furnishings.
   5. Combustible gaseous fuel.

10. Which of these is not a common standard for data transmission?

   1. Ethernet.
   2. RS484.
   3. RS232.
   4. RS124.
   5. C-bus.

11. Which is correct about the density of humid air?

   1. Decreases with increasing pressure.
   2. Increases with increasing air temperature.
   3. Varies with air temperature and pressure.
   4. Not affected by humidity.
   5. Increases as air velocity increases.

12. Which is correct about Kelvin?

   1. Name of the engineer who designed the first steam engine.
   2. Unit of heat.
   3. Measured in kJ/kg s.
   4. Temperature scale.
   5. Absolute temperature.

13. Which is correct about low-energy buildings?

   1. A low-energy building is one that requires the minimum amount of primary resource energy to build it.
   2. A low-energy building may consume more energy to construct.
   3. A low-energy building consumes less energy during its 100+ years of use than an equivalent building.
   4. We have no idea what an equivalent building is for a specific site.
   5. All buildings consume uncontrolled amounts of energy.

14. Which of these is where indoor odours, vapours and gases come from? More than one correct answer.

   1. Cleaning fluids used overnight.
   2. New furniture, carpets, floor coverings, sealants and adhesives.

3. Old furniture, carpets and floor coverings.
4. Personal hygiene products.
5. Cigarette smoke, diesel engine exhaust, road tar, painting work being done and creosote used on roofing.

15. We sense odours by:

1. Identifying smells.
2. Breathing onto others.
3. A measuring instrument.
4. Tasting them in our mouth.
5. Olfactory response.

16. Satisfactory air quality may be deemed when:

1. 100% of the full-time occupants are satisfied.
2. 85% of the full-time occupants are satisfied.
3. 50% of the full-time occupants are satisfied.
4. Complaints cease.
5. Odours have been eliminated.

17. Which is correct about low-energy building designs in the UK?

1. Are always modern and look impressive.
2. Are always found to be ideally comfortable by users.
3. Must have large windows and glazed walling.
4. Must have small windows and high levels of thermal insulation.
5. Should consume a minimum of primary energy when compared with similar types and sizes of buildings.

18. Thermal comfort *PPD* means?

1. Personal preferences determined.
2. Personal preferences determination.
3. Has no meaning.
4. Percentile people dissatisfied.
5. Predicted percentage of dissatisfied people.

19. Which of these is economic thickness of thermal insulation?

1. Found by calculation and intersecting lines on a graph for an application.
2. When graph of capital cost of additional thermal insulation becomes horizontal.
3. When a graph of energy savings value reaches a peak.
4. Always occurs when payback from energy cost savings reaches two and a half years.
5. Thickest amount the building designer can accommodate.

20. What does sustainability mean for low-energy buildings?

1. The mechanical and electrical services within this building all have a low maintenance requirement.
2. All the water, sewerage, paper and plastic waste output from this building go to recycling.
3. All the light bulbs and tubes from this building are recyclable.
4. Somebody has found a good argument why this design of building is less harmful to the global environment than competitive designs.

5. This building has consumed, and will continue to consume, more of the earth's physical resources than it can ever put back.

21. Which might be a means of reducing the refrigeration system energy usage in a small retail premises where food refrigeration, deep freezers and reverse cycle air conditioning are all needed?

1. Install smallest capacity compressors possible.
2. Carry out frequent maintenance checks and parts replacement.
3. Switch reciprocating compressors off for as long as possible and maintain wide temperature differentials.
4. Use same outdoor air-cooled condenser for all three systems.
5. Variable refrigerant volume scroll compressor with software-controlled digital operation programmable for all variations in year round duties.

22. What have I learnt from this study?

1. Nothing, it is all a fog to me!
2. Mechanical and electrical services within a building are not very important to the overall concept of the design and construction.
3. I can design or construct buildings; someone else must worry about the fiddly bits.
4. The building will work without the mechanical and electrical services anyway.
5. I now appreciate the importance and main features of the essential and desirable building services!

19 Understanding units

Learning objectives

Study of this chapter will enable the reader to:

1. practice answering short questions from a multiple choice of answers;
2. test own understanding of using units of measurement;
3. prepare for tests, assignments and written examinations where this form of questioning is provided;
4. evaluate why some answers are not entirely correct for the question;
5. have discussions with peers and instructors over the meaning of the incorrect answers;
6. lead into further study and investigation;
7. answer a range of questions relevant to the fifth edition and building services engineering work generally.

Key terms and concepts

acceleration due to gravity 398; atmospheric pressure 398; Celsius 403; density of water 399; electrical units 401; exponential 400; frequency 402; humid air 399; Joule 401; Kelvin 403; Newton 404; pressure 402; pressure drop rate 404; specific heat capacity 398; standard atmosphere 398; Stefan–Boltzmann 399; time 400; volume 404; Watt 401.

Introduction

This is a general knowledge section of questions covering measurement unit topics within the fifth edition and building services engineering generally. All questions should be understood by students of this topic area. There is only one absolutely correct answer to each question. Tackling these may stimulate additional study, discussion, questioning with peers or the instructor.

Questions

1. Which of these equals one standard atmosphere at sea level?

    1.   1.013 tonne/m^2.
    2.   1 bar.
    3.   10 000 N/m^2.
    4.   1013.25 mb.
    5.   10^6 N/m^2.

2. Which of these equals one standard atmosphere at sea level?

    1.   1013 tonne/m^2.
    2.   10^5 bar.
    3.   10^9 N/m^2.
    4.   14.7 lb/in^2.
    5.   10^6 N/m^2.

3. Which of these equals one standard atmosphere at sea level?

    1.   1 × 10^5 Pa.
    2.   1.01325 × 10^5 N/m^2.
    3.   1 × 10^4 N/m^2.
    4.   30 m H_2O.
    5.   1013.25 mm Hg.

4. Which of these equals one standard atmosphere at sea level?

    1.   9.807 m H_2O.
    2.   29.35 m H_2O.
    3.   10.3 m H_2O.
    4.   101 325 kJ/m^2.
    5.   1.205 kg/m^2.

5. Which of these is the acceleration due to gravity, g?

    1.   10 m/s^2.
    2.   30 ft/s^2.
    3.   186 000 miles/h.
    4.   Gravity is static.
    5.   9.807 m/s^2.

6. Which of these describes the acceleration due to gravity, g?

    1.   Calculated from a 1 kg weight free falling from a height.
    2.   Relative to distance from the Moon.
    3.   Constantly 9.807 m/s^2.
    4.   Varies with height above sea level.
    5.   Inversely proportional to depth below sea level.

7. Which is the specific heat capacity of air?

    1.   Sensible heat content kJ/kg.
    2.   Total heat content kJ/kg.
    3.   1.205 kJ/kg.
    4.   1.012 kJ/kg K.
    5.   4.186 kJ/kg K.

8. Which is the specific heat capacity of air?

    1. Ratio of C_p/C_v.
    2. Cannot be defined.
    3. Varies with atmospheric pressure.
    4. 1.012 kgK/W.
    5. 1.012 kJ/kgK.

9. Which is the specific heat capacity of water?

    1. 1.013 kW/kgK.
    2. 1.012 MJ/kgK.
    3. 4.186 kg K/kW.
    4. 4.186 kJ/kgK.
    5. 4.2 kg K/kJ.

10. Which is correct about the specific heat capacity of water?

    1. Varies with water pressure.
    2. 4.19 kW s/kgK.
    3. A ratio.
    4. 1.102 kJ/kgK.
    5. Used to calculate the flow rate of heating and cooling system water.

11. Which is correct about the Stefan–Boltzmann constant?

    1. Used to calculate convective heat transfer.
    2. 4.186 kJ/kgK.
    3. 1.012 kJ/kgK.
    4. 5.67×10^{-8} W/m^{2}K^4.
    5. Combines convective and radiant heat transfer.

12. Which is correct about the density of humid air?

    1. Decreases with increasing pressure.
    2. Increases with increasing air temperature.
    3. Varies with air temperature and pressure.
    4. Not affected by humidity.
    5. Increases as air velocity increases.

13. Which is correct about the density of humid air?

    1. 4.186 kg/m^3 at 21°C, 60% relative humidity.
    2. 1.013 kg/m^3 at 20°C, sea level.
    3. 0.802 m^3/kg.
    4. 1.205 kg/m^3 at 20°C, 1013.25 mb.
    5. 5.67 kg/m^3.

14. Which is correct about the density of water?

    1. Always 10^3 kg/m^3.
    2. 1013.25 m^3/kg.
    3. 101 325 kg/m^3.
    4. 1.205 MJ/m^{3}K.
    5. 1000 kg/m^3 at 4°C.

15. Which is correct about the density of water?

    1.   Cannot be measured.
    2.   Cannot be measured accurately.
    3.   Always relative to the specific gravity number.
    4.   1000 times that of air.
    5.   Specific gravity is 1.0.

16. Which is correct about the density of water?

    1. 1.205×10^5 kg/m^3.
    2.   1 tonne/m^3 at 4°C.
    3. 1.012×10^3 kJ/m^3.
    4.   1.27 kJ/kg K.
    5.   100 g/cm^3.

17. Which is correct about the density of water?

    1.   1 g/cm^3.
    2. 1.2×10^3 kg/m^3.
    3.   Specific gravity is 4.186.
    4.   1000 tonne/m^3 at 10°C.
    5.   1 kg/l.

18. What does the exponential e mean?

    1.   A logarithm.
    2.   A variable number.
    3.   Always 10^x, ten to the power x.
    4.   2.718.
    5.   Has no meaning.

19. What does the exponential e mean?

    1.   Something which is raised to a power.
    2. $e = 10^x$.
    3.   The sum of an infinite series.
    4. $e = \sqrt{-1}$.
    5. $e^1 = 2.718$.

20. Which of these has the correct units?

    1.   Mass is measured in kilograms.
    2.   1 tonne $= 10^6$ kg.
    3.   1 kg $= 10^9$ mg.
    4. 10^3 kg/m^3 $= 1$ kg/l.
    5. 10^6 m $= 1$ km.

21. Which of these are not the correct units?

    1.   1 h $= 3600$ s.
    2.   60 h $= 3600$ min.
    3. 3.6×10^3 s $= 1$ h.
    4.   1 year $= 8760$ h.
    5.   1 h $= 360$ s.

22. Which of these has the correct units?

    1. $1 N = 1 kg \times 1 m^2$.
    2. $1 J = 1 kg \times 1 m$.
    3. $1 W = 1 kg \times g\ m/s^2$.
    4. $10^3 J = 3600\ kN/m^2$.
    5. $1 J = 1\ N/m^2$.

23. Which of these has the correct units?

    1. $1 J = 1 N \times 1 m$.
    2. $1 J = 1\ W/s$.
    3. $10^3 J = 1\ kW/s$.
    4. $1 W = 10^3\ J/s$.
    5. $1 MJ = 10^3\ kW/s$.

24. Which of these has the correct units?

    1. $1 W = 1\ Nms$.
    2. $1 W = 1\ Js$.
    3. $1 W/s = 10^3\ J$.
    4. $1 W = 1\ Nm/s$.
    5. $1 kW/h = 10^3\ J/h$.

25. Which of these has the correct electrical units?

    1. $1 MW = 10^3\ W$.
    2. $10^3 kJ = 10^3\ kW/s$.
    3. $1 W = 1 V \times 1 A$.
    4. Electrical energy meters accumulate kW/h.
    5. $10^3 W = 10^3 V \times 10^3 A$.

26. Which of these has the correct electrical units?

    1. $10^3 kW/h = 10^3 \times 3600\ W/s$.
    2. $kWh = energy$.
    3. $1 GJ = 10^6 V \times 1 A$.
    4. $1 MJ = 10^6 V \times 1 A$.
    5. $1 kJ/s = 10^3 V \times 1 A$.

27. Which of these has the correct electrical units?

    1. $10^3 W = 10^3 V \times 1 A$.
    2. $1 MWh = 10^3\ Ws$.
    3. $1 kWh = 10^3\ Ws$.
    4. $1 kWh = 1000\ W/s$.
    5. $1 kWh = 1000\ W/h$.

28. Which of these has the correct units?

    1. $1\ atmosphere = 10^3\ b$.
    2. $1 Pa = 1\ N/m^2$.
    3. Pascal is a unit of radiation measurement.
    4. $1 kN/m^2 = 1\ b$.
    5. $1 mb = 10^3\ N/m^2$.

29. Which of these has the correct pressure units?

    1.   1.01325 mb = 1 atmosphere.
    2.   1 MN = 10^3 kN/m^2.
    3.   1 b = 1 kN/m^2.
    4.   13.6 mb = 13.6 N/m^2.
    5.   1 b = 10^5 N/m^2.

30. Which of these has the correct pressure units?

    1.   1 mb = 1 N/m^2.
    2.   1 b = 10^3 mb.
    3.   1 mb = 10^3 N/m^2.
    4. 10^3 kN/m^2 = 1 b.
    5.   1 mb = 10^6 b.

31. Which of these has the correct pressure units?

    1.   1 Nm = 1 Pa.
    2.   1000 Pa = 1 atmosphere.
    3.   1 kPa = 1 kN/m^2.
    4.   1 Pa = 1 mb.
    5.   1 Pa = 1 N/m^2.

32. Which has the correct meaning for frequency?

    1.   Number of times an event is repeated.
    2.   Cyclic repetition of an event.
    3.   Number of complete rotations per unit time.
    4.   Statistical correlation.
    5.   Occasional reoccurrence.

33. Which has the correct meaning for frequency?

    1.   Alternating current rate of increase.
    2.   Electrical single- or three-phase.
    3.   Torque of a motor.
    4.   Air changes per hour.
    5.   Revolutions per minute.

34. Which is not correct in relation to frequency?

    1.   3000 RPM = 50 Hz.
    2.   1 Hz = 1 Nm/s.
    3.   High-frequency fluorescent lamps work at 20 000 Hz.
    4.   VFD means variable frequency drive.
    5.   60 Hz = 3600 RPM.

35. Which is correct about Kelvin?

    1.   Name of the engineer who designed the first steam engine.
    2.   Unit of heat.
    3.   Measured in kJ/kgs.
    4.   Temperature scale.
    5.   Absolute temperature.

36. Which is correct about Kelvin?

    1. Where absolute zero gravity starts.
    2. Something to do with temperature.
    3. First name of Dr K. Celsius.
    4. °C + 273.
    5. Engineered the first closed-circuit piped heating system.

37. Which is correct about Kelvin degrees?

    1. Celsius scale plus 180.
    2. Are always negative values of Celsius degrees.
    3. Symbol K.
    4. Awarded by Kevin University, Peebles, Scotland.
    5. $K = C \times (9/5) + 32$.

38. Which is correct about Kelvin?

    1. Name of a famous Scottish scientist.
    2. Invented first bicycle in Scotland.
    3. $K = C + 273$.
    4. Kelvin McAdam invented tarmacadam for road surfacing.
    5. Degrees measured above absolute zero at $-180°F$.

39. Which is correct about Kelvin degrees?

    1. Measurement of room air temperature.
    2. Always used in heat transfer units.
    3. Used to specify absolute temperature and temperature difference.
    4. Fahrenheit plus 180.
    5. Zero scale commences at $-40°C$.

40. Which is correct about Celsius?

    1. Latin name of inventor of Roman hypocaust under floor heating system in 200 BC.
    2. Fahrenheit minus 32.
    3. $C = F \times (5/9) + 32$.
    4. $C = 32 - F \times (5/9)$.
    5. $C = (F - 32) \times (5/9)$.

41. Which is correct about Celsius?

    1. Called °C units.
    2. Kelvin degrees plus 273.13.
    3. $C = (F + 32) \times (5/9)$.
    4. Commonly used for cryogenic applications.
    5. $F = (C - 32) \times (5/9)$.

42. Which is correct about Celsius?

    1. Temperature scale in the centimetre, gram, second, CGS, metric system.
    2. Name of the Roman Senator in 35 AD who stabbed Caesar.
    3. $C = (F - 180) \times (5/9)$.
    4. Defines normal human body temperature of 98.4°.
    5. Temperature scale in the metre, kilogram, second, MKS, metric system.

43. Which is correct about volume?

    1.   1 cm^3 water occupies 1 l.
    2.   1 tonne water occupies 1000 m^3.
    3.   1 m^3 = 1000 l.
    4.   1 l water weighs 100 kg.
    5.   1 l water weighs 10 kg.

44. Which is correct about volume?

    1.   1 m^3 air weighs around 100 kg.
    2.   1 m^3 air weighs around 10 kg.
    3.   1 m^3 air weighs around 1 kg.
    4.   1 l occupies 1 m^2 area and 100 mm height.
    5.   1 l occupies 1 m^2 area and 10 mm height.

45. Which is correct about volume?

    1.   1 l water is contained in a cube of 100 mm sides.
    2.   1 l air is contained in a cube of 1000 mm sides.
    3.   There is such a thing as a volume sensor for a control system.
    4.   100 concrete blocks of 300 mm x 200 mm x 100 mm occupy a volume of 6 m^3.
    5.   1 tonne water occupies 10 m^3.

46. What is the volume of a room 12 m long, 8 m wide and having an average height of 4 m?

    1.   400 m^3.
    2.   62 m^3.
    3.   462 m^3.
    4.   384 m^3.
    5.   192 m^3.

47. Which is the correct length of a 1200 m^3 sports hall of average height 4 m and width 12 m?

    1.   25 m.
    2.   10 m.
    3.   250 m.
    4.   120 m.
    5.   12.5 m.

48. What are the units for pressure drop rate in a pipeline?

    1.   m head H_2O/m run.
    2. N/m^2.
    3.   mb/m.
    4. N/m^3.
    5. kN/m^3.

49. What does N/m^3 stand for?

    1.   Nanometres per m^2 pressure drop per metre run of pipe.
    2.   Neurons per cubic metre of room volume.
    3.   Newton's per square metre pressure drop per metre run of pipe or duct.

4. Newton per cubic metre is a density.
5. Nano-particles of Radon gas per cubic metre of air in a building.

50. What does N/m^3 stand for?

1. Normalized volumetric air change rate for a room.
2. Number of people in a building divided by building volume.
3. Volumetric coefficient.
4. Noise rating divided by room volume.
5. Pressure drop rate in a pipe or duct.

Appendix: answers to questions

Chapter 1 Built environment

10. t_r 15.0°C; t_{ei} 16.3°C. The room condition is below the comfort zone shown in Fig. 1.14. Surface temperatures are to be increased by adding thermal insulation, such as double glazing, and the air temperature should be raised.
11. 90 W.
12. 6211 W, work bench overheating.
13. 640 W.
14. 17.6°C.
15. 1.6°C.
16. 17.6°C.
17. 982, −11.2°C.
18. 923, −8.5°C.
19. 1139, −18.3°C.
20. Site A *WCI* 862, *EWCT* −5.8°C; Site B *WCI* 757, *EWCT* −1°C; site A has the more severe conditions.
21. *HSI* −26.8.
22. *HSI* 96.0; this is the maximum 8-h exposure for a fit, acclimatized young person; conditions vary through the day.
23. *HSI* 288, *AET* 11.7 min.
24. *HSI* 67.4, *AET* −12.7 min unrestricted.
25. *HSI* 648, *AET* 8.8 min.
26. t_r 14.3°C; t_{ei} 17.2°C. The room condition is outside the sedentary comfort zone.
27. 17.9°C, 17.2°C, 19.3°C, 20.3°C.
28. 13°C, 11.3°C, 15.7°C, 20°C.
29. 16.8°C, 15.4°C, 19.2°C, 21.4°C.
34. 2, 4
35. 1, 2, 4, 5
36. 3, 4
37. 2, 3, 5
38. 1, 2, 3, 4, 5
39. 2
40. 1, 2, 3, 4

45. 345 W.
47. C_i 0.977 decipol, C_o 0.3 decipol for vitiated outdoor air, Q 1770 l/s total, N 3.5 air changes/hour.
48. 1.47 l/s m^2.
49. 288 l/s m^2.
50. 0.33 l/s m^2.
51. 6.48 air changes/h.
52. 2.15 air changes/h.
53. $Q = 0.665$ air changes/h, no.
54. Answers in l/s. First figure is based on floor area, second is per person: Fanger 6250, 5000; ASHRAE 875 700; BS 1625, 1300; DIN 2375, 1900; CIBSE 1625, 1300; maximum N 5.1 air changes/hour.
55. C_1 0.612 decipol, G 0.517 olf/m^2, C_o 0.05 decipol, Q 9.2 l/s m^2, total Q 2760 l/s, N 6.6 air changes/hour.
61. 2, 4, 5
62. 3
63. 1, 4
64. 3
65. 1, 4
66. 4
67. 3
68. 2
69. 5
70. 3
71. 4
72. 5

Chapter 2 Energy economics

11. U_e 0.51 W/m^{2}K, L 53 mm.
12. A_f 2640 m^2, A_w 1776 m^2, B 1.035, T_T 8.46 W/m^2, T_E 24.85 W/m^2, T 33.3 W/m^2.
13. Gas 110, oil 158, coal 215, electric 224 tonnes C p.a.
14. 3
15. 2
16. 4
17. 1
18. 5
19. 4
20. 4
21. 1
22. 4
23. 1
24. 2
25. 2
26. 5
27. 2
28. 5
29. 5

Chapter 3 Heat loss calculations

4. 2330.5 W.
5. 20.112 kW.
6. 12.4°C.
7. Allowed heat loss per degree Celsius difference inside to outside is 3746.8 W/K; thus the proposal complies. Proposed heat loss 3407 W/K.
8. 83.14 kW.
9. 43%.
13. R_{si} 0.1 m²K/W, Q 50 W, U 2.78 W/m²K, R_n 6.67 m²K/W, 221 mm.
14. R_{si} 0.12 m²K/W, Q 19.2 W, 114, 120 mm used, U_n 0.29 W/m²K, 17.4°C.
15. R_{si}0.1 m²K/W, Q 20 W, U 1.82 W/m²K, extra R_a 0.18 m²K/W; 81.75, 90 mm used, U_n 0.23 W/m²K, new Q 4.83 W, 15.5°C.
16. 2, 3
17. 4
18. 2
19. 5
20. 4
21. 5
22. 5
23. 2
24. 2
25. 2
26. 3
27. 3
28. 5
29. 4
30. 4

Chapter 4 Heating

11. 0.95 l/s.
12. 2.4 m long × 700 mm high.
13. X 42 mm, Y 35 mm, Z 28 mm, radiator 1 22 mm, radiator 2 28 mm.
14. Expected internal temperature 26.5°C, system performance is satisfactory.
17. 3
18. 3
19. 4
20. 1
21. 2
22. 1
23. 3
24. 3, 5, 6, 7, 9
25. 2
26. 3
27. 3
28. 2
29. 5
30. 3

31. 3, 5
32. 3
33. 3
34. 4
35. 2
36. 2
37. 4
38. 4
39. 3
40. 5
41. 2
42. 5
43. 2
44. 3
45. 4
46. 3
47. 3
48. 2
49. 5
50. 3
51. 2
52. 4
53. 3
54. 5
55. 3
56. 5
57. 3

Chapter 5 Ventilation and air conditioning

1. From

$$Q = \frac{SH\,kW}{t_r - t_s} \times \frac{(273 + t_s)}{357} \; m^3/s$$

$$357Q\,(t_r - t_s) = SH(273 + t_3)$$

$$357Qt_r - 357Qt_s = 273SH + SH \times t_s$$

$$357Qt_r - 273SH = SH \times t_s + 357Qt_s$$

$$357Qt_r - 273SH = t_s(SH + 357Q)$$

and

$$t_s = \frac{357Qt_r - 273SH}{SH + 357Q}$$

$$= \frac{357 \times 5 \times 23 - 273 \times 50}{50 + 357 \times 5} = 14.94°C.$$

2. 0.793 m³/s.

4. 0.007469 kg H_2O/kg air.
5. (a) No; (b) 21.2°C w.b.; 0.877 m³/kg; (c) 6.186 kW.
6. (a) 4.25 m³/s; (b) 4.86 m³/s; (c) 87.45%; (d) 4.13 m³/s; (e) 0.61 m³/s; (f) 3.52 m³/s.
7. 20 air changes/h.
14. 1680 mm × 930 mm.
16. 2.68 m³/s, 10.72 air changes/h, 0.0076 kg H_2O/kg air.
17. t_s 28.6°C d.b., reduce supply air quantity to 1.7 m³/s and use t_s, 30°C d.b. if the room air change rate will not be less than 4 changes/h.
21. 14.45 air changes/h.
22. 15 air changes/h, 710 mm × 710 mm, 2 m³/s fresh air, 2 m³/s recirculated air, 3.6 m³/s extract air, 4 m³/s supply air duct, 0.4 m³/s natural exfiltration.
23. 4
24. 5
25. 2
26. 2
27. 1
28. 2
29. 5
30. 5
31. 3
32. 4
33. 2
34. 4
35. 1
36. 2
37. 4
38. 4
39. 5
40. 3
41. 1
42. 4
43. 1
44. 3
45. 2
46. 5
47. 3
48. 3, 4
49. 2, 4
50. 1, 5
51. 1, 3, 4
52. 3
53. 5
54. 2
55. 1
56. 5
57. 4
58. 5
59. 5
60. 4

61. 4
62. 5
63. 4
64. 4
65. 2
66. 3
67. 1
68. 3
69. 4
70. 4
71. 2
72. 1
73. 3
74. 3
75. 1
76. 3
77. 3
78. 3
79. 1
80. 5
81. 1
82. 4
83. 4
84. 3
85. 4
86. 3
87. 5
88. 4
89. 2
90. 4
91. 1
92. 5
93. 1

Chapter 6 Hot- and cold-water supplies

12. 13.44 kW.
15. 0.56 h.
16. 0.05 kg/s, 3.15 m head, pump C.
23. 3
24. 4
25. 3
26. 5
27. 5
28. 5
29. 5
30. 1
31. 1

32. 5
33. 5
34. 3
35. 5
36. 1
37. 5
38. 4
39. 5
40. 1
41. 5
42. 5
43. 3
44. 2
45. 5
46. 3
47. 1
48. 5
49. 4
50. 4
51. 4
52. 5
53. 5

Chapter 7 Soil and waste systems

  7. The furthest WC can only be 12.353 m from the stack.
  8. 53
15. 5
16. 5
17. 4
18. 3
19. 1
20. 2
21. 5

Chapter 8 Surface-water drainage

  2. 50 l/s, at least three.
  3. 5.04 l/s.
  4. 1.337 l/s.
  5. 1.45 l/s.
  6. 35.53 l/s.
  7. 2.921 l/s. Note that this is less than the figure given in Table 8.2 because of the simplifying assumptions made here.
  8. Either 125 mm with centre outlet or 150 mm with end outlet.
  9. Width W 120 mm, depth D 71.876 mm.
10. Yes, flow load 2.1 l/s, gutter capacity 2.936 l/s.
12. Storage volume 2.4 m^3, one pit diameter 1.25 m.

Chapter 9 Below-ground drainage

3. 100 mm.
4. 150 mm.
5. 1.235 m.
6. 225 mm, approximately 614.
13. 3
14. 5
15. 5

Chapter 10 Condensation in buildings

15. (a) 6.79°C, 6.11°C; (b) 13.89°C, 13.55°C, 2.89°C, 2.55°C; (c) 18.47°C, 17.83°C, 6.71°C, 2.91°C, 0.27°C; (d) 19.45°C, 19.06°C, −0.38°C.
16. 2.72°C.
17. −7.46°C.
18. 28.1 mm.
19. U value 0.46 W/m^{2}K, heat flow 9.26 W/m^2. Thermal temperature gradients are 22°C, 21.07°C, 20.38°C, 18.71°C, 11.76°C, 3.08°C, 2.9°C, 2.7°C, 2°C. Indoor dew-point 11.3°C, vapour pressure 1300 Pa, outdoor air −0.8°C and 568 Pa. Vapour resistance R_v 6.265 GN s/kg, mass flow of vapour G 1.168 × 10^{-7} kg/m^{2}s. Dew-points at the same interfaces as the thermal temperatures are 11.3°C, 11.3°C, 10.5°C, 10.5°C, 8.8°C, −0.7°C, −0.8°C, −0.8°C, −0.8°C. Condensation does not occur.
20. U value 0.6 W/m^{2}K, heat flow 7.74 W/m^2. Thermal temperature gradients are 14°C, 13.07°C, 12.84°C, 11.32°C, 3.58°C, 2.19°C, 1.22°C, 1°C. Indoor dew-point 6.5°C, vapour pressure 936 Pa, outdoor air −1.8°C and 531 Pa. Concrete block work resistivity taken as 200 GN s/kg m. Vapour resistance R_v 25.35 GN s/kg, mass flow of vapour G 1.6 x 10^{-8} kg/m^{2}s. Dew-points at the same interfaces as the thermal temperatures are 6.5°C, 6.5°C, 6.3°C, 0.09°C, −0.06°C, −0.06°C, −1.8°C, −1.8°C. Condensation does not occur.
21. 3
22. 4
23. 5
24. 3
25. 4
26. 1
27. 3
28. 4
29. 3
30. 1
31. 3
32. 4

Chapter 11 Lighting

9. 59%.
11. Room index 3, UF 0.73, MF 0.9, 36 luminaires in 3 rows of 12 along the 20 m dimension, 16.8 W/m^2, 21 A.
22. Lighting 3750 h/year, 3.81 × 10^6 lm, tungsten, 1814 lamps, replace 3401 per year, total annual cost £80 255 per year, fluorescent 569 lamps, replace 178 per year, total annual cost £14 403 per year, sodium 139 lamps, replace 22 per year, total annual cost £10 917 per year.

23. Lighting 1200 h/year, 35 0685 lm, tungsten 352 lamps, replace 211 per year, total annual cost £3559 per year, fluorescent 95 lamps, replace 16 per year, total annual cost £772 per year, halogen 37 lamps, replace 2 per year, total annual cost £423 per year.
24. 1
25. 4
26. 1
27. 1, 3, 5
28. 3
29. 4
30. 2
31. 3
32. 1
33. 3, 5
34. 4
35. 3
36. 1
37. 2
38. 3
39. 5
40. 4
41. 4
42. 2
43. 3
44. 3
45. 4
46. 2
47. 2
48. 3
49. 4
50. 5

Chapter 12 Gas

1. 1.026 l/s, 125 mm H_2O.
2. 5.394, 3.5, 0.75, 15, 1050 mb.
3. 3.333 Pa/m.
4. 27.17 m.
5. 1.47 l/s, 2.609 Pa/m, 32 mm.
6. 72 Pa.
11. 4, 5
12. 3
13. 1

Chapter 13 Electrical installations

5. 0.00172 Ω.
6. 0.2867 Ω.
7. 10.7%.
10. 12.6 kVA.
11. 19.2 Ω.

12. 0.2857 mA.
14. 9716.8 kW.
15. 28.6 m.
16. (i) 18.25 kVA, 25.4 A, (ii) £691.95.
26. Three earth rods give a total system resistance of 8.937 Ω.
27. 2, 4, 5
28. 5
29. 3
30. 1, 5
31. 4
32. 2, 4
33. 4, 5
34. 5
35. 1
36. 3
37. 5
38. 4
39. 1
40. 5
41. 2
42. 4
43. 1
44. 5
45. 2
46. 5
47. 3
48. 5
49. 5
50. 4
51. 4
52. 2
53. 2
54. 4
55. 2
56. 1

Chapter 14 Room acoustics

19. Reverberation time T 2.901 s at 125 Hz, 3.462 s at 250 Hz, 3.462 s at 500 Hz, 3.157 s at 1 kHz, 2.752 s at 2 kHz and 3.253 s at 4 kHz.
20. r 100 mm SPL 87 dB; r 1 m SPL 71 dB.
21. Directivity Q 2, r 0.5 m SPL 92 dB, reverberant SPL 79 dB.
22. 84 dBA.
23. Through the wall SPL_2 19 dB; through air vent 49 dB; open air vent causes noise to bypass the attenuation of the wall and may need acoustic louvres or an acoustic barrier.
24. Through the wall SPL_2 47 dB; through air vents in doors 59 dB; open air vent causes noise to bypass the attenuation of the wall; burner needs an acoustic enclosure.
25. SPL in roof is 32 dB; the large volume and short reverberation time assist in attenuating the plant room noise.
26. 33 dB.

27. 37 dB.
28. 39 dB.
29. See chapter explanation.
30. *NR* 40 is not exceeded in the room.
31. (a) *NR* 80; (b) *NR* 25, no intrusive noise from the chiller; (c) *NR* 45; (d) *NR* 20 when doors have equal sound reduction to the walls, have air-tight seals and are closed.
32. (a) *NR* 80; (b) 65 dB due to sound escape through door; (c) *NR* 35.
33. (a) *NR* 75; (b) *NR* 45, equivalent to the background noise level in a corridor; (c) *NR* 35; (d) *NR* 20, there is no intrusive noise.
34. (a) *NR* 60; (b) *NR* 40; (c) through the supply and return air ducts, noise radiation from the outer case of the fan coil unit, from the ceiling space through ceiling tiles, light fittings, noise break-in from the ceiling space into the supply and return air ducts and then into the office, structurally transmitted vibration from the fans, main air-handling plant noise through the outside air duct to the fan coil unit; (d) acoustic lining in the outdoor air, supply air and return air ducts, anti-vibration rubber mounts for the fan coil unit and the fan within it, acoustic lining within the fan coil unit, acoustic blanket above the recessed luminaires and above the ceiling tiles.
35. 3
36. 1
37. 3
38. 3
39. 5
40. 3
41. 3
42. 4
43. 5
44. 1
45. 4
46. 5
47. 5
48. 3
49. 5
50. 3
51. 5
52. 4
53. 4
54. 2
55. 1
56. 5
57. 3
58. 5
59. 2

Chapter 15 Fire protection

10. 2, 3, 5
11. 4
12. 2, 3
13. 1, 2, 3, 4, 5

14. 1, 2, 3, 4, 5
15. 3
16. 1, 2, 3, 4, 5
17. 5
18. 4
19. 3
20. 5
21. 1

Chapter 16 Plant and service areas

10. 5
11. 3
12. 1
13. 4
14. 2
15. 3
16. 2
17. 4
18. 2
19. 5
20. 2

Chapter 17 Mechanical transport

16. 1, 2, 4, 5
17. 2, 5
18. 3
19. 3, 4
20. 5
21. 3
22. 2

Chapter 18 Question bank

1. 4
2. 2
3. 4
4. 2
5. 5
6. 3
7. 3
8. 4
9. 3
10. 4
11. 3
12. 5
13. 3
14. 1, 2, 4, 5

15. 5
16. 2
17. 5
18. 5
19. 1
20. 4
21. 5
22. 5

Chapter 19 Understanding units

1. 4
2. 4
3. 2
4. 3
5. 5
6. 3
7. 4
8. 5
9. 4
10. 5
11. 4
12. 3
13. 4
14. 5
15. 5
16. 2
17. 5
18. 4
19. 5
20. 4
21. 5
22. 1
23. 1
24. 4
25. 3
26. 2
27. 1
28. 2
29. 5
30. 2
31. 3
32. 3
33. 5
34. 2
35. 5
36. 4
37. 3
38. 3

39. 3
40. 5
41. 1
42. 5
43. 3
44. 3
45. 1
46. 4
47. 1
48. 4
49. 3
50. 5

References

Action Energy, Department of the Environment, Transport and the Regions Best Practice Programme, Introduction to Energy Efficiency in Publication Offices EEB006 (2002).

ASHRAE (1985) *ASHRAE Fundamentals Handbook*, American Society of Heating, Refrigeration and Air Conditioning Engineers, Atlanta, GA.

Awbi, H.B. (1991) *Ventilation of Buildings*, E & FN Spon, London.

Belding, H.S. and Hatch, T.F. (1955) Index for evaluating heat in terms of resulting physiological strains. *Heating, Piping and Air Conditioning*, **27** (8), 129.

Bowyer, A. (1979) Space allowances for building services – outline design stage, *Technical Note TN4/79*, Building Services Research and Information Association, BSRIA, Bracknell, ISBN 0860220850, UK.

BRE Digest 205: 1977. *Domestic Water Heating by Solar Energy*, September, Building Research Establishment, Garston, Watford, UK.

BRE Digest 206: 1977. *Ventilation Requirements*, Building Research Establishment, Garston, Watford, UK.

BRE Digest 369: February 1992. *Interstitial Condensation and Fabric Degradation*, Building Research Establishment, Garston, Watford, UK.

British Gas (1980) *The Effect of Thermal Insulation in Houses on Gas Consumption*, Watson House, London.

BS 848: 1980. *Fans for General Purposes, Part 1: Methods of Testing Performance*.

BS 5572: 1978. *Sanitary Pipework*.

BS Code of Practice CP 326: 1965. *The Protection of Structures Against Lightning*.

BS Code of Practice CP 1013: 1965. *Earthing*.

BS Code of Practice 413: 1973. *Ducts for Building Services*.

Butler, H. (1979a) Lightning protection of buildings, *Technical Note TN1/79*, Building Services Research and Information Association, BSRIA, Bracknell, ISBN 0860220745, UK.

Butler, H. (1979b) Space requirement for building services distribution systems – detail design stage, *Technical Note TN3/79*, Building Services Research and Information Association, BSRIA, Bracknell, ISBN 0860220842, UK.

Chadderton, D.V. (1997a) *Air Conditioning, A Practical Introduction*, 2nd edn, F & FN Spon, London.

Chadderton, D.V. (1997b) *Building Services Engineering Spreadsheets*, E & Spon, London.

Chartered Institution of Building Services Engineers (1973) *Practice Notes, Number I, Combustion and Ventilation Air for Boilers and Other Heat Producing Appliances: Installations Not Exceeding 45 kW*, London.

Chartered Institution of Building Services Engineers (1981) *Building Energy Code*, Parts 1–3, London.

Chartered Institution of Building Services Engineers (1986) *CIBSE Guide A*, London.

Chartered Institution Building Services Engineers (1995) Building services. *CIBSE Journal*, **17** (4), 39.

Coffin, M.J. (1992) *Direct Digital Control for Building HVAC Systems*, Chapman & Hall, London.

Courtney, R.G. (1976a) An appraisal of solar water heating in the UK, *Building Research Establishment Current Paper CP 7/76*, January, Watford, UK.

Courtney, R.G. (1976b) Solar energy utilisation in the UK: current research and future prospects, *Building Research Establishment Current Paper 64/76*, October, Watford, UK.

Crawshaw, D.T. (1976) Coordinated working drawings, *Building Research Establishment Current Paper CF 60/76*, Watford, UK.

Diamant, R.M.E. (1977) *Insulation Deskbook*, Heating and Ventilating Publications, London.

Ellis, P. (1981) Listening to the user. *Journal of the Chartered Institution of Building Services*, **3** (4), 40.

Fanger, P.O. (1972) *Thermal Comfort*, McGraw-Hill, New York.

Fanger, P.O. (1988) Introduction of the olf and decipol units to quantify air pollution perceived by humans indoors and outdoors. *Energy in Buildings*, **12**, 1–6.

Haines, R.W. and Hittle, D.C. (1983) *Control Systems for Heating, Ventilating and Air Conditioning*, Chapman & Hall, London.

Horlock, J.H. (1987) *Cogeneration: Combined Heat and Power Thermodynamics and Economics*, Pergamon, Oxford.

Jenkins, B.D. (1991) *Electrical Installation Calculations*, Blackwell Scientific, Oxford.

Jones, W.P. (1985) *Air Conditioning Engineering*, 3rd edn, Edward Arnold (Publishers) Limited, London.

Kut, D. (1993) *Illustrated Encyclopedia of Building Services*, E & FN Spon, London.

Levermore, G.J. (1992) *Building Energy Management Systems*, E & FN Spon, London.

McKenna, G.T., Parry, C.M. and Tilic, M. (1981) What future for task lighting? *Journal of The Chartered Institution of Building Services*, **3**, 39.

Moss, K.J. (1996) *Heating and Water Services Design in Buildings*, E & FN Spon, London.

Moss, K.J. (1997) *Energy Management and Operating Costs in Buildings*, E & FN Spon, London.

National Joint Utilities Group (1979) *Provision of Mains and Services by Public Utilities on Residential Estates*, Publication Number 2, November, London.

Payne, G.A. (1978) *The Energy Manager's Handbook*, IPC Science and Technology Press, London.

Saunders, L.H. (1981) Thermal insulation and condensation. *Building Research Establishment News*, 55.

Smith, B.J., Peters, R.J. and Owen, S. (1985) *Acoustics and Noise Control*, Longman, Harlow, UK.

Southampton Geothermal booklet. (1992) Southampton Geothermal Heating Company, Garret House, Manor Royal, Crawley, West Sussex, RH10 9UT.

Swaffield, J.A. and Wakelin, R.H.M. (1976) Observation and analysis of the parameters affecting the transport of waste solids in internal drainage systems. *Public Health Engineer*, November, 165–171.

Threlkeld, J.L. (1962) *Thermal Environmental Engineering*, Prentice-Hall International, New Jersey, USA.

Uglow, C.E. (1981) The calculation of energy use in dwellings. *Building Services Engineering Research and Technology*, **2** (1), 1–14.

Wise, A.F.E. (1979) *Water, Sanitary and Waste Services for Buildings*, Batsford, London.

Index